Handbook of Research on Teacher Education in History and Geography

Cosme Jesús Gómez Carrasco,
Pedro Miralles Martínez and
Ramón López Facal

Handbook of Research on Teacher Education in History and Geography

PETER LANG

Bibliographic Information published by the Deutsche Nationalbibliothek

The Deutsche Nationalbibliothek lists this publication in the Deutsche Nationalbibliografie; detailed bibliographic data is available online at http://dnb.d-nb.de.

Library of Congress Cataloging-in-Publication Data

A CIP catalog record for this book has been applied for at the Library of Congress.

This book was funded by some research projects granted by Spanish Ministry of Science and Innovation, Spanish Agency of Research, and FEDER EU funds, grant numbers PGC2018-094491-B-C31, PGC2018-094491-B-C32, PGC2018-094491-B-C33; and "Seneca Foundation. Regional Agency for Science and Technology", grant numbers 20638/JLI/18 and 20874/PI/18.

ISBN 978-3-631-81897-8 (Print)
E-ISBN 978-3-631-84838-8 (E-PDF)
E-ISBN 978-3-631-84839-5 (EPUB)
E-ISBN 978-3-631-84840-1 (MOBI)
DOI 10.3726/ b18091

Peter Lang – Berlin · Bern · Bruxelles · New York · Oxford · Warszawa · Wien

This publication has been peer reviewed.

www.peterlang.com

Acknowledgement

This book was funded by some research projects granted by Spanish Ministry of Science and Innovation, Spanish Agency of Research, and FEDER EU funds, grant numbers PGC2018-094491-B-C31, PGC2018-094491-B-C32, PGC2018-094491-B-C33; and "Seneca Foundation. Regional Agency for Science and Technology", grant numbers 20638/JLI/18 and 20874/PI/18.

Contents

Contributors 11

Cosme J. Gómez-Carrasco, Pedro Miralles-Martínez, and Ramón López-Facal
Introduction 15

Part I Research approaches in the training of geography and history teachers

Victor Arias, Benito Arias, and Jairo Rodríguez-Medina
Chapter 1 Quantitative research in education 35

Jairo Rodríguez Medina and Bartolomé Rubia Avi
Chapter 2 Mixed methods in education research 53

Stéphane Lévesque and Jean-Philippe Croteau
Chapter 3 Teaching the narrative competence in history education: Lessons from Canada 71

Fredrik Alvén
Chapter 4 Please let it be a history test with a content I can handle: How girls and boys perform in the national test in history in Sweden 87

Part II Researches about opinions and perceptions of teachers in initial training and in practising

David Parra Monserrat, Xosé Manuel Souto González, and Nancy Palacios Mena
Chapter 5 The representation of geographical and historical education among teachers-in-training: An Ibero-American perspective ... 115

María de la Encarnación Cambil, Esther Jiménez, Mª Montserrat Pastor and Ander Delgado
Chapter 6 An analysis of the perceptions of trainee Secondary Education teachers on current historiographical trends 137

Juan Carlos Colomer, Jorge Sáiz, and Antonio J. Morales
Chapter 7 History and geographical thinking skills: between disciplinary knowledge and knowledge for teaching in Teacher Education 159

Cosme J. Gómez Carrasco, Álvaro Chaparro Sáiz, María del Mar Felices de la Fuente and Juan Antonio Inarejos Muñoz
Chapter 8 Perceptions regarding historical competences in trainee geography and history teachers 177

Francisco Javier Trigueros Cano, José María Campillo Ferrer, Pedro Miralles-Martínez, and Jesús Molina Saorín
Chapter 9 The acquisition of historical competences in Spain: Perceptions of primary and secondary education teachers 201

Helena Rausell, Álvaro-Francisco Morote, Rafael Valls, and Moisés Domingos
Chapter 10 Social representations and teaching of cross-cutting topics for history and geography teachers in training: Gender and climate change 221

Carlos Fuertes, Elvira Asensi, Carlos Fuster, and Sérgio Claudino
Chapter 11 Socially acute questions and critical citizenship in trainee geography and history teachers: From theory to classroom observation 241

José A. Armas Castro, Jorge Conde Miguélez, Luis Alberto Marques Alves, and Cristina María Maia
Chapter 12 Learning to teach history in secondary education: Preservice teachers' attitudes when faced with emotional and controversial issues 259

Belén Castro-Fernández, Ramón López Facal, Guadalupe Jiménez-Esquinas, and Daniel Schugurensky
Chapter 13 What can be learned from decontextualised heritage? 283

Part III Heritage and ICT in the training of geography and history teachers

Olaia Fontal Merillas and Silvia García-Ceballos
Chapter 14 Heritage education in teacher education and training 309

Andres Domínguez-Almansa, Manuela Costa-Casais and Tania Riveiro-Rodríguez
Chapter 15 Cliffs with memory: Historical and social teacher training via climatic and environmental transformations 331

Francisco-Javier Hinojo-Lucena, Inmaculada Aznar-Díaz, María-Pilar Cáceres-Reche, and José-María Romero-Rodríguez
Chapter 16 Mobile learning methodology: Analysis of the use of mobile devices in teaching 355

Janire Castrillo, Iratxe Gillate, Ursula Luna, and Alex Ibañez-Etxeberria
Chapter 17 Developing social and civic competence via apps: The role of historical memory in the initial teacher training. 367

José Monteagudo-Fernández, Juan Ramón Moreno-Vera, Raimundo A. Rodríguez-Pérez, and Ramón Cózar-Gutiérrez
Chapter 18 Motivation strategies for gamified flipped classrooms in Social Sciences education 391

Contributors

Fredrik Alvén
University of Malmö

Benito Arias
University of Valladolid (Spain)

Victor Arias
University of Salamanca (Spain)

José A. Armas Castro
University of Santiago de Compostela (Spain)

Elvira Asensi
University of Valencia (Spain)

Inmaculada Aznar-Díaz
University of Granada (Spain)

María-Pilar Cáceres-Reche
University of Granada (Spain)

María de la Encarnación Cambil
University of Granada (Spain)

José María Campillo Ferrer
University of Murcia (Spain)

Janire Castrillo
University of the Basque Country (Spain)

Belén Castro-Fernández
University of Santiago de Compostela (Spain)

Álvaro Chaparro Sáiz
University of Almería (Spain)

Sérgio Claudino
University of Lisbon (Portugal)

Juan Carlos Colomer
University of Valencia (Spain)

Manuela Costa-Casais
University of Santiago de Compostela (Spain)

Ramón Cózar-Gutiérrez
University of Castilla-La Mancha (Spain)

Jean-Philippe Croteau
Sichuan University (China)

Ander Delgado
University of the Basque Country (Spain)

Moisés Domingos
Federal University of Rio Grande do Norte (Brazil)

Andrés Domínguez-Almansa
University of Santiago de Compostela (Spain)

María del Mar Felices de la Fuente
University of Almeria (Spain)

Olaia Fontal Merillas
University of Valladolid (Spain)

Carlos Fuertes
University of Valencia (Spain)

Carlos Fuster
University of Valencia (Spain)

Silvia García-Ceballos
University of Zaragoza (Spain)

Iratxe Gillate
University of the Basque Country (Spain)

Cosme J. Gómez-Carrasco
University of Murcia (Spain)

Francisco-Javier Hinojo-Lucena
University of Granada (Spain)

Alex Ibañez-Etxeberria
University of the Basque Country (Spain)

Juan Antonio Inarejos Muñoz
University of Castilla-La Mancha (Spain)

Esther Jiménez
Complutense University of Madrid (Spain)

Guadalupe Jiménez-Esquinas
University of Santiago de Compostela (Spain)

Stéphane Lévesque
University of Ottawa (Canada)

Ramón López-Facal
University of Santiago de Compostela (Spain)

Ursula Luna
University of the Basque Country (Spain)

Cristina María Maia
Polytechnic Institute of Porto (Portugal)

Luis Alberto Marques Alves
Polytechnic Institute of Porto (Portugal)

Jorge Conde Miguélez
University of Santiago de Compostela (Spain)

Pedro Miralles-Martínez
University of Murcia (Spain)

José Monteagudo-Fernández
University of Murcia (Spain)

Antonio J. Morales
University of Valencia (Spain)

Juan Ramón Moreno-Vera
University of Murcia (Spain)

Álvaro-Francisco Morote
University of Valencia (Spain)

Nancy Palacios Mena
University of Los Andes (Colombia)

David Parra Monserrat
University of Valencia (Spain)

Mª Montserrat Pastor
Autonomous University of Madrid (Spain)

Helena Rausell
University of Valencia (Spain)

Tania Riveiro-Rodríguez
University of Santiago de Compostela (Spain)

Jairo Rodríguez-Medina
National University of Distance Education (Spain)

Raimundo A. Rodríguez-Pérez
University of Murcia (Spain)

José-María Romero-Rodríguez
University of Granada (Spain)
Bartolomé Rubia Avi
University of Valladolid (Spain)

Jorge Sáiz
University of Valencia (Spain)

Jesús Molina Saorín
University of Murcia (Spain)

Daniel Schugurensky
Arizona State University (USA)

Xosé Manuel Souto González
University of Valencia (Spain)

Francisco Javier Trigueros Cano
University of Murcia (Spain)

Rafael Valls
University of Valencia (Spain)

Cosme J. Gómez-Carrasco, Pedro Miralles-Martínez, and Ramón López-Facal

Introduction Research challenges regarding trainee geography and history teachers

Knowledge of the opinions, practices and expectations of trainee and practising teachers is necessary in order to be able to improve teacher training and teaching practice. This Handbook addresses the challenges faced by history and geography teachers, who, in several European countries, such as Spain and France, share their initial training and teaching in both subjects. It includes contributions by researchers from eight countries. The majority of them (eleven) are from Spanish universities and have participated in a wide-ranging research project, enjoying the collaboration and participation of researchers from seven other countries (Brazil, Canada, China, Colombia, Portugal, Sweden and the United States). Rather than being the sum of individual contributions, this work is the fruit of collaboration by a network which incorporates the main Spanish research groups in the teaching of the social sciences: history, geography, heritage and other social sciences (Red 14).

An emerging academic knowledge domain

In recent decades, the initial and ongoing training of teachers has become a key issue (González & Skultety, 2018). Studies carried out in several countries have agreed on the need to update teacher training programmes in order to improve teaching and learning processes in compulsory education (Barnes, Fives, & Dacey, 2017; König & Blömeke, 2012; Korthagen, 2010). The research carried out points towards the need for comparative studies, thereby making it possible to transfer these findings into practical teaching (Darling-Hammond & Bransford, 2005; König, Ligtvoet, Klemenz, & Rothland, 2017; Schmidt, Blömeke, &Tatto, 2011).

Researchers in the field of teacher training agree on the fact that it is of prime importance to analyse the knowledge and conceptions of teachers in order to guide initial teacher training programmes (Darling-Hammond, 2006; Fives & Buehl, 2012; Virta, 2002). Their studies highlight research which aims to calibrate the different types of professional knowledge of teachers, placing emphasis on the mastery of everyday tasks in the classroom (König & Pflanz, 2016; Oliveira, Lopes, & Spear-Swerling, 2019).

Shulman's (1989) proposals have been influential in the definition of the categories designed for research in the field of teacher training. Based on his analysis, when evaluating teaching competencies, researchers tend to differentiate between Content Knowledge (CK), Pedagogical Content Knowledge (PCK) and General Pedagogical Knowledge (GPK) (Kleickmann, Richter, Kunter, Elsner, Besser, Krauss, & Baumert, 2012). CK refers to the knowledge of a specific subject and is related with the contents which teachers are obliged to teach. GPK implies the broad principles and strategies of management and organisation within the classroom (Blömeke, Busse, Kaiser, König, & Sühl, 2016). PCK implies the capacity of relating the specific material of the subject with teaching aims (Monte-Sano, 2011). The latter is a type of knowledge which explores the social representations of those learning a specific subject, how the learner understands this knowledge, the methods and resources required for teaching that subject and the selection and organisation of the specific contents in order to make them appropriate to the reality of the classroom (Meschede, Fiebranz, Möller, & Steffensky, 2017).

Several decades ago, studies on the teaching of history underwent a cognitive change, which did not occur at the same time or at the same pace in all countries. Wilschut (2011) has situated the origin of this trend in the 1970s, when Bruner's theories and Bloom & Krathwohl's taxonomies of educational objectives began to have a decisive influence on proposals relating to history education. A landmark in this change took place in the United Kingdom in 1972 with the setting up of the History Project 13–16, which later became known as the Schools History Project (SHP). The idea of this project was that pupils should "make" history rather than simply memorising events of the past. This project was the origin of the Concepts of History and Teaching Approaches Project. Thus, a research line was initiated linked to what is known as historical thinking (Martínez-Hita & Gómez, 2018). This approach attempts to provide students with the necessary intellectual tools to be able to analyse the past and relate it with understanding the problems of the present (Chapman, 2011; Counsell, 2011; Lee, 2005; Lee & Ashby, 2000).

In the United States, this line of research has given rise to studies influenced by cognitive psychology and the expert-novice analytical technique (VanSledright, 2011, 2014; Wineburg, 2001). Such research has generated studies in which the use of historical sources and the work of the historian occupy a prime role (Levstik & Barton, 2008; Monte-Sano, De la Paz, & Felton, 2014; Reisman, 2012; Wineburg, Martin, & Monte-Sano, 2013). In Canada, the work carried out at the Centre for the Study of Historical Consciousness, led by Peter Seixas, stands out. This centre has worked hard to define historical consciousness and historical thinking and to adapt these ideas to the reality of the classroom in a practical way via projects such as the Historical Thinking Project and the Assessment of Historical Thinking (Lévesque, 2008; Seixas & Morton, 2013). In addition to the work of this group in Canada, there has been a significant increase in studies on history education which attempt to combine the two concepts mentioned above (Létourneau, 2014; Zanazanian, 2015). These studies have had a great impact on the academic production of other countries, such as Australia (Parkes & Sharp, 2014) and the Netherlands (Bjorn, Sanne, Itzél, & Theo, 2018; Grever, Peltzer, & Haydn, 2011; Van Boxtel & Van Drie, 2012; Van Boxtel, Grever, & Klein, 2015; Van Drie & Van Boxtel, 2008).

In the Ibero-American sphere, the work by Barca (2005), Cerri and Amézola (2010), Domínguez (2015), Gómez and Miralles (2015, 2016), López-Facal (2014), Mora and Ortiz (2012), Sáiz and López-Facal (2015), and Schmidt (2017) shows how the afore-mentioned Anglo-Saxon proposals have been adapted to research on history education, including contributions influenced by German educational science, which place emphasis on the ethical dimension of history teaching.

This increase in research in the area of history education has, in recent years, led to the publication of many monographs. Among these publications, those by Counsell, Burn and Chapman (2016), Carretero, Berger and Grever (2017) and Metzger and Harris (2018a) stand out for examining key methodological concepts, current research lines, teaching praxis and the uses and objectives of the teaching of history. These reviews agree on the fact that there has been a significant increase in research in this area since the 1990s (Metzger & Harris, 2018b). They show that historical thinking and historical consciousness are two fundamental axes of research in this area in recent decades (Seixas, 2017), and that these studies have primarily focused on the curriculum, textbooks and, to a lesser degree, on interviews, pupils' perceptions, observation records for the assessment of intervention proposals and case studies (Epstein & Salinas, 2018). In recent years, validated questionnaires have begun to have a greater impact in this field of knowledge, in the same way as other data collection tools and observation scales (De Groot-Reuvekamp, Anje, & Van Boxtel, 2017; De Groot-Reuvekamp, Ros, & Van Boxtel, 2018a, 2018b; Van Straaten, Wilschut, & Oostdam, 2018).

In spite of the increase in research over recent decades, there is still a limited number of studies dealing with the issue of PCK in the teaching of history and other social sciences, such as geography, from a global perspective and with a solid empirical basis (Pollock, 2014). Banks & Parker (1990) and Adler (2008) have pointed out that the majority of research on teacher training in the social sciences has been carried out on the basis of specific experiments which cannot easily be generalised. They highlight several classic studies in this area, such as those by Wilson and Wineburg (1993), McDiarmid (1994), Seixas (1998), Fragnoli (2005) and Bain and Mirel (2006), which made it possible to verify the epistemological and methodological conceptions of trainee teachers (Sáiz, Gómez, & López-Facal, 2018). The studies by Korthagen (2010), Lévesque (2014), Monte-Sano (2011) and Westhoff and Polman (2008) have examined in depth the interrelationship between the epistemological conceptions of teachers in relation to the concepts of historical thinking and the objective of the teaching of history (Pollock, 2014). In recent years, the studies carried out in the Netherlands, particularly by Van Boxtel's group, have made it possible to make progress in robust empirical analyses, both in terms of statistics and in conceptual depth (Stoel, Logtenbergb, Wansink, Huijgend, Van Boxtel, & Van Drie, 2017).

Later, Tuithof et al. (2019) carried out a systematic analysis of the research on Pedagogical Content Knowledge in history education. They highlighted the great number of qualitative studies with small samples and also indicated that a considerable amount of the existing research focuses on disciplinary strategies of history, such as argumentation and the use of primary sources in the classroom (Ledman, 2015; Monte-Sano & Budano, 2013). Only a few studies deal with issues of teaching methodology and, when they do so, they focus on some specific aspect of the subject, such as critical pedagogy in the classroom (Blevins, Magins, & Salinas, 2020).

The majority of the research which evaluates training programmes focuses on the analysis of exercises carried out by teachers, direct observation and perception questionnaires (De Groot-Reuvekamp, Ros, & Van Boxtel, 2018b; Gómez, Monteagudo, Moreno, & Sáinz, 2019). Few are the studies which evaluate the improvement of these teachers' skills in initial training when they begin their teaching practice in order to verify the effectiveness of the programmed activities. When this has been done, it has been with small samples and with qualitative techniques (Domínguez-Almansa & López Facal, 2017).

It is necessary to approach PCK from a holistic point of view in order to improve the teaching of history and for trainee teachers to abandon the epistemological baggage which conceives of history as a closed set of knowledge. Carrying out research on the training of future teachers may provide a diagnosis of what must be changed and the tools which should be employed. Therefore, it is considered necessary to perform a diagnostic analysis of the opinions and perceptions of trainee teachers with the aim of bringing about an improvement in teaching skills in terms of Pedagogical Content Knowledge.

The proposal of this handbook

In the first part of this book, the approaches of research in history and geography teaching are analysed, along with their relationship with teacher training. The first two chapters examine research methods. In the first chapter, Víctor Arias, Benito Arias and Jairo Rodríguez (from the Universities of Valladolid and Salamanca) explore in depth the quantitative approaches of research, both in terms of experimental designs and those which use greater internal controls in the research processes. Measurement and statistics are fundamental for quantitative research, as they constitute the link between the data observed and explanatory theoretical models. The chapter begins with an introduction to research trends in teacher training and a review of the main lines of research in this field. They conclude by describing the main designs in quantitative research in the context of education, which they illustrate with examples.

The second chapter addresses mixed methods. The techniques and procedures of quantitative data analysis have a great presence in high impact journals in education. Researchers in education sciences increasingly assume the complementarity of the paradigms and both conventional models are considered significant and valuable. Working with mixed methods seems to fit well with educational research in history education and in social studies. The research questions posed in the field of geography and history teaching and learning obtain more precise answers via the use of mixed methods, surpassing those obtained with quantitative and qualitative approaches when these are applied separately. Rodríguez-Medina and Rubia-Avi (the University of Valladolid) present the main research designs proposed from the mixed methods approach, along with some recent examples which illustrate their practical application in the field of education and also in geography and history teacher training.

The third chapter deals with narrative analysis in the area of history education and its potentiality. Levésque (the University of Ottawa) and Croteau (the University of Sichuan) highlight the importance of the theories of historical consciousness for the teaching of history, as they focus on the role played by narration in the attempts of contemporary peoples to define and orient themselves in time. History is a narrative act of (re)construction of the past. It uses cognitive means to interpret past realities connecting the dimensions of time (past, present, foreseen future) into coherent histories which create meaning. Far from being an oversimplification of realities, narration is a realisation of how we translate knowing how to narrate. The ability to provide meaning to the past is unavoidably related with narrative acts, or what Jürgen Straub calls "narrative competence."

In the fourth chapter, Fredrik Alvén (the University of Malmö) addresses an issue of great interest for teacher training, namely the difference in the conceptions of historical knowledge according to the sex of the pupils. He contributes the results of a study carried out in the context of the national history test in Sweden between 2013 and 2019. At the beginning of the study period, the national test mainly required methodological skills (source analysis, argumentation, etc.) but, later, factual knowledge was added. According to the results of the test, girls were more successful in the first type of knowledge and boys in the second. The test writers could decide whether they wanted boys or girls to perform better, thus conferring upon them a moral responsibility. What principles could they base themselves upon when writing the test?

The second part of the book brings together contributions on the opinions and perceptions of trainee geography and history teachers. Their knowledge, prejudices and attitudes regarding history, geography and heritage education and the teaching of socially conflictive issues are analysed. Schools are not currently the main sphere in which knowledge is produced, reproduced and socialised. Neither are teachers the only or main agents of its transmission, nor is education a mere accumulation of knowledge. The role and functions of teachers and the concept of education itself are questioned and relativised as the result of social, cultural and technological changes: the omnipresence of social networks and the diversity of the mass media; the immediate access offered by technology and devices; the advancements in knowledge and the vastness of its scope; the modification of the role of the family in its responsibility in children's education; and the need to educate free and autonomous citizens who are able to fulfil their potential to bring about a democratic, multicultural, changing and diverse society.

These social and cultural changes imply the need to train teachers to diagnose the learning situation of their pupils, to design more appropriate teaching and learning strategies and to intervene in the classroom in the most appropriate way for each situation.

Teachers should be trained to carry out their job in a context of accelerated and unpredictable changes. They need to acquire autonomy and, at the same time, work in collaboration with other colleagues and assume a curricular approach which is open, flexible and adaptable to its environment and a society which is ever more diverse and mixed. They must have rigorous and updated knowledge of the subjects of geography and history and also of innovative theories, methodologies and teaching practice. Being a teacher implies a commitment to society and to one's surroundings.

Chapter 5 presents an Ibero-American perspective, with the contributions of Drs Parra and Souto (the University of Valencia) and Palacios (the University of The Andes). Their research was carried out by way of study reports reflecting the conceptions of 200 trainee teachers studying a primary education degree and a master's degree in secondary education. Tradition and individual memory, both based on life experiences and personal memories, are remodelled and mixed with the collective memory of school. This research aims to demonstrate that the influence of this phenomenon has created a consensual common sense which hinders change, normalises codes and guarantees the continuity of certain narratives, practices and uses which are not useful for a civic and democratic education. The authors evaluate the degree of influence of school experiences on the participants, the impact of pedagogical and disciplinary education and the effect of a system and curricular framework in which the configuration and reproduction of these representations are produced.

In Chapter 6, Cambil (the University of Granada), Jiménez (the Complutense University of Madrid), Delgado (the University of the Basque Country) and Pastor (the Autonomous University of Madrid) analyse the perceptions of trainee secondary teachers regarding current historiographical trends. Historiography and the teaching of history have not mixed well in spite of the significant influence of the former on the latter. The authors study the perceptions regarding traditional and current historiographical approaches of these future teachers once they have finished their period of teaching practice. The aim is to analyse which historiographical trends currently have greater presence in the teaching and learning of history, how they influence teaching methodology and the focus of activities carried out in the classroom. The difficulties and advantages of their use are also examined, along with the challenge supposed by methodological innovation from current historiographical approaches, such as the history of the present, the history of everyday life, gender perspective, etc., and how the capacity for historical thinking in secondary education can be improved based on these aspects.

In Chapter 7, Colomer, Sáiz and Morales (the University of Valencia) analyse the historical and geographical knowledge and the thinking aptitudes and abilities of training teachers and to what extent they incorporate disciplinary and didactic knowledge. Their study aims to verify whether trainee teachers only acquire these aptitudes in a theoretical way or whether they are capable, or not, of putting them into practice when necessary.

Chapter 8 deals with the perceptions of trainee geography and history teachers regarding historical competences. For Gómez-Carrasco (University of Murcia), Felices and Chaparro (the University of Almería) and Inarejos (the University of Castilla-La Mancha), the changes which have taken place in our society require new teaching strategies and new objectives to build society in a broad and multicultural world. The data revealed by the questionnaire employed in the research show the unanimous preference for skills of a procedural and methodological nature. A clear preference among women for second-order contents and methodological concepts is also shown.

In Chapter 9, Trigueros, Campillo, Miralles and Molina (the University of Murcia) approach the acquisition of historical skills in Spanish teachers. They analyse the perceptions of teachers on history education with the aim of promoting significant practices, competency-based learning and mutual consciousness and cooperation. They note the lack of agreement among teachers regarding progress which should be made and the necessary fundamental changes in direction. They conclude that both teachers and students should advance towards a form of history education which places specific skills in the foreground. In their opinion, the nucleus of history teaching in the classroom must ensure that the central concepts of historical thinking are put into practice.

Chapter 10 focuses on social representations and the teaching of cross-cutting issues to trainee geography and history teachers. Rausell, Morote and Valls (the University of Valencia) and Domingos (the University of Rio Grande do Norte) evaluate the representations of trainee teachers regarding socially relevant cross-cutting issues: questions of gender and environment. Their aim is to evaluate how trainee teachers incorporate these issues into teaching and learning processes. The authors analyse and classify discourse and representations relating to gender identity and environmental issues via initial surveys, self-evaluation surveys and narratives. The results indicate, along the same lines as the study by Armas, Conde, Alves and Maia (Chapter 10), that the prior training received by the students generates significant differences when defining and, consequently, understanding the concept of gender. Greater limitations are identified in the activities which they proposed for their future pupils, in spite of the fact that all of them claimed to have received specific training on this issue.

Fuertes, Asensi and Fuster (the University of Valencia), in collaboration with Claudino (the University of Lisbon), deal, in Chapter 11, with the relationship between the theoretical knowledge of socially relevant issues which university education provides to trainee teachers and the subsequent approach to these issues in the trainee teachers' initial experiences in the geography and history classroom. The authors aim to identify the complex relationship which will bring about changes and continuities in the perceptions and praxis of trainee teachers and identify their difficulties and possibilities, making it possible to promote the education of a critical citizenship and to improve teacher training itself. This is an initial study which will be complemented with data currently being processed in Portugal in order to obtain comparative results.

In Chapter 12, Drs Armas and Conde (the University of Santiago de Compostela), Alves (the University of Porto) and Maia (the Polytechnic Institute of Porto) present a study on the attitudes of trainee secondary education history teachers towards the teaching of sensitive and controversial issues. The results indicate that trainee teachers are, in general, in favour of teachers not expressing their own opinions when debating such topics in class, or expressing, but not imposing, their opinions whilst promoting the manifestation of other opinions. The majority reject the idea of the teacher imposing a single interpretation or excluding burning issues from school debate. In the analysis of the data, the authors consider that the prior training of the trainee teachers and the training programme received have a greater influence on the results than other socio-demographic variables.

In Chapter 13, Drs Castro, López-Facal and Jiménez-Esquinas (the University of Santiago de Compostela) and Schugurensky (Arizona State University) analyse the perceptions of trainee primary education teachers regarding the inclusion of relevant and socially conflictive issues from the perspective of heritage education. The results obtained from a study on a town flooded by the construction of a dam and the trauma brought about by the forced relocation of the population are presented. The data of the learning are evaluated in order to understand the social problems and analyse to what extent trainee primary teachers consider that such issues can contribute towards the construction of a socially committed citizenship. The results reveal that trainee teachers modify their perceptions of how to approach heritage education when research methodology is incorporated into their training process.

The third part of the Handbook brings together two sets of issues; on the one hand the results of incorporating ICT into the training of history and geography teachers and, on the other, the incorporation of heritage education from a different and complementary perspective to that included in the previous chapter (Castro et al.).

In Chapter 14, Fontal (the University of Valladolid) and García-Ceballos (the University of Zaragoza) advocate a proposal for incorporating a heritage education programme for teacher training based on consciousness and the connection with its material context. This proposal is based on an exploratory study which reveals fragmented, disjointed and reductionist knowledge among future teachers. The authors propose the definition of skills to be developed in heritage education, which should be oriented towards an integral and humanistic definition of heritage, providing tools and experiential learning for future practice.

Domínguez-Almansa, Costa-Casais and Riveiro (the University of Santiago de Compostela) present, in Chapter 15, the results of a quite unusual experiment in heritage education. They analyse the learning outcomes of 191 trainee teachers based on an experiment grounded in the theory of Place-Based Education. In this experiment, the participants related unique geomorphological formations with the historical changes associated to them. The results and conclusions demonstrate the potentiality of interdisciplinary work.

The final chapters deal with ICT resources in teacher training and in practical teaching. Researchers from the University of Granada, Hinojo-Lucena, Aznar, Cáceres-Reche and Romero, analyse the use of mobile devices in teaching (mobile learning methodology). In Chapter 16, these authors analyse the degree of implementation of mobile learning in several universities in the region of Andalusia (Spain) and determine the reasons why some teachers do not apply this method in their teaching. They conclude that the majority do use this methodology and that those who do not employ this technique allege that they lack sufficient training and also fear that their pupils may be distracted.

Furthermore, in Chapter 17, researchers from the University of the Basque Country, Castrillo, Gillate, Luna and Ibáñez-Etxeberria, deal with the development of social and citizen competence via the use of applications: the role of historical memory in initial teacher training. To do so, they have analysed results deriving from the use of an educational resource designed for the study of the Spanish Civil War (1936 to 1939) in the town of Eibar. In the opinion of the authors, the results confirm that this type of resource contributes to increasing motivation by incorporating the direct testimony of those who experienced the tragic events and improving understanding of the past thanks to the use of primary sources. However, they also highlight the importance of the role of the teacher as a mediator when using applications of this kind, which, in general, are not designed to be used in formal education.

Chapter 18 provides information on different motivational strategies for gamified flipped classrooms. Drs Monteagudo, Moreno and Rodríguez-Péres (the University of Murcia) and Cózar (the University of Castilla-La Mancha) reveal the perceptions of students and trainee teachers regarding gamification and the flipped classroom strategy as motivational resources. Flipped classroom and video games have occupied an increasingly important position in the teaching of the social sciences and constitute a phenomenon which has come to stay. However, they cannot replace face-to-face teaching or more traditional methods, which help to establish skills such as the habit of reading and knowing how to work in a team. ICT resources do not arise with educational aims, but they are a first-level resource for facilitating the teaching process and providing examples. They can contribute towards an increase in participation among pupils and the construction of their conceptual knowledge, making use of a reward-based assessment system and the use of cooperative working strategies during lessons. However, education must assume the priority of educational objectives relating to values such as tolerance, respect towards others, ethics and the preservation of heritage and the environment.

References

Adler, S. A. (2008). The Education of Social Studies Teachers. In. L. S. Levstik & C. A. Tyson (Eds.). *Handbook of Research on Social Studies Education* (pp. 329–351). New York: Routledge.

Bain, R. & Mirel, J. (2006). Setting Up Camp at the Great Instructional Divide: Educating Beginning History Teachers. *Journal of Teacher Education*, 57(3), 212–219.

Banks, J. A. & Parker, W. C. (1990). Social Studies Teacher Education. In W. R. Houston (Ed.). *Handbook of Research on Teacher Education* (pp. 674–686). New York: Macmillan.

Barca, I. (2005). "Till New Facts Are Discovered": Students' Ideas about Objectivity in History. In R. Ashby, P. Gordon & P. Lee (Eds.). *Understanding History: Recent Research in History Education* (pp. 68–82). New York: Routledge Falmer.

Barnes, N., Fives, H. & Dacey, C. (2017). U.S. Teachers' Conceptions of the Purposes of Assessment. *Teaching and Teacher Education*, 65, 107–116.

Bjorn, W., Sanne, A., Itzél, Z. & Theo, W. (2018). Where Does Teaching Multiperspectivity in History Education Begin and End? An Analysis of the Uses of Temporality. *Theory & Research in Social Education*, 46(4), 495–527.

Blevins, B., Magins, K. & Salinas, C. (2020). Critical Historical Inquiry: The Intersection of Ideological Clarity and Pedagogical Content Knowledge. *The Journal of Social Studies Research*, 44, 35–50.

Blömeke, S., Busse, A., Kaiser, G., König, J. & Sühl, U. (2016). The Relation between Content-Specific and General Teacher Knowledge and Skills. *Teaching and Teacher Education*, 56, 35–46.

Carretero, M., Berger, S. & Grever, M. (eds.) (2017). *Palgrave Handbook of Research in Historical Culture and Education*. London: Palgrave-Macmillan.

Cerri, L. F. & Amézola, G. (2010). El estudio empírico de la conciencia histórica en jóvenes de Brasil, Argentina y Uruguay. *Didáctica de las Ciencias Experimentales y Sociales*, 24, 3–23.

Chapman, A. (2011). Taking the Perspective of the Other Seriously? Understanding Historical Argument. *Educar em Revista*, 42, 95–106.

Counsell, C. (2011). Disciplinary Knowledge for All, the Secondary History Curriculum and History Teachers' Achievement. *The Curriculum Journal*, 22(2), 201–225.

Counsell, C., Burn, K. & Chapman, A. (2016). *MasterClass in History Education. Transforming Teaching and Learning*. London: Bloomsbury.

Darling-Hammond, L. (2006). Assessing Teacher Education. The Usefulness of Multiple Measures for Assessing Program Outcomes. *Journal of Teacher Education*, 57(2), 120–138.

Darling-Hammond, L. & Bransford, J. D. (2005). *Preparing Teachers for a Changing World: What Teachers Should Learn and Be Able to Do*. San Francisco, CA: Jossey-Bass.

De Groot-Reuvekamp, M., Anje, R. & Van Boxtel, C. (2017). "Everything Was Black and White…": Primary School Pupils' Naive Reasoning While Situating Historical Phenomena in Time. *Education* 3(13), 1–16.

De Groot-Reuvekamp, M., Ros, A. & Van Boxtel, C. (2018b). A successful Professional Development Program in History: What Matters? *Teaching and Teacher Education*, 75, 290–301.

De Groot-Reuvekamp, M., Ros, A. & Van Boxtel, C. (2018a). Improving Elementary School Students' Understanding of Historical Time: Effects of Teaching With "Timewise". *Theory and Research in Social Education*, 46(1), 35–67.

Domínguez, J. (2015). Evaluación del pensamiento histórico y conceptos metodológicos de la historia. In *Pensamiento histórico y evaluación de competencias* (pp. 35–64). Barcelona: Graó.

Domínguez-Almansa, A. & López Facal, R. (2017). Patrimonios en conflictos. Competencias cívicas y formación profesional en educación primaria. *Revista de Educación*, 375, 86–109. https://doi.org/10.4438/1988-592X-RE-2016-375-336.

Epstein, T. & Salinas, C. S. (2018). Research Methodologies in History Education. In S. A. Metzger & L. M. Harris (Eds.). *The Wiley International Handbook of History Teaching and Learning* (pp. 61–92). Arizona: Wiley.

Fragnoli, K. (2005). Historical Inquiry in a Methods Classroom: Examining Our Beliefs and Shedding our Old Ways. *The Social Studies*, 96(6), 247–255.

Gómez, C. J. & Miralles, P. (2015). ¿Pensar históricamente o memorizar el pasado? La evaluación de los contenidos históricos en la educación obligatoria en España. *Revista de Estudios Sociales*, 52, 52–68.

Gómez, C. J. & Miralles, P. (2016). Historical Skills in Compulsory Education: Assessment, Inquiry Based Strategies and Argumentation. *New Approaches in Educational Research*, 5(2), 139–146.

González, G. & Skultety, L. (2018). Teacher Learning in a Combined Professional Development Intervention. *Teaching and Teacher Education*, 71, 341–354.

Grever, M., Peltzer, B. & Haydn, T. (2011). High School Students' Views on History. *Journal of Curriculum Studies*, 43(2), 207–229.

Kleickmann, T., Richter, D., Kunter, M., Elsner, J., Besser, M., Krauss, S. & Baumert, J. (2012). Teachers' Content Knowledge and Pedagogical Content Knowledge: The Role of Structural Differences in Teacher Education. *Journal of Teacher Education*, XX(X), 1–17.

König, J. & Blömeke, S. (2012). Future Teachers' General Pedagogical Knowledge from a Comparative Perspective: Does School Experience Matter?. *ZDM – The International Journal on Mathematics Education*, 44(3), 341–354.

König, J., Ligtvoet, R., Klemenz, S. & Rothland, M. (2017). Effects of Opportunities to Learn in Teacher Preparation on Future Teachers' General Pedagogical Knowledge: Analyzing Program Characteristics and Outcomes. *Studies in Educational Evaluation*, 53, 122–133.

König J. & Pflanz, B. (2016). Teacher Knowledge Associated with Performance? On the Relationship between Teachers' General Pedagogical Knowledge and Instructional Quality. *European Journal of Teacher Education*, 39(4), 419–436.

Korthagen, A. J. (2010). Situated Learning Theory and the Pedagogy of Teacher Education: Towards an Integrative View of Teacher Behavior and Teaching Learning. *Teaching and Teacher Education*, 26(1), 98–106.

Ledman, K. (2015). Navigating Historical Thinking in a Vocational Setting: Teachers Interpreting a History Curriculum for Students in Vocational Secondary Education. *Journal of Curriculum Studies*, 57, 77.93.

Lee, P. (2005). Putting Principles into Practice: Understanding History. In M. S. Donovan & J. D. Bransford (Eds.). *How Students Learn: History in the Classroom* (pp. 29–78). Washington, DC: National Academies Press.

Lee, P. & Ashby, R. (2000). Progression in Historical Understanding among Students Ages 7–14. In P. N. Stearns, P. Seixas & Wineburg, S. (Eds.).

Knowing, Teaching, and Learning History: National and International Perspectives (pp. 199–222). New York, NY: University Press.

Létourneau, J. (2014). *Je me souviens? Le passé du Québec dans la conscience de sa jeunesse*. Quebec: Fides.

Lévesque, S. (2008). *Thinking Historically. Educating Students for the 21st Century*. Toronto: University of Toronto Press.

Lévesque, S. (2014). What Is the Use of the Past for Future Teachers? A Snapshot of Francophone Student Teachers in Ontario and Québec Universities. In R. Sandwell & A. V. Heyking (Eds.). *Becoming a History Teacher* (pp. 115–138). Toronto: University of Toronto Press.

Levstik, L. & Barton, K. (2008). *Doing History. Investigating with Children in Elementary and Middle Schools*. New York: Routledge.

López-Facal, R. (2014). La LOMCE y la competencia histórica. *Ayer*, 94, 273–285.

Martínez-Hita, M. & Gómez, C. J. (2018). Nivel cognitivo y competencias de pensamiento histórico en los libros de texto de Historia de España e Inglaterra. Un estudio comparativo. *Revista de Educación*, 379, 145–169.

McDiarmid, G. W. (1994). Understanding History for Teaching: A Study of the Historical Understanding of Prospective Teachers. In M. Carretero & J. F. Voss (Eds.). *Cognitive and Instructional Process in History and the Social Sciences* (pp. 159–185). Hillsdale: Lawrence Erlbaum Associates Publishers.

Meschede, N., Fiebranz, A., Möller, K. & Steffensky, M. (2017). Teachers' Professional Vision, Pedagogical Content Knowledge and Beliefs: On Its Relation and Differences between Pre-Service and In-Service Teacher. *Teaching and Teacher Education*, 66, 158–170.

Metzger, S. A. & Harris, L. M. (Eds.). (2018a). *The Wiley International Handbook of History Teaching and Learning*. Arizona: Wiley.

Metzger, S. A. & Harris, L. M. (2018b). Introduction: History Education in (and for) a Changing World. In S. A. Metzger & L. M. Harris (Eds.). *The Wiley International Handbook of History Teaching and Learning* (pp. 1–10). Arizona: Wiley.

Miralles, P., Gómez, C. J., Arias, V. B. & Fontal, O. (2019). Digital Resources and Didactic Methodology in the Initial Training of History Teachers. *Comunicar*, 61, 45-56. https://doi.org/10.3916/C61-2019-04.

Monte-Sano, C. (2011). Learning to Open Up History for Students: Preservice Teachers' Emerging Pedagogical Content Knowledge. *Journal of Teacher Education*, 62(3), 260–272.

Monte-Sano, C. & Budano, C. (2013). Developing and Enacting Pedagogical Content Knowledge for Teaching History: An Exploration of Two Novice

Teachers' Growth Over Three Years. *Journal of the Learning Sciences*, 22, 171–211. https://doi.org/10.1080/10508406.2012.742016

Monte-Sano, C., De la Paz, S. & Felton, M. (2014). *Reading, Thinking and Writing about History. Teaching Argument Writing to Diverse Learners in the Common Core Classroom, Grades 6–12.* New York: Teacher College Press.

Mora, G. & Ortiz, R. (2012). El modelo de educación histórica. Experiencias de innovación en la Educación Básica de México. Enseñanza de las Ciencias Sociales. *Revista de investigación*, 11, 87–98.

Parkes, R. J. & Sharp, H. (2014). Nietzschean Perspectives on Representations of National History in Australian School Textbooks: What Should We Do with Gallipoli? Ensayos. *Revista de la Facultad de Educación de Albacete*, 29(1), 159–181.

Pollock, S. A. (2014). The Poverty and Possibility of Historical Thinking: An Overview of Recent Research into History Teacher Education. In R. Sandwell & A. V. Heyking (Eds.). *Becoming a History Teacher* (pp. 60–74). Toronto: University of Toronto Press.

Reisman, A. (2012). "Reading Like a Historian": A Document-Based History Curriculum Intervention in Urban High Schools. *Cognition and Instruction*, 30(1), 86–112.

Sáiz, J. & López-Facal, R. (2015). Competencias y narrativas históricas: el pensamiento histórico de estudiantes y futuros profesores españoles de educación secundaria. *Revista de Estudios Sociales*, 52, 87–101.

Sáiz, J., Gómez, C. J. & López-Facal, R. (2018). Historical Thinking, Causal Explanation and Narrative Discourse in Trainee Teachers in Spain. *Historical Encounters: A Journal of Historical Consciousness, Historical Cultures, and History Education*, 5(1), 16–30.

Schmidt, M. A. (2017). ¿Qué hacen los historiadores cuando enseñan la Historia? Contribuciones de la teoría de Jörn Rüsen para el aprendizaje y el método de enseñanza de la Historia. *Clío & Asociados. La historia enseñada*, 24, 26–37.

Schmidt, W. H., Blömeke, S. & Tatto, M. T. (2011). *Teacher Education Matters. A Study of the Mathematics Teacher Preparation from Six Countries.* New York: Teacher College Press.

Seixas, P. (1998). Student Teacher Thinking Historically. *Theory and Research in Social Education*, 26(2), 310–341.

Seixas, P. (2017). Historical Consciousness and Historical Thinking. In M. Carretero, M. Grever & S. Berger (Eds.). *Palgrave Handbook of Research in Historical Culture and Education* (pp. 59–72). London: Palgrave McMillan.

Seixas, P. & Morton, T. (2013). *The Big Six Historical Thinking Concepts.* Toronto: Nelson Education Ltd.

Shulman, L. S. (1989). Paradigmas y programas de investigación en el estudio de la enseñanza: una perspectiva contemporánea. In M. C. Wittrock (Ed.). *La investigación de la enseñanza, I. Enfoques, teorías y métodos* (pp. 9–91). Barcelona: Paidós/MEC

Stoel, G., Logtenbergb, A., Wansink, B., Huijgend, T., Van Boxtel, C. & Van Drie, J. (2017). Measuring Epistemological Beliefs in History Education: An Exploration of Naïve and Nuanced Beliefs. *International Journal of Educational Research*, 83, 120–134.

Tuithof, H., Logtenberg, A., Bronkhorst, L., Van Drie, J., Dorsman, L. & Van Tartwijk, J. (2019). What Do We Know about the Pedagogical Content Knowledge of History Teachers: A Review of Empirical Research. *Historical Encounters: A Journal of Historical Consciousness, Historical Cultures, and History Education*, *6*(1), 72–95.

Van Boxtel, C. & Van Drie, J. (2012). "That's in the Time of the Romans!" Knowledge and Strategies Students Use to Contextualize Historical Images and Documents. *Cognition & Instruction*, 30(2), 113–145.

Van Drie, J. & Van Boxtel, C. (2008). Historical Reasoning: Towards a Framework for Analyzing Student's Reasoning about the Past. *Educational Psychology Review*, 20, 87–110.

Van Straaten, D., Wilschut, A. & Oostdam, R. (2018). Measuring Students' Appraisals of the Relevance of History: The Construction and Validation of the Relevance of History Measurement Scale (RHMS). *Studies in Educational Evaluation*, 56, 102–111.

VanSledright, B. A. (2011). *The Challenge of Rethinking History Education. On Practice, Theories, and Policy*. New York: Routledge.

Virta, A. (2002). Becoming a History Teacher: Observations on the Beliefs and Growth of Student Teachers. *Teaching and Teacher Education*, 18(6), 687–698.

Westhoff, L. M. & Polman, J. L. (2008). Developing Preservice Teachers' Pedagogical Content Knowledge about Historical Thinking. *International Journal of Social Education*, 22(2), 26–38.

Wilschut, A. (2011). *Images of Times. The Role of a Historical Consciousness of Time in Learning History*. Charlotte: IAP.

Wilson, S. & Wineburg, S. (1993). Wrinkles in Time and Place: Using Performance Assessment to Understand the Knowledge of History Teachers. *American Educational Research Journal*, 30(4), 729–769.

Wineburg, S. (2001). *Historical Thinking and Other Unnatural Acts: Charting the Future of Teaching the Past*. Philadelphia: Temple University Press.

Wineburg, S., Martin, D. & Monte-Sano, C. (2013). *Reading Like a Historian. Teaching Literacy in Middle & High School History Classrooms.* New York: Teacher College Press.

Zanazanian, P. (2015). Historical Consciousness and Metaphor: Charting New Directions for Grasping Human Historical Sense-Making Patterns for Knowing and Acting in Time. *Historical Encounters Journal*, 2(1), 16–33.

Part I Research approaches in the training of geography and history teachers

Victor Arias, Benito Arias, and Jairo Rodríguez-Medina

Chapter 1 Quantitative research in education

Abstract: Taking into account that a research method must answer and adapt to the research question or questions, quantitative research includes both experimental designs and other designs that emphasize the internal control of the research process. The research process begins with the posing of a question or series of questions related to a problem to which the researcher will try to answer by applying the scientific method. Measurement and statistics are fundamental for quantitative research as they constitute the link between observed (empirical) data and explanatory theoretical models. The chapter begins with an introduction on trends in research in teacher training and a review of the main lines of research in this field. In addition, some of the most widespread quantitative approaches applied to the area of education are illustrated with examples of recent research. Finally, the main designs in quantitative research in the educational context are described and also illustrated with examples.

Keywords: Quantitative researchs, Research tools, Methods, Research designs

Introduction

Starting from the basic premise that the research method must be subject to the research questions and the purpose of the study, it is possible to identify some of the main purposes for the application of the quantitative approach in research in teacher training. These include hypothesis testing, model development, theory testing or the generalization possibilities of the results in various populations. From the quantitative approach, research questions are posed that can and should be studied empirically on the basis of verified facts linked to relevant theories. In addition, methods are used that allow rigorous observation of these facts and that can be replicated by other researchers.

The research process begins with a question or series of questions that are usually posed with the intention of solving a research problem. In the recent history of teacher education research, research questions have typically been derived from practice. In this sense, one of the purposes of research on teacher training has been to identify effective interventions that could be generalized in various contexts (Phelan, 2011). Despite having received much criticism, the quantitative approach has been the dominant perspective in teacher education research for years (Popkewitz, Tabachnick, & Zeichner, 1979; Zeichner, 1999). Although it is true that both the themes and the methodological approaches have been modified over the years, at present, an important current of quantitative approach continues to persist in research on teacher training.

Although each approach has strengths and weaknesses, the key issue is to develop a research process that addresses the wide variety of possible questions about teacher education. As Zeichner (2009) points out, in general terms, a multidisciplinary and multi-methodological approach is necessary to respond to this diversity and complexity. However, the specific questions and problems require different research approaches and specific training in each of them. In this chapter, first, the main current lines of research in teacher training are identified that are approached from a quantitative approach, then the methodology and characteristics of the research designs applied in these works are analyzed.

Trends in teacher training research

Research on teacher training is "an emerging, complex and multifaceted field" (Cochran-Smith et al., 2016, p. 439), which has reached a priority place in educational research. In the past five years, since 2015, it has become a central topic in education research (Huang et al., 2020). Cochran-Smith and Villegas (2015) and Cochran-Smith, Villegas, Abrams et al. (2015) propose the presence of three main trends in research in teacher training, of a political-ideological, intellectual and demographic nature, which derive and converge in what they have called research programs in teacher training. In their review of more than 1500 empirical works, Cochran-Smith and Villegas (2015) identify three programs: (a) research on the responsibility, efficacy and policies of teacher preparation; (b) research on teacher preparation for the knowledge society; and (c) research on teacher training in equality and diversity; would make up the bulk of research in this area.

The works of the first research program are related to the preparation of teachers, the effectiveness and the impact of educational policies. Cochran-Smith and Villegas (2015) identify four groups of studies in this program: (a) policies regarding the professional certification of teachers; (b) institutional responses to policies and analysis of trends and discourses; (c) testing and evaluation of teacher or training program candidates and new evaluation tools;

and (d) program evaluation studies. Regarding the second research program, the studies are oriented to the analysis of the training of future teachers. In this program, Cochran-Smith, Villegas, Abrams et al. (2015) identify six subgroups: (a) teacher training in specific subjects; (b) the influence of work in the classroom for learning the profession; (c) the influence of practice on learning the profession; (d) studies that analyze the contents, structures, and pedagogical models of the various teacher training programs; (e) studies that analyze the characteristics and work of teacher educators; and (f) studies focused on the role of teacher education according "to the assumption that learning to teach is a process that begins before formal teacher preparation and continues beyond it" (Cochran-Smith & Villegas, 2015, p. 66). Regarding the last program, on teacher training in equality and diversity, Cochran-Smith et al. (2016) differentiate four groups of work with their respective lines of research: (a) studies that analyze the influence and opportunities offered by training and internships to future teachers to teach in a context of diversity; (b) strategies to attract and train a diverse teaching force; (c) analysis of the contents, structures and pedagogical models to prepare future teachers to develop their work in contexts of diversity; and (d) analysis of the learning and experience of teacher educators with diversity.

Considering this classification, although it is possible to find examples of the application of the quantitative approach in any of the programs and their subgroups, according to Cochran-Smith et al. (2016), this approach is applied more frequently in the subgroup oriented to the analysis of professional certification policies and especially in the subgroups of studies oriented to analyze the tests and evaluation of candidates for teachers or training programs and new tools for the evaluation of programs. Many of these studies address questions about the validity of tests for selecting teachers or evaluating teacher performance and rely on large-sample data sources and advanced analytical techniques. Therefore, we can place the methodology, designs and analysis techniques presented in this chapter within the groups and lines of research proposed by Cochran-Smith et al. (2016) in this program. Throughout the chapter, an attempt will be made to locate each of the examples in the subgroup and research line corresponding to classification, taking into account that sometimes the boundaries between groups can be fuzzy and may overlap. For example, Ruales, Agirdag and Van Petegem (2020) designed an instrument to evaluate the multicultural sensitivity of teachers in training, this work, therefore, would fit both the line of research related to the development and validation of tests and evaluation tools as in that related to the experience of future teachers in diverse environments.

Research process

As noted above, the method, techniques and tools for data collection and analysis must be at the service of the research question or questions, so it seems appropriate to first clarify the approach and purposes of the quantitative approach in research in teacher training. In this way, it is possible to identify different related research lines, demarcated by the research questions posed. These types of studies generally address questions about the validity and reliability of the tests for teacher selection (Klassen, Kim, Rushby, & Bardach, 2020), the evaluation of their performance (Gitomer, Martínez, Battey, & Hyland, 2019; Krammer, Pflanzl, & Mayr, 2019), or the correlation between the scores in the selection tests and the performance of the teachers, among others. The focus of interest is on the development, application and evaluation of assessment instruments for training programs and the personal and professional characteristics of future teachers (beliefs, attitudes, self-efficacy, skills) (Voet & De Wever, 2019). The questions posed in these works are, in general, related to the analysis of the psychometric properties of the measurement instruments (reliability, convergent, discriminant, predictive validity, factorial structure, etc.).

Specifically, in the field of research in historical education, of the 10 works identified by Epstein and Salinas (2018) in their review of the literature, two of them evaluate the beliefs of teachers regarding historical knowledge and the teaching strategies used (Fogo, 2014; Hicks, Doolittle, & Lee, 2004). Other works in this same area are oriented to the design, development and application of evaluation instruments of teacher training programs or to evaluate the impact of different teaching strategies in teacher training programs (Gómez-Carrasco, Monteagudo-Fernández, Moreno-Vera, & Sainz-Gómez, 2020). Miralles-Martínez, Gómez-Carrasco, Arias, and Fontal-Merillas (2019) analyze the relationship between the perception of the use of digital resources and the methodological and epistemological conceptions of teachers in training.

The approach and formulation of the research questions constitute the basis on which the research will be based. In general terms, from a quantitative approach, research questions tend to be transformed into hypotheses that guide the research work and its design. In this sense, the literature review is a crucial step that will make it possible to outline the hypotheses and locate the research within the field of study. Once the literature has been reviewed, it is convenient to identify the specific objectives of the research and the variables that will be analyzed. The purposes of this first phase of the process are to delimit the problem precisely and obtain sufficient information to make the appropriate decisions regarding the design. In addition, during this phase, both the objectives and the data collection strategy will be defined.

The selection of instruments for data collection is not an arbitrary question but depends again on the purpose, objectives and research questions. Among the main data collection techniques used from the quantitative approach in research in teacher training are questionnaires, objective tests and tests, scales, systematic observation or interviews.

Research methodology and designs

Design is understood as the structured plan that facilitates the chain of decision-making that must be carried out throughout the study, and which essentially concern the selection and assignment of participants to groups, control of extraneous variables, and the collection, management and analysis of data (Arnau, Anguera, & Gómez, 1990). The objectives of the design, therefore, are aimed at maximizing the rigor of the research, optimizing the resources and effort invested in the research, facilitating the replicability of the study and controlling the variance. The quality of the research design in the quantitative approach will be determined by its internal validity or capacity to control extraneous variables that could affect the results and its external validity or design capacity to generalize the results to other contexts.

Research designs

Research designs from a quantitative approach can be experimental, quasi-experimental, correlational or descriptive. Experimental designs are one of the most powerful research methods and allow establishing causal relationships between variables (Shadish, Cook, & Campbell, 2002). In experimental designs, participants are randomly assigned to the experimental (participating in the intervention) and control (not participating in the intervention) group. This allows comparison between groups to determine whether the intervention was effective because the differences in outcomes can be attributed to the treatment and not to other possible factors. It is therefore essential to rule out possible alternative explanations for the observed result. The distinguishing feature of experimental designs is the manipulation of the independent variable. Independent variables that are frequently manipulated in educational research include training programs, tasks, strategies, resources, or materials used in training. The dependent variables that are analyzed are usually related to performance, motivation, attitudes, beliefs or perception.

However, they are difficult to apply in educational and social contexts where participants are already part of groups (Berliner, 2002). As an alternative, the researcher can apply quasi-experimental designs. In recent years, there has been an increase in the application of quasi-experimental research designs in educational research (Gopalan, Rosinger, & Ahn, 2020). These research designs resemble the experimental ones to the extent that one group participates in the intervention program and another group does not.

Depending on the control of threats to internal validity, it is possible to distinguish between pre-experimental, experimental and quasi-experimental designs. Pre-experimental designs do not allow to establish causal relationships. The experimental designs are characterized by using randomization in the selection of the participants and in the assignment of these to the experimental and control groups. Quasi-experimental designs lack the feature of random selection and allocation of participants but employ other strategies to try to control for the effect of extraneous variables.

Pre-experimental designs

Pre-experimental designs provide little or no control for extraneous variables. Despite this, although these designs are not recommended, they are used with some frequency in research on teacher training. In these types of designs, participants undergo a pre-evaluation, are exposed to an intervention, and are evaluated again after the intervention. However, studies using this design do not provide sufficient evidence on the causal relationship between the dependent and independent variables and may be subject to a number of threats to internal validity such as history, maturation, test effects, and the regression effect to the mean (Marsden & Torgerson, 2012).

Pretest-posttest design with only one group

The single group pretest-posttest design usually takes place in three phases: (1) administering a pretest that measures the dependent variable; (2) participation in the intervention of the experimental group; and (3) evaluation of the dependent variable by applying a posttest. Differences are analyzed by comparing the pretest and posttest scores.

Two strange variables that are not controlled in this design are history and maturation. All the events that occur between the pretest and the posttest (not controlled by the researcher), regardless of the intervention, could affect learning. Without a control group to allow comparison, the results obtained using these types of single-group designs cannot be exclusively attributed to the intervention. However, this type of design can be useful for exploring hypotheses and suggesting further research.

Through this design, Lee, Tice, Collins, Brown, and Smith, (2012), investigated the change in the perception of their preparation of future teachers after participation in teaching practices. 130 teacher candidates participated who responded to a questionnaire before and after doing the practices. The results indicated a significant increase in the perception of their preparation in five categories: (a) knowledge of pedagogical content, (b) planning and preparation for teaching, (c) classroom management, (d) promotion of family participation, and (e) professionalism.

Similarly, Yazici and Yildirim (2017), investigate the perceived self-efficacy of history teachers in training and the effect of a pedagogical training program on this perception. Using a single-group pre-experimental pre-post-test design, they analyzed the perception of self-efficacy of 178 history teachers in training. The results indicated that the pedagogical training did not make a significant difference in the perceived self-efficacy of the history teachers in training, except in the classroom management subscale, in which the self-efficacy scores of the history teachers in training decreased after completion of pedagogical training.

Posttest design with non-equivalent control group

In the posttest only design with a non-equivalent control group, the group comparison is applied to two pre-established groups, of which only one is exposed to the intervention. Although this design uses two groups for comparison, the participants are not randomly assigned to the groups and no pretest measurements are taken. Both groups are assumed to be comparable before the start of the intervention. To assess the effect of the intervention, the scores of both groups are compared.

Hsieh, Lin, Liu, and Tsai (2019) apply this design to analyze the effect of peer training on the practice of science teachers and on the scientific competencies of their students. 132 students and four teachers participated and were assigned to the control and experimental groups. The results showed that the students in the experimental group had "more positive perceptions of their teachers' teaching practice than those in the comparison group" (Hsieh et al., 2019, p. 1). In addition, students in the experimental group classes outperformed their peers in the control group in their score in scientific competencies. The results indicate that the peer training model can improve the teaching practice of science teachers and the scientific competencies of their students.

Quasi-experimental designs

As noted above, in educational research, it is often not possible to select and assign participants to groups at random. In quasi-experimental designs, random assignment to treatment groups does not occur so participants are not randomly selected or assigned to groups. To compensate for this lack, some control techniques have been developed and proposed, such as measures before and after the application of the intervention or program, or the inclusion of multiple control groups. Two types of quasi-experimental designs are usually differentiated, the first one is characterized by the presence of a measure prior to the intervention (pre-test) and another after it (post-test) and the second by the presence of multiple measures. before and after the intervention.

Pretest-posttest design with non-equivalent control group

This is one of the most used designs, the experimental group receives the intervention or participates in a program while the control group does not. Both groups are evaluated with the same measuring instrument at two moments, before and after the intervention. By not using a rule of random assignment of participants to groups, it is not possible to know if the groups are equivalent before starting the investigation, but by means of the pretest it is possible to verify this equivalence. If no significant differences are found in the pretest, selection bias can be ruled out. If differences are found between groups, it is possible to apply statistical procedures to adjust the final results once these initial differences have been considered. Since the data are collected simultaneously in both groups, the potential problems of maturation, instrumentation, preliminary testing, history and regression should be controlled.

This design has been used by Saini and Abraham (2019) to evaluate the effect of a teacher training program based on Facebook on the performance and involvement of teachers in training. The results showed a significant improvement in the performance and levels of commitment of the teachers in training of the experimental group, in which the formative approach based on Facebook was applied, compared to those of the control group in which a conventional formative approach was applied. For their part, Huijgen, Holthuis, van Boxtel, van De Grift, and Suhre (2019), evaluate the effectiveness of a didactic unit related to the cold war on the capacity of students for historical contextualization (the ability to place facts in its historical context) also by means of a quasi-experimental design with a non-equivalent control group. In this case, they use an evaluation instrument built on the basis of open questions whose answers are coded by two independent observers. The results showed that the students in the experimental group gave fewer responses supported by presenteeism and used greater historical contextualization compared to the students in the control group.

The analysis of the data obtained through this type of design is complex since it requires adjusting for the lack of equivalence between the groups. The options used usually go through models of analysis of variance (ANOVA), or models of covariance (ANCOVA) in the scores obtained in the pretest. In the first case (ANOVA), it is convenient to carry out the procedure in two stages (Reichardt, 1979), first on the pretest scores and only if no significant differences are found, apply it to the posttest scores. In the second case (ANCOVA), baseline scores are included as a covariate to try to adjust for baseline differences between experimental and control groups. Saini and Abraham (2019) use ANCOVA to control the effect of knowledge and the involvement of teachers in training in the pretest, so that these scores were considered covariates in the data analysis. Subsequently, they use the multivariate analysis of covariance (MANCOVA), an extension of ANCOVA, which allows two or more dependent variables to be included in the same analysis.

Interrupted time series designs

Time series designs involve the periodic measurement of a variable or series of variables in a group and participation in an intervention program. So the dependent variable is measured chronologically on successive occasions, the intervention is introduced, and measurements of the variable are made again. The objective is to evaluate the effect of the intervention by observing the change or stability of the level and trend of the data. Several modalities of time series designs can be differentiated, the simple design of time series of a single group can be understood as a pre-experimental pretest-posttest design with only one improved group. The greatest weakness of this design is its inability to control the history; it cannot be ruled out that other events or circumstances other than the intervention may have influenced the observed change.

Another modality, which represents an improvement over the previous design, is the design of time series with a non-equivalent control group. In this case, a control group is incorporated that will be evaluated at the same time as the experimental group but is not subjected to the intervention. This design allows greater control over extraneous variables that may affect its internal validity than the simple single-group design. However, this type of design is used infrequently in teacher education research due to the large amount of time and data required.

Markowitz (2018) uses a time series design to estimate the association between the implementation of the No Child Left Behind (NCLB) regulation in the USA and the involvement of students in school. Their results reveal that some of the features of the implementation of NCLB may have had unintended consequences for student engagement with the school. The level and trend changes suggest that student engagement with the school increased slightly during the first years of implementation but decreased over time.

The analysis of the data obtained through these designs can be carried out both visually and statistically. Visual analysis requires the graphical representation of the data and its subsequent analysis regarding the possible change in the level and trend of the data, as well as the immediacy or delay in the appearance of the effect. The statistical analysis must take into consideration that the data obtained through this design are not independent. This violation of the independence assumption requires the use of statistical techniques that allow controlling autocorrelation, such as time series analysis. In the cases in which there is no autocorrelation, it is possible to apply the classical linear model. In those cases in which autocorrelation is present, it is possible to transform the scores in a first stage of the analysis to eliminate the autocorrelation and apply a regression analysis on the transformed scores.

Experimental designs

The essential component of experimental designs is that participants are randomly assigned to treatment groups. Randomization is a powerful technique for controlling threats to internal validity. The experimental designs can be classified according to various criteria: the number of independent variables, the degree of manipulation of the variables, the temporal sequence in the measurement of the variables, the level of control of the experimental situation, level of control of the threats to the internal validity, the representativeness of the variables or the representativeness of the sample (Nesselroade & Cattell, 1988). Considering the number of independent variables, it is possible to differentiate between single-variable or unifactorial designs and factorial designs. Factorial designs combine the effect of several independent variables of which at least one is manipulated.

Control group with pretest and posttest

It is the basic model, it includes a pre-test and a post-test, in which both the experimental group and the control group participate. The purpose of the pretest is to ensure that the two groups are comparable, while the posttest allows us to check the effect of the intervention. The control group allows researchers to verify whether the observed changes are a consequence of the treatment or whether other external variables may have affected the results. Strange variables related to time are minimized since both the experimental and control groups perform the tests simultaneously. Initial randomization ensures statistical equivalence between groups before the intervention.

Barr et al. (2015) apply this design to study the effect of a training program, aimed at active secondary education history teachers, with the purpose of improving the perception of self-efficacy, involvement and satisfaction of teachers on the one hand, and on the other. another, to develop skills for citizenship and for historical thinking in students. 84 schools participated that were randomly assigned to the control and experimental groups. Various measurements were made for each of the dependent variables studied before and after the intervention. The results revealed positive effects in the eight areas of the teachers' perception of self-efficacy and in two of the four professional involvement results. Likewise, the students in the experimental group outperformed the students in the control group in their perception of the classroom climate, tolerance for the rights of those who have beliefs that are contrary to their own, and the civic self-efficacy of the students. Regarding the historical understanding of the students, the participation of the teachers in the training program improved the historical thinking skills of the students. These results provide evidence that "an approach that integrates intellectual rigor, engaged academic discussion, and ethical reflection processes to learn history – and then connect that history with current social and civic concerns, commitments, and participation - is effective for foster the fundamental skills of historical thinking that students can apply to understand new content" (Barr et al., 2015, p. 35).

Whitford and Emerson (2019) use this design to analyze the efficacy of a brief intervention for the development of empathy with the aim of reducing unconscious bias and racial and cultural stereotypes of female teachers in training. The selection and assignment of the participants to the control and experimental groups was carried out randomly. All of them answered the implicit association test before and after the intervention. The results showed a significant reduction in unconscious bias in the experimental group compared to the control group after having participated in the intervention.

Repeated measures designs

In these designs, the participants of the experimental groups are evaluated in two or more experimental conditions. One of the advantages of repeated measures designs is that they require fewer participants, and also allow you to control for individual differences. Randomization allows controlling for threats to internal validity related to participant characteristics, maturation, and statistical regression.

As in the interrupted time series designs, in the analysis of data obtained through repeated measures designs it is necessary to take into account the violation of the assumption of independence of the observations. The three main approaches to analyzing this type of data are univariate, multivariate, and mixed. If the scores are normally distributed, the repeated measures analysis of variance models allow studying the effect of the factors when at least one of them is an within-subjects factor. However, when the sphericity assumption (equality of variances of differences between pairs of repeated measures) is violated, it is possible to apply the multivariate analysis of variance (O'Brien & Kaiser, 1985). Lastly, the mixed approach allows to reproduce the results of both the univariate and multivariate approaches and model the covariance structures, thus offering greater versatility and flexibility (Kowalchuk, Keselman, Algina, & Wolfinger, 2004).

Peebles and Mendaglio (2014) use this design to analyze the impact of a training course in inclusive education and a practical experience on the perception of self-efficacy of teachers in training to work in inclusive settings. One hundred forty one Canadian trainees participated and attended a mandatory ten-week inclusion course and a three-week post-course practical experience. The researchers collected data on the perception of self-efficacy of the pre-service teachers before participating in the course, after participating in the course, and after practical experience. The results of the analysis of variance of repeated measures indicated that both the course and the practical experience produced a significant increase in the perception of self-efficacy in the teachers in training.

Factorial designs When the intention is to analyze the effect of two or more independent variables that influence the dependent variable, we are faced with factorial designs. Each independent variable is analyzed at each of its levels. These designs can present different levels of complexity depending on the number of independent variables (factors) and the number of levels of each variable. The simplest factorial design (2×2), has two factors with two levels each, and therefore allows to define four groups. These designs can be more complex, for example, in a $3 \times 3 \times 4$ factorial design, there are three independent variables with three, three and four levels, respectively, resulting in 36 experimental groups. Participants are randomly assigned to groups that cover all possible combinations of levels of each independent variable, thus enabling the analysis of the effect of both each variable by itself (main effects) and its combination (interaction effects). However, when too many factors are analyzed simultaneously, statistical analysis becomes increasingly complex.

Wenz and Hoenig (2020) analyze discrimination based on ethnicity and social class by primary school teachers in Germany. To do this, they asked 237 randomly selected elementary school teachers to evaluate fourth-grade students' written tests. They applied a 2 × 2 × 3 factorial design, in which they take into account the quality of the text (high or low); sex (girl or boy); and the ethnic and socioeconomic origin (German upper class, German middle class, immigrant) inferred by the teachers based on the fictitious names assigned to the authors of the written test. Although they found no evidence of discrimination in the scores given to the jobs, the expectations of teachers about the future performance of children suggest a discriminatory bias based on ethnicity and social class. The effect is conditional on the quality of the test evaluated (interaction effects) and was only observed when the text evaluated was of high quality.

Conclusions

Research designs are indispensable in educational research in general and in teacher education research in particular because they help to plan the process and to control unwanted extraneous variables. As Zientek, Capraro and Capraro (2008) warn, some of the deficiencies detected in the research reports on teacher training with a quantitative approach are related to the lack of (a) effect sizes, (b) confidence intervals and (c) reliability and validity coefficients. The quantitative approach can contribute to improving the design of teacher training programs since, as Cochran-Smith and Zeichner (2009) have pointed out, "There are few quantitative studies examining the impact of methods course or field experiences on immediate practice and virtually none that examine impact on long practice over time or on students' learning once the novice teachers are in charge of their own classrooms " (p. 16).

Zeichner (2009) highlights eleven aspects related to the design and research methodology aimed at strengthening research in teacher training. Their recommendations include the need to describe in detail the data collection and analysis methods as well as the contexts in which the research is carried out. It also points out that greater attention must be paid to the validity and reliability of the data collection instruments used, since a large part of the research reports on teacher training do not provide this information. This lack of information affects the validity of the inferences and also makes it difficult to replicate the research work. Furthermore, it is necessary to relate empirical data to relevant theoretical frameworks that explain the research results and to develop more reliable data collection instruments.

This chapter has reviewed some of the main concepts related to the application of research designs that are used in the field of teacher training. By way of conclusion, it is advisable for researchers to use validated measurement instruments and robust designs whenever possible, however, as Zeichner (2009) points out, "given the complexity of research in teacher training, no single methodological approach can provide everything you need to understand how and why teacher education influences educational outcomes " (p. 743).

References

Arnau, Jaime/Anguera, Teresa/Gómez, Juana: *Metodología de la investigación en ciencias del comportamiento*. Murcia: Publicaciones de la Universidad de Murcia, 1990.

Barr, Dennis et al.: "A randomized controlled trial of professional development for interdisciplinary civic education: Impacts on humanities teachers and their students". *Teachers College Record*, 117(2), 2015, pp. 1–52.

Berliner, David Charles: "Educational research: The hardest science of all". *Educational Researcher*, 31(8), 2002, pp. 18–20. DOI: 10.3102/0013189X031008018.

Cochran-Smith, Marilyn/Villegas, Ana María: "Framing teacher preparation research: An overview of the field, Part 1". *Journal of Teacher Education*, 66(1), 2015, pp. 7–20. DOI: 10.1177/0022487114549072.

Cochran-Smith, Marilyn/Zeichner, Kenneth M.: *Studying teacher education: The report of the AERA panel on research and teacher education*. Lawrence Erlbaum: Mahwah, NJ, 2009.

Cochran-Smith, Marilyn et al.: "Critiquing teacher preparation research: An overview of the field, Part II". *Journal of Teacher Education*, 66(2), 2015, pp. 109–121. DOI: 10.1177/0022487114558268.

Cochran-Smith, Marilyn et al.: "Research on teacher preparation: Charting the landscape of a sprawling field". In: D. H. Gitomer/C. A. Bell (eds.): *Handbook of research on teaching (Fifth Edition)*. American Educational Research Association: Washington, DC, 2016, pp. 439–547.

Epstein, Terrie/Salinas, Cinthia: "Research methodologies in history education". In: Alan Metzger,/Lauren McArthur Harris (eds.): *The Wiley international handbook of history teaching and learning*. Wiley-Blackwell: Hoboken, NJ, 2018, pp. 61–91.

Fogo, Bradley: "Core practices for teaching history: The results of a Delphi panel survey". *Theory & Research in Social Education*, 42(2), 2014, pp. 151–196. DOI: 10.1080/00933104.2014.902781.

Gitomer, Drew/Martínez José Felipe/Battey, Dan/Hyland Nora E.: "Assessing the assessment: Evidence of reliability and validity in the edTPA". *American Educational Research Journal*, 2019. DOI: 10.3102/0002831219890608.

Gómez-Carrasco, Cosme J./Monteagudo-Fernández, José/Moreno-Vera, Juan Ramón/Sainz-Gómez, Marta: "Evaluation of a gamification and flipped-classroom program used in teacher training: Perception of learning and outcome". *PLoS ONE*, 15(7), 2020, e0236083, DOI: 10.1371/journal.pone.0236083.

Gopalan, Maithreyi/Rosinger, Kelly/Ahn, Jee Bin: "Use of quasi-experimental research designs in education research: Growth, promise, and challenges". *Review of Research in Education*, 44(1), 2020, pp. 218–243. DOI: 10.3102/0091732X20903302.

Hicks, David/Doolittle, Peter/Lee, John: "Social studies teachers' use of classroom based and web based historical primary sources". *Theory & Research in Social Education*, 32(2), 2004, pp. 213–247. DOI: 10.1080/00933104.2004.10473253.

Hsieh, Fu-Pei/Lin, Huann-Shyang/Liu, Shu-Chiu/Tsai, Chun-Yen: "Effect of peer coaching on teachers' practice and their students' scientific competencies". *Research in Science Education*, 70, 2019. DOI: 10.1007/s11165-019-9839-7.

Huang, Cui/Yang, Chao/Wang, Shutao/Wu, Wei/Su, Jun/Liang, Chuying: "Evolution of topics in education research: A systematic review using bibliometric analysis". *Educational Review*, 72(3), 2020, pp. 281–297. DOI: 10.1080/00131911.2019.1566212.

Huijgen, Tim/Holthuis, Paul/van Boxtel, Carla/van De Grift, Wim/Suhre, Cor: "Students' historical contextualization and the cold war". *British Journal of Educational Studies*, 67(4), 2019, pp. 439–468. DOI: 10.1080/00071005.2018.1518512.

Klassen, Robert M./Kim, Lisa E./Rushby, Jade V./Bardach, Lisa: "Can we improve how we screen applicants for initial teacher education?" *Teaching and Teacher Education*, 87, 2020, 102949. DOI: 10.1016/j.tate.2019.102949.

Kowalchuk, Rhonda/Keselman, Harvey/Algina, James/Wolfinger, Russell D: "The analysis of repeated measurements with mixed-model adjusted F tests". *Educational and Psychological Measurement*, 64(2), 2004, pp. 224–242. DOI: 10.1177/0013164403260196.

Krammer, Gerog/Pflanzl, Barbara/Mayr, Johannes: "Using students' feedback for teacher education: Measurement invariance across pre-service teacher-rated and student-rated aspects of quality of teaching". *Assessment and Evaluation in Higher Education*, 44(4), 2019, pp. 596–609. DOI: 10.1080/02602938.2018.1525338.

Lee, Joohi/Tice, Kathleen/Collins, Denise/Brown, Amber/Smith, Cleta/Fox, Jill: "Assessing student teaching experiences: Teacher candidates' perceptions of preparedness". *Educational Research Quarterly*, 36(2), 2012, pp. 3–19.

Markowitz, Anna: "Changes in school engagement as a function of No Child Left Behind: A comparative interrupted time series analysis". *American Educational Research Journal*, 55(4), 2018, pp. 721–760. DOI: 10.3102/0002831218755668.

Marsden, Emma/Torgerson, Carole J.: "Single group, pre- and post-test research designs: Some methodological concerns". *Oxford Review of Education*, 38, 2012, pp. 583–616. DOI: 10.1080/03054985.2012.731208.

Miralles-Martínez, Pedro/Gómez-Carrasco, Cosme J./Arias, Víctor B./Fontal-Merillas, Olaia: "Digital resources and didactic methodology in the initial training of history teachers". *Comunicar*, 27(61), 2019, pp. 41–51. DOI: 10.3916/C61-2019-04.

Nesselroade, John R./Cattell, Raymond B.: *Handbook of multivariate experimental psychology (Second edition)*. Plenum Press: New York, 1988.

O'Brien, Ralph G./Kaiser, Mary Kister: "MANOVA for analyzing repeated measurement design: An extensive primer". *Psychological Bulletin*, 92(2), 1985, pp. 316–333.

Peebles, Jodi L./Mendaglio, Sal: "The impact of direct experience on preservice teachers self-efficacy for teaching in inclusive classrooms". *International Journal of Inclusive Education*, 18(12), 2014, pp. 1321–1336. DOI: 10.1080/13603116.2014.899635.

Phelan, Anne M.: "Towards a complicated conversation: Teacher education and the curriculum turn". *Pedagogy, Culture and Society*, 19(2), 2011, pp. 207–220. DOI: 10.1080/14681366.2011.582257.

Popkewitz, Thomas S./Tabachnick, B. Robert/Zeichner, Kenneth M.: "Dulling the senses: research in teacher education". *Journal of Teacher Education*, *30*(5), 1979, pp. 52–60. DOI: 10.1177/002248717903000527.

Reichardt, Charles: "The statistical analysis from non-equivalent group designs". In: Thomas D. Cook/Donald T. Campbell (eds.): *Quasi-experimentation: Design and analysis issues for field settings*. Rand McNally: Chicago, IL, 1979, pp. 147–205.

Ruales, Shelanee Theresa P./Agirdag, Orhan/Van Petegem, Wim: "Development and validation of the multicultural sensitivity scale for pre-service teachers". *Multicultural Education Review.* Advance online publication, 2020. DOI: 10.1080/2005615X.2020.1808926.

Saini, Charu/Abraham, Jessy: "Implementing Facebook-based instructional approach in pre-service teacher education: An empirical investigation". *Computers and Education*, 128, 2019, pp. 243–255. DOI: 10.1016/j.compedu.2018.09.025.

Shadish, William R./Cook, Thomas D./Campbell, Donald T.: *Experimental and quasi-experimental designs for generalized causal inference.* Houghton Mifflin: Boston, MA, 2002.

Voet, Michiel/De Wever, Bram: "Teachers' adoption of inquiry-based learning activities: The importance of beliefs about education, the self, and the context". *Journal of Teacher Education*, 70(5), 2019, pp. 423–440. DOI: 10.1177/0022487117751399.

Wenz, Sebastian E./Hoenig, Kerstin: "Ethnic and social class discrimination in education: Experimental evidence from Germany". *Research in Social Stratification and Mobility*, 65, 2020, 100461. DOI: 10.1016/j.rssm.2019.100461.

Whitford, Denise K./Emerson, Andrea M.: "Empathy intervention to reduce implicit bias in pre-service teachers". *Psychological Reports*, 122(2), 2019, pp. 670–688. DOI: 10.1177/0033294118767435.

Yazici, Fatih/Yildirim, Tercam: "Effect of pedagogical formation program on pre-service history teachers' perceived self-efficacy". *European Journal of Educational Research*, 6(3), 2017, pp. 357–366. DOI: 10.12973/eu-jer.6.3.357.

Zeichner, Kenneth M.: "The new scholarship in teacher education". *Educational Researcher*, 28(9), 1999, pp. 4–15. DOI: 10.3102/0013189X028009004.

Zeichner, Kenneth M.: "A research agenda for teacher education". In: Marilyn Cochran-Smith/Kenneth M. Zeichner (eds.): *Studying teacher education: The report of the AERA panel on research and teacher education.* Lawrence Erlbaum: Mahwah, NJ, 2009, pp. 737–759.

Zientek, Linda/Capraro, Mary Margaret/Capraro, Robert M.: "Reporting practices in quantitative teacher education research: One look at the evidence cited in the AERA Panel Report". *Educational Researcher*, 37(4), 2008, pp. 208–216. DOI: 10.3102/0013189x08319762.

Jairo Rodríguez Medina and Bartolomé Rubia Avi

Chapter 2 Mixed methods in education research

Abstract: Over the last three decades, the mixed methods approach has become the third methodological movement in educational research. While quantitative data analysis techniques and procedures seem to occupy an ever more relevant position in educational research, researchers in educational sciences are increasingly moving towards accepting the complementarity of paradigms, with both conventional models being considered important and valuable. This may be due to the fact that research questions raised in the field of education obtain more precise answers through mixed methods, and their advantages outweigh those of quantitative and qualitative approaches when each is applied separately. This chapter presents the main research designs proposed from the mixed methods approach together with some recent examples illustrating their practical application in the field of education. In addition, the main research sources and evidences of validity are discussed. Ultimately, the claim is made that mixed methods are necessary to ensure the interpretative rigor of the data and the quality of the design. This is further illustrated with some examples.

Keywords: Mixed methods, Qualitizing, Quantitizing, Research design, Validity

Introduction

Mixed-method research requires the collection, analysis and integration of quantitative and qualitative data (Greene, Caracelli, & Graham, 1989) and has been defined as a research approach in which "(...) the researcher collects and analyzes data, integrates findings and draws inferences using qualitative and quantitative approaches or methods in a single study or research programme" (Tashakkori & Creswell, 2007, p. 4). This approach emerges as an alternative to the qualitative-quantitative confrontation and allows capturing the complexity of the teaching-learning process and the contexts in which it takes place. In the last 20 years there has been an important growth in research with mixed methods, so that this approach has come to be called the third methodological movement (Teddlie & Tashakkori, 2009), or the third research paradigm (Johnson & Onwuegbuzie, 2004). Even so, in the words of Maxwell (2015), "a substantial amount of current research combining the qualitative and the quantitative in terms of methods and data, in both the natural and social sciences, is ignored, despite its potential relevance to the field of mixed-method research" (p. 13).

Several recent systematic reviews coincide in pointing out the increase in scientific production with mixed methods that are being applied in almost all areas of research, but especially in the fields of health, education, psychology and management. In the last 20 years, a growing group of education researchers have adopted this approach, so that between 2000 and 2008 the number of studies using mixed methods in education research represented approximately 25 % of all publications on mixed methods (Teddlie & Tashakkori, 2009). The studies by López-Fernández and Molina-Azorín (2011) and Ross and Onwuegbuzie (2010) have analyzed the prevalence of publications featuring mixed methods in education journals, noting a gradual increase in recent years, especially in the *Journal of Mixed Methods Research*. Similarly, in historical education research, mixed methods are becoming increasingly relevant in a general context of growth in empirical research in this area (Epstein & Salinas, 2018) and, more specifically, in teacher training (De La Paz, Malkus, Monte Sano, & Montanaro, 2011; Marcus, Levine, & Grenier, 2012).

These results show that mixed methods are of significant value in solving problems in different educational contexts (Teddlie & Tashakkori, 2009). The main premise is that by combining and integrating quantitative and qualitative approaches, it is possible to achieve a better understanding and a deeper insight into research problems than by using either approach alone. As Greene (2007) points out, one approach alone will lead to a partial understanding of the phenomenon under investigation. Research with mixed methods, therefore, allows for a wider and more complete, as well as complementary, understanding of educational phenomena and can provide explanations that include different perspectives, increasing the usefulness of the results obtained. Furthermore, mixed methods do not simply combine types of data, but they encompass as well different ways of understanding the world, reality and access to knowledge, so as to achieve, as proposed by Greene (2007), the construction of a mental model of understanding the phenomenon being researched.

Ontological, epistemological and axiological points of view

The foundations of mixed methods come from multiple sources and cover different views of the world (worldview), reality (ontology), knowledge (epistemology) and values (axiology). These views are combined in different paradigms. Mertens (2015) suggests that paradigms are philosophical frameworks on ethics, reality, knowledge and research. Paradigms, therefore, represent the conceptual framework for understanding reality and define how we research social phenomena and how we validate that knowledge. For Kuhn (1962), a paradigm is defined as "the set of practices that define a scientific discipline in a particular period of time" (p. 175). A paradigm, therefore, represents the values and beliefs of a group.

Creswell and Plano Clark (2017) and Mertens (2015) agree on identifying four main paradigms, four models or ways of understanding reality: positivist and post-positivist; constructivist; transformative; and pragmatic. While positivism and postpositivism are associated with the quantitative approach, constructivism is usually associated with the qualitative approach. The transformative paradigm focuses on social justice and social transformation supported by qualitative research processes. Both the philosophical pragmatism advocated by William James, Herbert and Dewey, and the neo-pragmatism advanced in the 1960s by authors such as Kaplan and Cornel West (Mertens, 2010), are associated with the use of multiple data collection methods to solve research problems. Post-positivist and constructivist paradigms are often included in methodology manuals, while the more recent transformative paradigm has been gaining momentum in recent years (Creswell, 2009). The pragmatic paradigm provides a philosophical framework for advocates of mixed-method research (Teddlie & Tashakkori, 2009).

For more than half of the 20th century, the dominant paradigms in educational research were positivism and postpositivism. They argue that objectivity, based on statistical test results, should be the standard in research; therefore, the researcher should remain neutral to avoid biases that distort the results. Positivism borrows methods from the experimental sciences. It recognizes that in education research, and social science research in general, the assumptions required for the rigorous application of these methods are difficult to achieve, and so develops quasi-experimental methods.

One of the basic principles of constructivism is the idea that reality is socially constructed (Brunner, 1960; Vygostky, 1981). All knowledge, in this sense, is developed within a pre-existing social environment, which is constantly being interpreted and reinterpreted. One of the basic premises of the constructivist paradigm is that knowledge is socially constructed and that researchers should try to understand reality taking into account the context and the point of view of the participants in the research. At the same time, it states that the objects of analysis are also based on linguistic representations of the subjects' thought.

Qualitative methods such as interviews, observation, content analysis, or grounded theory are the most common in this paradigm. Research can only be conducted through interaction between the researcher and the participants

(Denzin & Lincoln, 2005; Lincoln & Guba, 2000). This approach is often described as hermeneutic and dialectical, since it is about "obtaining multiple perspectives that produce better interpretations of the meanings that are compared and contrasted through a dialectical exchange that involves the juxtaposition of conflicting ideas, thus forcing a reconsideration of previous positions'" (Mertens, 2015, p. 68). In addition, the perceptions of a various types of people must be sought.

The transformative paradigm, following from the positions of the Frankfurt School (Habermas, 1984), points out the need to put the focus of research on the field of social justice, so that these researchers position themselves, consciously and explicitly, alongside the less powerful, in an effort to achieve social transformation. The transformative paradigm is grounded on a wealth of academic research literature including mixed methods, qualitative, participatory action research, feminist research, critical ethnography, indigenous researchers, disability researchers and researchers from the international development community. Transformative researchers are plural and eclectic in their methodologies, tending to use qualitative methods such as interviews, observations and life stories, but emphasizing the need for greater rigor.

Finally, pragmatism provides a philosophical framework for mixed-method research (Tashakkori & Teddlie, 2010). Effectiveness is used as a criterion for judging the value of research. The criterion for judging the appropriateness of a method is whether it achieves its purpose. The method should be chosen on the basis of the purpose, questions and goals of the research. For the proponents of this paradigm consider that mixed methods offer a practical solution, or one that is closer to the overall reality regarding the quantitative vs. qualitative trade-off. Pragmatism provides flexibility for researchers to select the method, or combination of methods, that best suits their research questions (Mertens, 2015). From a pragmatic perspective, the research question is more important than the method and so is the worldview with respect to the method. These researchers rely on the criterion of what works, or what best represents the model of interpretation of that reality, so as to determine which method or methods to use in order to answer a particular research question.

The mixed methods research process

Mixed designs

Research design is defined as the plan devised so as to address the problem and attempt to answer the research questions, select the approach and perspective, identify the type of data, how they will be collected and who will participate, how the data will be analyzed, interpreted and reported, and the degree of confidence that can be placed in the validity and reliability of each element of the research. In short, it is the plan that facilitates the chain of decision-making that must be carried out throughout the empirical study, always subordinated to the definition of the latter's objectives and essentially concerned with the collection, processing and analysis of data.

When considering mixed designs and their typologies, there is a wide variety of proposals, but in essence the basic characteristics that define them is the sequencing or the concurrence of the quantitative and qualitative phases (Creswell, 2003, 2014; Hesse-Biber, 2010; Morse, 2010). Therefore, the designs are organized from these two perspectives of analysis, techniques or intentions in the research process. Usually the designs are in turn divided into sequential or concurrent, while their perspective regarding their emphasis on either the qualitative or the quantitative approach is the subject of discussion. Tab. 1 shows some of the main typologies, which rest on criteria like the designs' main focus and the interfaces across methods, techniques and data. They also address the main priorities that shape the orientation of the several types, as explained at the beginning of the chapter, as well as whether they start from a specific theoretical perspective or not.

Teddlie and Tashakkori (2009) suggest different research designs with mixed methods and propose the coexistence of five typologies: parallel or concurrent, sequential, conversion, multilevel and integrated. In turn, Creswell and Plano Clark (2019) place the designs on a continuum with two poles: fixed designs and emergent designs. Fixed designs would be those in which the use of qualitative and quantitative approaches has been pre-established and planned before starting the research process, while emerging designs would be characterized by the inclusion of a new approach (qualitative or quantitative) after the research process has started. In addition, they propose a framework with three designs that they consider basic: convergent, sequential explanatory and sequential exploratory.

The three basic design models featuring mixed methods are the concurrent or convergent parallel, the sequential explanatory, and the sequential exploratory. These three types of design differ from one other, either in terms of the concurrence or sequencing of methods (i.e., the difference between concurrent and sequential models), or as a result of priority given to the methods employed (the explanatory-sequential approach prioritizes the quantitative part of the study, while the exploratory-sequential approach prioritizes the qualitative part).

Tab. 1: Examples of classifications of mixed method designs

Authors	Starting point	Prioritized methods	Data use and integration
Creswell (2003)	Concurrent	At the same time.	Along the process of data collection
	Sequential-Qualitative first	Qualitative	During the analysis and with a theoretical basis
	Sequential-Quantitative first	Quantitative	Along the process of data interpretation and with a theoretical basis. Combined data integration and with a theoretical basis
Morse (2003, 2010)	Sensitivity	Accumulation of strategies	Dependent on the study's pertinence and depth
	Sequentiality	Perspective check	Comparing information between one method and the next
	Triangulation	Comparison and combination of information	Analyzing the information from different perspectives
	Complementarity	Enriching information	Integrating the information during the process of analysis of each step in the investigation
	Theoretical support	Theoretical orientation	Starting from a theoretical perspective that indicates the concepts researched beforehand

Parallel mixed designs, also called concurrent designs, are those in which the qualitative and quantitative approaches are implemented simultaneously but independently so that conclusions are based on the results of the analyzes of both types of data. In the notation proposed by Morse (2003) the parallel or concurrent design is denoted by the symbol [+]. Materechera (2018) uses a parallel design to study teachers' perceptions of educational inclusion by conducting in-depth interviews (qualitative instrument) and applying a questionnaire (quantitative instrument). Her study attempts to address the research questions by integrating the inferences derived from the analysis of the data obtained through both instruments. Using the same type of design, Yücel and Usluel (2016) analyze the participation and interaction of students in an online collaborative learning environment by recording and analyzing the logs (records of interaction on the online platform, quantitative) and carrying out an analysis of the content of the posts generated on the platform (qualitative).

As pointed out by Teddlie and Tashakkori (2009), although parallel designs have great potential for obtaining meta-inferences (conclusions that integrate qualitative and quantitative results), they are difficult to carry out due to the complexity of simultaneously performing multiple lines of research. In general, they often require the intervention of different teams of researchers, experienced in each of the approaches and in the several techniques of data analysis, as well as in examining the same phenomenon by using two different approaches in parallel. Parallel analysis and subsequent integration can be difficult, especially when the results differ.

Sequential mixed designs are those in which one or the other of the approaches (quantitative and qualitative or vice versa) are executed successively, depending on the initial or emerging approach of the research. Therefore, at least two phases of data collection are required so that the results of the first phase provide the basis for the data collection of the next phase. The final inferences are based on the results of both strands of the study. The second and subsequent phases of the study are developed in order to confirm or refute the inferences of the previous phases or to enlarge the previous results. Finally, the final conclusions are based on the results obtained throughout the various phases. In the notation proposed by Morse (2003) the sequential design is denoted by the symbol [→].

The character of this type of design and its purpose can be either explanatory or exploratory. In the first case, the process begins with a phase of quantitative data collection and analysis followed by a phase of qualitative data collection and analysis, with the intention of explaining or deepening the results of the previous phase. In the second case, the process starts with a qualitative data collection and analysis phase to continue with a quantitative data collection and analysis phase.

Creswell and Plano Clark (2019) consider the sequential explanatory design to be the simplest to implement and highlight some of its advantages. It is an easy design to implement because the researcher carries out the two phases in succession and on each occasion collects only one type of data so that one single researcher can carry it out successfully. In addition, the final report can be drawn up so as to include a quantitative section followed by a qualitative one, which ultimately facilitates the reporting process. However, despite its greater simplicity, this type of design generally requires more time to complete all the phases than concurrent or parallel designs.

Zumbrunn, McKim, Buhs and Hawley (2014) set up an explanatory-sequential design in order to examine how the perception of belonging, academic motivation and engagement can mediate the relationship between contextual academic characteristics and performance. In the first phase of the (quantitative) study, two structural equation models are contrasted in relation to the perception of belonging and motivation. After contrasting the goodness of fit of both models, in a second phase (qualitative), they conduct interviews with a sub-sample of students to try to understand how they describe their perception of belonging in relation to classroom experiences.

Lacka and Wong (2019) examine the impact of virtual learning environments and social networks on the academic outcomes of university students through an exploratory sequential design. To do this, they first interview 17 higher education students (qualitative phase) with the aim of identifying categories relating to learning outcomes. In the second phase (quantitative), they design a questionnaire so as to analyze the use of these technologies by university students and their impact on these outcomes. The findings of the qualitative phase informed the quantitative research stage, during which a descriptive analysis was carried out to reveal the consequences of the use of digital technologies on these outcome categories.

Bernhard (2019) applies this design to analyze the epistemological position of Austrian history teachers and their views on history teaching itself. In a first qualitative phase he collects data from 26 Austrian schools through observation of the classes taught and interviews with 50 teachers. The content analysis of both the observations and the interviews allows him to construct a questionnaire that was used in the second phase of his study, which had a quantitative nature. The qualitative phase of the study helped to identify the main problems and define the hypotheses that were tested in the quantitative phase of the study.

Mixed conversion designs are those in which data are transformed: either from qualitative to quantitative ("quantitizing") or vice versa ("qualitizing"). In these designs, data are collected by using one method and then transformed and analyzed using the other method. In the first case ("quantitizing"), qualitative data are generally converted into quantitative data by counting the occurrences of each of the categories under examination, i.e. the frequencies. A special case of this type of design is that of observational designs, in which the categorical data collected through systematic observation are transformed into quantitative data taking as a reference the order and duration parameters (Anguera, Blanco-Villaseñor, Losada, & Sánchez-Algarra, 2020). This transformation, therefore, results in a more robust type of quantification, which allows sequential analysis of delays, analysis of polar coordinates or detection of time patterns (T-Patterns) that reveal the structure underlying the recorded data.

Escolano-Pérez et al. (2019) employ an observational conversion design to assess the metacognitive skills employed by 5-year-olds in solving a puzzle. To do this, they designed an observational instrument which they used to code children's behavior during the resolution of the proposed task. The transformation of categorical data, obtained through observation and recording, into quantitative data ("quantitizing") starting from the parameters of order and duration of the behaviors observed led to the detection of temporal patterns, sequential analysis of delays and analysis of polar coordinates. The results revealed differences in metacognitive skills between children who solved the task and those who did not.

Multilevel mixed designs are those in which different types of data (both quantitative and qualitative) are integrated and/or used at different levels of the research (groups, schools, districts, regions). Numerical data can be used at one level (students) and qualitative data can be used at another level (groups). Both types of data can be collected both in parallel and sequentially. Despite the possibilities they offer in the field of education research, as pointed out by Headley and Plano Clark (2020), there is not yet a clear consensus on the implications of this type of design.

Papadimitriou (2010) analyzes quality management in Greek universities at the macro, meso and micro levels. At the macro level he analyzes how the print media addressed the issue quality management in higher education during a particular period. He does so by analyzing the quantitative and qualitative content of publications in the print media during 2005. At the meso level, taking into account the inferences of the macro level, his study carries out a qualitative content analysis of the reports by eight external evaluators and uses a questionnaire to understand how Greek university rectors and vice-rectors had implemented quality management in their universities. At the micro level, the study conducts a survey on the perceptions and concerns of department heads about quality management. Finally, he integrates the results from each level to identify the factors influencing the adoption of quality management in Greek universities.

Fully integrated mixed designs are those in which there is a constant integration of qualitative and quantitative methods at each stage of the process. Therefore, qualitative and quantitative strategies are combined continually: planning and design, data collection, sampling, analysis and inference drawing (Creamer, 2018). This integration across the several stages of the inquiry is the distinctive feature of this type of design.

Love and Wells (2018) use this design to analyze the extent of the relationship between technology and the science education experiences of technology and engineering teachers as well as their teaching strategies. This research began with a content analysis of the curricula that identified behaviors that would subsequently be observed in the classroom. This was followed by a questionnaire to which 55 teachers responded. Respondents were classified according to their answers into three categories: novice, intermediate and veteran. The results of the analysis of the questionnaire data guided the selection of eight teachers to carry out the classroom observation in the next phase. Qualitative records were extracted from the observation and quantified, and the correlations between these records and the other instruments were analyzed. Finally, the teachers who had been observed were interviewed to allow the researchers to ask specific questions about teaching strategies with respect to the observed scientific content. The interview responses helped to corroborate and validate the teaching strategies that had been observed.

Data analysis

Mixed analysis involves the application of qualitative and quantitative data analysis techniques that can be combined, connected, or integrated into an investigation. This combination, connection or integration of data and techniques is probably one of the most complex phases of mixed-method research (Onwuegbuzie & Combs, 2011). Teddlie and Tashakkori's (2009) proposal makes it possible to organize mixed analysis strategies in relation to basic research designs. Thus, according to this typology, parallel, sequential, conversion, multilevel and fully integrated analyses can be differentiated.

The first consideration to be made is the justification and purpose of mixed analysis. In this sense, Greene et al. (1989) propose five possible purposes: triangulation, complementarity, development, initiation and expansion. Triangulation refers to the comparison of various data sources, analysis procedures, research methods, or inferences. It requires that data sets (qualitative and quantitative) evaluate the same phenomenon; therefore they must be collected simultaneously and independently. In case the purpose of the study with mixed methods is complementarity, the analysis should be targeted at contrasting the degree to which both data sources (qualitative and quantitative) produce complementary results. Complementarity is usually assessed through parallel mixed designs and analyses. In this case, the results of one type of analysis are interpreted in order to improve, extend, illustrate or clarify the results obtained through the analysis of the other type of data (Greene et al., 1989). Sometimes, analysis of data obtained through one of the approaches (quantitative or qualitative) can contribute to developing the research process through the complementary approach. For example, grouping on the basis of qualitative observations and then comparing these groups with quantitative techniques, or conversely grouping on the basis of the results of quantitative data analysis and then deepening the analysis of representatives of these groups through interviews, observation or focus groups (Bae & Lai, 2020). Vangrieken and Kyndt (2020) selected a group of 17 teachers representing extreme cases of individualistic and collaborative attitudes to conduct in-depth interviews based on their responses to several questionnaires to which 3018 participants initially responded.

Initiation involves looking for contradictions or contrasts between the results of the analysis of both types of data. It involves assessing the extent to which the results diverge, which makes it possible to highlight the paradoxes and contradictions that arise when the results are compared. Expansion aims to broaden the scope of the research by selecting the most appropriate analysis techniques for each of the research components.

A second consideration before tackling the mixed analysis concerns the number of data types to be analyzed. Traditionally, data analysis in mixed-method research has consisted in analyzing quantitative data with quantitative techniques and qualitative data with qualitative techniques (Creswell & Plano Clark, 2017). Fetters and Freshwater (2015) suggest that the combination of quantitative and qualitative elements affords more than the sum of the individual components (1 + 1 = 3), however, other authors have opted for a deeper integration, trying to eradicate the qualitative vs. quantitative dichotomy, and have proposed the formula "1 + 1 = 1" (Onwuegbuzie, Hitchcock, Natesanc, & Newman, 2018) to represent this integration at the level of data transformation. The process, in this case, would involve the transformation of qualitative data into quantitative data ("quantitizing") (Escolano-Pérez et al., 2019) or of quantitative data into qualitative data ("qualitizing") (Taylor & Tashakkori, 1997). This type of data transformation, typical of mixed methods, implies a complete integration of both methodological approaches.

A third consideration prior to data analysis concerns the priority or preponderance of qualitative or quantitative components. If the quantitative component is given greater priority, the analysis is essentially quantitative-dominant. Conversely, if the qualitative analysis component is given higher priority, then the analysis is essentially a mixed qualitative-dominant analysis. The time sequence of the mixed analysis should be another aspect considered before conducting the analysis itself, as it can be carried out in a chronological sequence or in a parallel or concurrent manner. If the two sets of data are analyzed independently, it will be a parallel mixed analysis. Conversely, if there is a defined order for the analyses, it will be a sequential mixed analysis. Finally, it is necessary to consider the phases of analysis to be carried out. Mixed analyses may involve several phases, which will need to be identified and delimited beforehand. Onwuegbuzie, and Combs, (2010) and Greene (2007) suggest some of these possible analytical phases: data transformation, reduction, comparison, correlation, graphical representation of data, integration, etc.

After considering the above questions, the researcher is ready to start the analysis of the data. These preliminary considerations should lead to a series of decisions that will allow the selection of the most appropriate analysis strategy or strategies to answer the research questions posed. The following are the main possibilities for analysis according to the research design adopted.

Parallel mixed analysis

In the view of Teddlie and Tashakkori (2009), parallel mixed analysis is probably the most widely used data analysis strategy in mixed-method research. It is generally associated with purposes such as triangulation and complementarity. It involves carrying out both a qualitative and a quantitative analysis of the data, using the strategies, techniques and tools specific to each approach. The results obtained will be linked, combined or integrated in meta-inferences in the final phase. These analyses can lead to convergent or divergent results, but they facilitate a common interpretation.

Mixed conversion analysis

Mixed data conversion analysis occurs either when the qualitative data collected are quantified ("quantitizing") or when the quantitative data are converted to narrative or other types of qualitative data ("qualitizing"). In conversion analysis there is only one source of data. Quantification allows statistical analyzes to be applied to data sets that were originally categorical in nature. Usually, qualitative categorical data are transformed into frequencies. As mentioned above, special mention should be made of the observational methodology that transforms qualitative data into quantitative data, based not only on the frequency parameter but also on the order and duration parameters, which makes it possible to carry out a wide variety of analyses (Anguera et al., 2020).

Sequential mixed analysis

Sequential mixed analysis occurs when the analysis of qualitative and quantitative data is done in chronological order. As well as differentiating between exploratory and explanatory-sequential designs, it is possible to establish these same differences in the analysis. A third type of mixed-sequential analysis is iterative mixed-sequential analysis, which occurs in sequential studies involving more than two phases.

The question of validity in mixed designs

As Burke Johnson and Onwuegbuzie (2006) point out, due to the complexity arising from the combination of qualitative and quantitative approaches, mixed research gives rise to what has been called the problem of integration. A better and deeper understanding of the phenomenon under examination is only possible if the results of both approaches are linked or integrated through inferences and meta-inferences. These inferences are conclusions and interpretations made on the basis of the analysis of data obtained in one study (Teddlie & Tashakkori, 2009). They use the term inference to refer to three related concepts: (a) the process of inference, which begins with data collection; (b) the quality of inferences, which encompasses the quantitative concepts of internal validity and statistical conclusion validity and the qualitative concepts of credibility and reliability; and (c) the transferability of inferences, which includes the concepts of generalizability and external validity.

To assess the appropriateness and rigor of inferences, Teddlie and Tashakkori (2009) develop a set of criteria relating to design quality and interpretive rigor. With regard to design quality they propose four criteria: (a) adequacy of design: does the design fit the questions and purpose of the research? (b) Fidelity of the design: was each phase of the design correctly implemented? (c) Internal consistency of the design: are the design phases consistent with the problem, the research questions posed and the sampling? (d) Adequacy of analysis: are the data analysis techniques employed appropriate and adequate to answer the research questions? In terms of interpretive rigor, on the other hand, they pose six criteria: (a) Interpretive consistency: are the conclusions consistent with the results obtained from the data analysis? (b) Theoretical consistency: are the inferences consistent with theoretical approaches and with the empirical results of other work? (c) Interpretive agreement: do other experts support the interpretation of the results (e.g. through peer review)? (d) Interpretive distinction: is it possible to exclude other alternative conclusions that explain the results? (e) Integrative effectiveness: are the inferences drawn from each approach effectively and uniformly integrated into consistent meta-interferences? (f) Interpretative correspondence: do the meta-inferences respond to the initial design approach and purpose?

Another aspect identified by Greene (2007) to support the validity of mixed methods in terms of their quality focuses on the concept of "legitimacy" (Onwuegbuzie, 2003). Thus, attending to the nine types of legitimacy proposed by Onwuegbuzie and Johnson (2006) we can identify aspects that serve as examples to understand the concept and the process to be followed in research with mixed methods:

1. Integration of the sample, where the sample designs (quantitative and qualitative) produce meta-inferential analyses. Thus, if the selection of the sample or the space where the research is to be carried out is really exemplary, we can have sufficient legitimacy for a study to be valid.
2. Inclusion of internal and external perspectives in a study, where the researcher presents the internal perspective of what is observed and relates it to the perspective from external observation, in order to construct descriptions supported by the dialogue between perspectives.
3. Compensation of the methods in terms of the weaknesses of one towards the other. When we use different methods, they have strengths and weaknesses. Therefore, the weaknesses of each method must be compensated by the other in order to achieve a solid global vision.
4. Inferences extracted from the evolution of the research design itself. In many cases this occurs when an erroneous sequencing occurs and has to be changed. Therefore, modification is the best example of legitimacy.
5. The conversion or transformation of data that helps to quantify the qualitative or the reverse (Anguera et al., 2020).
6. The use of research assumptions from different paradigmatic perspectives. The mixture of qualitative and quantitative perspectives from ontology, axiology, methodology and research practice can be used to verify, from different perspectives, the value of research.
7. The commensurability or comparison process, which seeks the existence of a common factor that can be expressed in verifiable terms through the replication of qualitative and quantitative measures/indicators.
8. The methodological soundness of each method used. The legitimacy of the quantitative and qualitative components of the study resulting from the use of mixed quantitative, qualitative and validity types of data, which produce high-quality meta-inferences, is likewise addressed.
9. The political connection within the mix of methods in a given context, a question that can be checked when transferring into the context what has been extracted from it as well as the relationships that have been established with the types of quantitative and qualitative data.

We can say that this construction of legitimacy-related aspects entails a form of mixing and combining data arrived at by using several epistemologies. All in all, it involves a complementary process of comparison, compensation and double perspectives when it comes to analyzing the value of a study.

Mixed methods enable us to capture the complexity inherent in the teaching-learning processes by incorporating and combining the quantitative and qualitative approaches. Fetters and Moliza-Azorín (2020) defend the need to apply mixed methods in the field of evaluation of educational intervention programmes. However, the successful implementation of mixed designs in education requires knowledge and mastery of quantitative and qualitative research designs, techniques and practices, which is usually time-consuming, and the coordinated and collaborative work of teams of researchers (Ponce & Pagán-Maldonado, 2015).

References

Anguera, Teresa/Blanco-Villaseñor, Ángel/Losada José Luis/Sánchez-Algarra, Pedro: "Integration of qualitative and quantitative elements in observational methodology". *Ámbitos. Revista Internacional de Comunicación*, 49, 2020, pp. 49–70. DOI: 10.12795/Ambitos.2020.i49.04.

Bae, Christine Lee/Lai, Mark H. C.: "Opportunities to participate in science learning and student engagement: A mixed methods approach to examining person and context factors". *Journal of Educational Psychology*, 112(6), 2020, pp. 1128–1153. DOI: 10.1037/edu0000410.

Bernhard, Roland: "Using mixed methods to capture complexity in an empirical project about teachers' beliefs and history education in Austria". *History Education Research Journal*, 16(1), 2019, pp. 63–73. DOI: 10.18546/herj.16.1.06.

Bruner, Jereome S.: *The process of education*. Harvard University Press: Cambridge, 1960.

Creamer, Elizabeth G.: *An introduction to fully integrated mixed methods research*. Sage Publications: Thousand Oaks, CA, 2018.

Creswell, John W.: *Research design: Qualitative, quantitative, and mixed methods approaches* (2nd ed.). Sage: Thousand Oaks, CA, 2003.

Creswell, John W.: *Research design: Qualitative, quantitative, and mixed methods approaches* (3rd ed.). Sage: Thousand Oaks, CA, 2009.

Creswell, John. W./Plano Clark, Vicki L.: *Designing and conducting mixed methods research*. Sage: Los Angeles, CA, 2017.

De La Paz, Susan/Malkus, Nathaniel/Monte Sano, Chauncey/Montanaro, Elizabeth: "Evaluating American history teachers' professional development: Effects on student learning". *Theory & Research in Social Education*, 39(4), 2011, pp. 494–540. DOI: 10.1080/00933104.2011.10473465.

Denzin, Norman K./Lincoln, Yvonna S.: *The SAGE handbook of qualitative research*. Sage: New York, 2005.

Epstein, Terrie/Salinas, Cinthia: "Research methodologies in history education". In: Alan Metzger/Lauren McArthur Harris (eds.): *The Wiley international handbook of history teaching and learning*. Wiley-Blackwell: Hoboken, NJ, 2018, pp. 61–91.

Escolano-Pérez, Elena/Herrero-Nivela, María Luisa/Anguera, Teresa: "Preschool metacognitive skill assessment in order to promote educational sensitive response from mixed-methods approach: Complementarity of data analysis". *Frontiers in Psychology*, 10, 2019, 1298. DOI: 10.3389/fpsyg.2019.01298.

Greene, Jennifer C.: *Mixed methods in social inquiry*. Jossey-Bass: San Francisco, CA, 2007.

Greene, Jennifer C./Caracelli, Valerie J./Graham, Wendy F.: "Toward a conceptual framework for mixed-method evaluation designs". *Educational Evaluation and Policy Analysis*, 11, 1989, pp. 255–274. DOI: 10.2307/1163620.

Habermas, Jürgen: *The theory of communicative action, Vol. 1: Reason and the rationalization of society*. Beacon Press: Boston, MA, 1984.

Headley, Marcia Gail/Plano Clark, Vicki L.: "Multilevel mixed methods research designs: Advancing a refined definition". *Journal of Mixed Methods Research*, 14(2), 2020, pp. 145–163. DOI: 10.1177/1558689819844417.

Hesse-Biber, Sharlene: "Qualitative approaches to mixed methods practice". *Qualitative Inquiry*, 16(6), 2010, pp. 455–468. DOI: 10.1177/1077800410364611.

Lincoln, Yvonna S./Guba, Egon G.: "Paradigmatic controversies, contradictions, and emerging confluences". In: Norman K. Denzin/Yvonna S. Lincoln (eds.): *Handbook of qualitative research* (2nd ed.). Sage: Thousand Oaks, CA, 2000, pp. 163–188.

Lopez-Fernandez, Olazt/Molina-Azorín, José F.: "The use of mixed methods research in interdisciplinary educational journals". *International Journal of Multiple Research Approaches*, 5(2), 2011, pp. 269–283. DOI: 10.5172/mra.2011.5.2.269.

Love, Tyler S./Wells, John G.: "Examining correlations between preparation experiences of US technology and engineering educators and their teaching of science content and practices". *International Journal of Technology and Design Education*, 28(2), 2018, pp. 395–416. DOI: 10.1007/s10798-017-9395-2.

Marcus, Alan S./Levine, Thomas H./Grenier, Robin S.: "How secondary history teachers use and think about museums: Current practices and untapped promise for promoting historical understanding". *Theory & Research in Social Education*, 40(1), 2012, pp. 66–97. DOI: 10.1080/00933104.2012.649466.

Materechera, Ellen K.: "Inclusive education: Why it poses a dilemma to some teachers". *International Journal of Inclusive Education*, 24(7), 2018, pp. 771–786. DOI: 10.1080/13603116.2018.1492640.

Mertens, Donna. M.: *Transformative research and evaluation.* Guilford: New York, 2009.

Mertens, Donna M.: *Research and evaluation in education and psychology: Integrating diversity with quantitative, qualitative, and mixed methods.* (3rd ed.) Sage: Thousand Oaks, CA, 2010.

Mertens, Donna M.: *Research and evaluation in education and psychology: Integrating diversity with quantitative, qualitative, and mixed methods.* (4ª ed.) Sage: Thousand Oaks, CA, 2015.

Morse, Janice M.: "Principles of mixed methods and multimethod research design". In: Abbas Tashakkori/Charles Teddlie (eds.): *Handbook of mixed methods in social and behavioral research* (8). Sage: Thousand Oaks, CA, 2003, pp. 189–120.

Onwuegbuzie, Anthony J./Burke Johnson, Robert: "The validity issue in mixed research". *Research in the Schools*, 13(1), 2006, pp. 48–63. DOI: 10.1016/j.jsis.2005.08.002.

Onwuegbuzie, Anthony J./Combs, Julie P.: "Emergent data analysis techniques in mixed methods research: A synthesis". In: Abbas Tashakkori/Charles Teddlie (eds.): *Handbook of mixed methods in social and behavioral research* (2nd ed.). Sage: Thousand Oaks, CA, 2010, pp. 397–430.

Onwuegbuzie, Anthony J./Combs, Julie P.: "Data analysis in mixed research: A Primer". *International Journal of Education*, 3(1), 2011, E13. DOI: 10.5296/ije.v3i1.618.

Onwuegbuzie, Anthony J./Hitchcock, John/Natesan, Prathiba/Newman, Isadore: "Using fully integrated Bayesian thinking to address the 1 + 1 = 1 integration challenge". *International Journal of Multiple Research Approaches*, 10(1), 2018, pp. 666–678. DOI: 10.29034/ijmra.v10n1a43.

Papadimitriou, Antigoni: "Looking for clues about quality: A multilevel mixed design on quality management in Greek universities". *Electronic Journal of Business Research Methods*, 8, 2010, pp. 85–94.

Ponce, Omar A./Pagán-Maldonado, Nellie: "Mixed methods research in education: Capturing the complexity of the profession". *International Journal of Educational Excellence*, 1(1), 2015, pp. 111–135. DOI: 10.18562/ijee.2015.0005.

Tashakkori, Abbas/Creswell, J. W: "The new era of mixed methods." *Journal of Mixed Methods Research*, 1(1), 2007, pp. 3–7. DOI: 10.1177/2345678906293042.

Tashakkori, Abbas/Teddlie, Charles: *Mixed Methods in Social and Behavioral Research* (2nd ed.). Sage: Thousand Oaks, CA, 2009.

Tashakkori, Abbas/Teddlie, Charles: *Handbook of mixed methods in social and behavioral research* (2nd ed.). Sage: Thousand Oaks, CA, 2010.

Taylor, Dianne/Tashakkori, Abbas: "Toward an understanding of teachers' desire for participation in decision making". *Journal of School Leadership*, 7(6), 1997, pp. 609–628. DOI: 10.1177/105268469700700604.

Teddlie, Charles/Tashakkori, Abbas: *Foundations of mixed methods research*. Sage: Thousand Oaks, CA, 2009.

Vangrieken, Katrien/Kyndt, Eva: "The teacher as an island? A mixed method study on the relationship between autonomy and collaboration". *European Journal of Psychology of Education*, 35, 2020, pp. 177–204. DOI: 10.1007/s10212-019-00420-0.

Vygotsky, Lev S: *Thought and language*. The MIT Press: Cambridge, 1986.

Yücel, Ümmühan A. I./Usluel, Yasemin K.: "Knowledge building and the quantity, content and quality of the interaction and participation of students in an online collaborative learning environment". *Computers and Education*, 97, 2016, pp. 31–48. DOI: 10.1016/j.compedu.2016.02.015.

Zumbrunn, Sharon/McKim, Courtney/Buhs, Eric/Hawley, Leslie R.: "Support, belonging, motivation, and engagement in the college classroom: A mixed method study". *Instructional Science*, 42(5), 2014, pp. 661–684. DOI: 10.1007/s11251-014-9310-0.

Stéphane Lévesque and Jean-Philippe Croteau

Chapter 3 Teaching the narrative competence in history education: Lessons from Canada[1]

Abstract: Narration is considered a structural feature of historical thinking and learning. However the conceptualization of narrative competence remains vague in history didactics literature. Relying on an empirical study with Canadian students, this chapter discusses how history education needs to prepare teachers to be more conscious of and proactive in providing learners with the ability to construct and deconstruct historical narratives.

Keywords: Narration, Narrative competence, Historical consciousness, Storytelling

Setting the stage

In our most recent book, *Beyond History for Historical Consciousness*, we claim that theories of historical consciousness are central to history education because they focus on the role that narration plays in contemporary peoples' attempts to define and orient themselves in time (Reich et al., 2015; Rüsen, 2005; Seixas, 2004). Leading historian Jörn Rüsen (2005) contends that the fundamental form within which historical consciousness realizes its practical function of life orientation is that of the narrative. The purpose of narrative is to transform the past into meaningful history (or histories) by giving a direction, sense and coherence to otherwise disorganized past events. As he putsit, "for more than four decades narrativity has gradually become the most convincing answer to the question for the distinctive nature of history. Historians tell stories, and story-telling follows the rules of narration" (Rüsen, 2005, p. 3). History is thus a narrative act of (re)construction of the past. It uses cognitive means of interpreting past actualities by connecting the dimensions of time (past, present, envisioned future) into coherent, meaning-making stories. Far

1 This chapter is based on arguments presented more extensively in our most recent book *Beyond History for Historical Consciousness: Students, Narrative, and Memory.* Toronto: University of Toronto Press, 2020. This paper si result of research project granted by Seneca Foundation. Regional Agency for Science and Technology, grant number 20638/JLI/18.

from being an oversimplification of realities, narrative would be a solution to how we translate knowing into telling. From this perspective, our ability to make sense of the past would inevitably be connected to narrative acts, or what Jürgen Straub calls "narrative competence" (Straub, 2005, p. 54).

In recent years, history educators in the United States (U.S.), Germany, Sweden, France, and other European nations have explored the importance of narrative competence, looking at learners' ability to develop historical narratives and their ability to critically analyse the deep structure of narratives (Calder, 2013; Eliasson et al., 2015; Grever & Adriaansen, 2019; Kölbl & Konrad, 2015; Lantheaume & Létourneau, 2016; Reich et al., 2015; Waldis et al., 2015). Canadian educators are only starting to investigate these aspects of historical consciousness. Until recently, the focus was instead placed on questions of national belonging, citizenship skills, and grand narratives of the nation (Granatstein, 1998; Osborne, 2000; Sandwell, 2012). But things are now gradually changing, at least in our scholarly understanding of students' historical thinking (Duquette, 2015; Gibson, 2017; Létourneau, 2014; Lévesque & Croteau, 2020; Lévesque et al., 2013; Zanazanian, 2015). As leading Canadian educator Peter Seixas contends, the [Canadian] model of historical thinking was developed – pragmatically, like the American and British – in order to be communicable and intelligible to teachers and their students, and yet rich enough (like the German) to lead them into explorations of fundamental epistemological and ontological problems of history. (Seixas, 2015, p. 5)

The Canadian historical thinking model has resulted in a variety of standards and practices among Canadian faculties of education and provincial school systems that specifically address students' historical learning and assessment using key concepts: evidence, significance, continuity and change, causation, perspective-taking, and ethical judgment. But this model has also been criticized for paying too little attention to the interpretative, deconstructionist nature of history, as well as for ignoring narration as an essential meta-historical concept – and ultimately as a competence. As Seixas himself recognizes, the Canadian historical thinking model "has omitted discussion on how these historical thinking concepts... can contribute to the construction and critique of narrative interpretations" (Seixas, 2017, p. 262).

This state of affairs may seem paradoxical in today's context of Canadian school reforms that aim to place greater emphasis on inquiry, critical thinking skills, and multiple perspectives. Indeed, how are students supposed to evaluate competing narrative perspectives if they have not learned how narratives are constructed? For Québec historian Jean-François Cardin (2015), the current situation is rooted in a certain misunderstanding of new pedagogies that place students at the heart of learning and that dispute the role of narratives in education. Such narrations are perceived to be harmful to critical thinking and active learning of the history method. As Marc-André Éthier (2014) puts it, "if history is to be taught as a science, then schools should respect its disciplinary nature and insist not on reinstating one or more stories, but on the practice of historical inquiry."

Narration and history

We believe that this opposition between "narration" and the "history discipline" is sterile. It creates an imaginary gap between the historian and his or her work. In many ways, the controversy over the nature of history generated by postmodern theorists such as Hayden White and Keith Jenkins is largely over. For years now, the act of narration has been studied and recognized as an epistemological foundation of history knowledge and represents a proven construction which, as Paul Ricoeur contends, "exposes the discourse of the historian" (as cited in Cardin, 2015, p. 3). Indeed, the need to make sense of and give meaning to the world (past and present) implies a process by which historians, no less than citizens, transform past realities and makes their actualization possible through history. As Rüsen argues, "the linguistic form within which historical consciousness realizes its function of [temporal]orientation is that of the narrative" (Rüsen, 2005, p. 26). It is through narration that it becomes possible to inscribe the dimensions of time (the temporality) in a continuous process and thus to offer a synthetic vision of the human experience.

In any given society, there are always conflicting interpretations of the past, and some of these are purely driven by ideological or political agendas. They are, in the words of Jocelyn Létourneau, "mythistories" in the sense that they embody collective references, clichés, stereotypes, and representations through which the past, the present and the future are not only decoded but anticipated (Létourneau, 2006, p. 80). Narrative competence is about learning to understand why it is that narratives are plural, multifaceted, and never definitive. It also involves coming to appreciate the intellectual tools emerging from the history discipline to evaluate these representations, how they are constructed and what power and limit they have. Without such disciplinary competence, as Sam Wineburg contends, people are like passengers on board a collective ship with no rudder to navigate "the shoals of the competing narratives that vie for [their] allegiance" (Wineburg, 2000, p. 311).

If we are interested in teaching students to construct and deconstruct historical narratives, then efforts should be made in history education to conceptualize what we mean by *historical narrative*. As Monika Waldis and her colleagues observe, "the conceptualization of narrative competence remains vague" precisely because we lack effective understandings of narration in history (Waldis et al., 2015, p. 117). For Seixas (2017), recent debates in the field has led to a consensus on three propositions about historical narration.

First, *narrative representations and the past are conceptually and ontologically different*. The historical narrative is indeed a construct that, from a retrospective vision, gives a sense of coherence and direction to the past. It does not simply show what actually happened as Leopold von Ranke once proclaimed. This realistic view of history, as an objective reality out there that can be recovered and truthfully represented in the right story of the past, has gradually given way to more deconstructionist approaches which reveal the presumptive and contingent nature of historical knowledge. Most historians –at least in the academic world –now accept that their narratives are not direct copies of the past but historical representations of a particular linguistic mode or discourse. As such, they cannot be cross-referenced with the past itself (as it is gone), only with other representations of it.

Second, there can be *more than one historical narrative for any set of events from the past*, centring on different questions, perspectives, ethical choices, and narrative focalisations. Unlike the naive view of history sometimes held by students, there is no such thing as a complete, definitive grand narrative that encompasses all aspects of the past. In addition, narratives intend to answer particular questions about the past and these questions are vast and informed by ongoing, contemporary realities. In other words, new events trigger additional questions about the past that were not initially envisioned in previous narrative works. "The past," as Arthur Chapman (2016) puts it, "has a future and the future keeps rewriting the past" (p. 4).

Finally, *all narratives can be criticized for their plausibility*. Unlike fictional narratives, historical narratives aim at representing a coherent reality of the past that conveys true, verifiable statements. But since history is a mode of narrative representation, it has no absolutist epistemological meaning in itself, only a sense of correspondence as plausibility. History is thus a complex activity which demands strategies for explanation of argumentation and justification of narrative arrangement, and these disciplinary strategies differentiate historical narratives from other stories of the past generated by the practices of collective memory such as mythistories.

Narrative competence: narration, content and expression

Following these three prepositions, it becomes possible to envision a model of narrative competence that would inform teacher education, and ultimately the development of historical consciousness among learners. Recently, historian Alun Munslow (2007) has proposed a simplified, yet intelligible model that in our view characterizes narrative competence in history while making it possible to integrate Canadian concepts of historical thinking. This model, inspired by the influential work of Gérard Genette (1980), is represented in the form of a triangle with *narration* (authorial function) at the apex, and *content/story* (the past), and *expression* (mode of representation) located at the other two vertex points of the triangle.

Content/story

First, the past has to be organized in story-form. As we noted earlier, the past is not initially structured in a coherent story. While historical agents had narrative visions of their own, these representations are contextually-situated in the past and created for purposes of their own time and, as such, do not provide us with present-day stories we generate about this past. The content/story element is the location for the narrator's creation of the story space which delimits the beginning, the end, and the orderly arrangement of events.

The concept of story space requires a more detailed understanding of the explanatory strategies for content/story that historians use as way to present narratives. After all, historical narratives are not simply summaries of established facts. When historians write narratives, they offer constructed arguments which are open to discussion and debate. These strategies might be best considered in terms of key questions that implicitly drive the work of history narrators: How was the story created? What is the orderly arrangement of the discrete elements? How is the argumentation historically plausible? What ethical choices were made? How are these choices culturally-bound?

What follows from these questions is a more specific discussion on the role of historical thinking in structuring an argument in story-form. Interestingly, Seixas (2017) has recently proposed to transpose historical thinking concepts in terms of "narrative plausibility," that is the disciplinary patterns of narrative construction. Seixas' transposition helps understand how historical narratives present particular arguments about the past using key concepts and related dimensions, as well as in reference to the logic of the claims made by their authors. The following presents historical thinking concepts as adapted by Seixas to the dimensions of narrative plausibility (Seixas, 2017).

Tab. 1: Narrative plausibility in reference to historical thinking concepts

Evidence	Evidentiary plausibility
Continuity and change	Temporal plausibility
Cause and consequence	Causal plausibility
Perspective taking	Empathetic plausibility
Ethical dimension	Normative plausibility

- Evidentiary plausibility refers not only to established facts (like dates, events, actors) but also to "the body of evidence whose interpretation support a particular narrative" (Seixas, 2017, p. 260).
- Temporal plausibility serves to establish the temporal experience of past times as expressed in the story-form, that is the means of understanding the past and its specific temporal quality as well as its translation into our understanding of the present.
- Causal plausibility is necessary to structure an argumentation so that events assembled as causes in the narrative "convincingly determine the events and the conditions identified as consequences" (Seixas, 2017, p. 262).
- Empathetic plausibility makes it possible to understand human experiences and their complexity. Through perspective-taking, it becomes possible to appreciate the minds and feeling of people, past and present.
- Normative plausibility refers to the moral and ethical values that underline the narrator's own positionality and perspective or moral compass of the story.

The transposition of historical thinking concepts onto elements of narrative plausibility provides useful elements for the realization of the content/story of narration. It shows that the organization of the past into story-form by the historian requires important narrative decisions that cannot be left to intuitive storytelling practices. But narrative theorists like Gérard Genette, Roland Barthes, Tzvetan Todorov, and Vladimir Propp have argued that the story-arrangement is also governed by distinctive narrative choices or logical constraints that any series of events must respect in order to be coherent and intelligible. Propp, for instance, was the first one to identify schematic narrative patterns based on a series of recurrent constant functions intended to structure the content/story of folk tales. Applied to history, scholars like James Wertsch (2008) have also discussed the intrinsic structure of historical narratives suggesting, among other things, that schematic templates (generalized functions that characterize narratives) are powerful cognitive tools that provide distinctive ways of emploting events for historical memory and identity purposes. In our works, we have also demonstrated how young and adult Canadians construct narratives of the nation using schematic narrative templates (e.g., *la survivance* of French Canada, European colonial progress) supplied by collective memory.

Once learning history is thought as coming to understand how narratives are structured in story-form, with distinctive procedures and principles of narrative plausibility, it becomes easier to promote narrative competence. But narration requires another key element for Munslow, the telling (the narrative mode of representation), for every content/story there is always a history-narrator.

The expression

Narrative is not merely a medium of report designed to show what actually happened; it is a discourse that uses aesthetic principles to turn the past into "the-past-as-history" (Munslow, 2007, p. 182). Following Aristotle's rhetoric, Victor Ferry (2011) has recently explained how historical narratives are dependent upon both "extratechnical" evidence from the dimensions of narrative plausibility and rhetorical "technical" evidence generated by the narrator. For Ferry, the discursive nature of narrative makes it impossible to separate the two sets of proofs even if historians tend to ignore how the narrative act "goes all the way through the process" of their historical investigation (Munslow, 2007, p. 24). As he observes, the persuasive power of historical narrative is partly based on "processes that are part of the very structure of the text" (Ferry, 2011, p. 128).

But how does technical evidence, that is the rhetorical act of narrating, makes historical narratives more plausible? In order to answer this question, we have to look at three interrelated elements of narration as a mode of representation: the historian as narrator, voice, and focalisation.

First, *historians use narratives as a way to present their arguments*. Their role is not to rescue stories from the past but to offer representations, convincing representations of it –reference to the past, explanation of the past, and meaning for the past. These representations are neither fixed nor universal; they reveal the historian's own argument and positionality – his or her historical perspectives, emotions, and beliefs. Talking more broadly about narratology, Barthes asks us to always consider the invariable question when approaching narratives: who is speaking? Answers to this question helps reveal the diegetic universe that historians create for arranging their narratives. This universe represents the spatiotemporal environment within which the story unfolds. Variations in narrative diegetic universe occur for different reasons: the perspective of the narrator (his/her viewpoint), the purpose of the narrative (the goal set by the narrator), the relation between the subject and the narrator (how he/she relates to and engage with sources), and the context in which the narrative is produced (the historical culture in which the narrator lives). Thus, on any given history subject, there can be more than one plausible narrative due to the fact that each narrator has a distinctive positionality. This positionality shapes the discourse of the narrators or how they narrate the past-as-history.

Narrators can also increase the plausibility of their narratives by generating arguments within a diegetic universe that better reveals their own subjectivity. Unlike fictional stories, historical narratives must be framed in discursive ways that comply with canons of historical argumentation. As Chapman observes, histories are written by narrators (hence subjective) but are written in ways that are open to scrutiny by an interpersonal audience (the narratee(s)). The plausibility of their narrative argument is thus linked "to the degree to which the patterns of explanation (and of sense-making) are made explicit and acceptable to the audience" (Körber, 2015, p. 18).

Looking at the historian as narrator leads us to consider the twin concepts of voice and focalisation of the narrative. The *narrator has a particular voice which is usually made clear through narrative choices*. This voice can be personal and explicit with references to the "I" form in text (first-person narrator) but it can also be external or implied without direct reference to the narrator, as it is often the case in history textbooks – thus leading to the students' naive illusion that there is no narrator. This position is referred to as the omniscient narrator because he or she tells the story without direct involvement in the content/story but knows everything about the characters and the events described.

Focalisation refers to the focal choices of the narrator in the organization and presentation of the diegetic universe. It is defined in terms of the point of view from which the story is being told: the narrator, the characters or the historical agents. If the voice deals with "who speaks" in the story, focalisation is concerned with "who sees" within the diegetic universe (Munslow, 2007, p. 48). To tell a story from a character's point of view means to present the historical events as they are perceived, felt, interpreted and evaluated by him or her at a particular moment, and with the information available to the character.

The historian as narrator has the power over both the voice and the focalisation of the narrative discourse. In fact, historians often resort to various types and combinations of voice and/or focalisation to make their narratives more intelligible and plausible. As a matter of fact, the forms that historians use in their writing often reveal their distinctive purpose and own identity as narrators. He or she makes sense of the past through those "narrative-making decisions and turns" (Munslow, 2007, p. 138). As French historian Arlette Farge confesses, "For me, it's not fiction. I am convinced that words, the place of words, the syntax, the wording, the expression, the writing process, makes it possible for a story to present sensitive, unrevealed movements, opinions, crowds, and feelings. Words can do that" (Farge as quoted in Ferry, 2011, p. 130).

Teaching the narrative competence: instructional implications

A growing body of research indicates that students are not only capable of creating historical narratives of the collective past but, when provided with adequate tools and strategies, they can also engage in active reconstruction of the past using principles of historical thinking. The German FUER model is one such example showing that purposeful didactical approaches for constructing narratives offer evidence of students' ability and active role in "making history" (Körber & Meyer-Hamme, 2015, p. 98). Yet, current teacher education programs, at least in Canada, rarely engage pre-service teachers in the practice of making history. As leading Canadian educator Alan Sears puts it, in speaking with the pre-service teachers I meet today, little seem to have changed. Many, even those with majors in history, have little or no first-hand experience with the processes of doing history.... They haven't, in other words, had to think historically, but rather have been relatively passive observers of others' attempts to do so (Sears, 2014, p. 12).

The results of such passive strategies are consequential. If teachers do not know how to engage in the construction and deconstruction of historical narratives, how are they supposed to teach their students to do so? Our own research results indicate that Canadian students' historical narratives are highly simplified, selective stories of the collective past with linear patterns of historical development and limited focalisation and voices from a diversity of experiences that characterize Canada's past. In many ways, their narratives have more to do with collective memory than with recent developments in Canadian historiography. It is evident from our studies that current teacher education approaches have not solicited the intellectual tools of the narrative competence. As Robert Martineau points out, "rarely are the resources of 'narrative thinking' considered to reflect on situations or realities of the past, to generate representations based on historical sources, and to offer plausible interpretations expressed in the form of a narrative" (Martineau, 2010, p. 178–179).

We believe like Sears that if we want to change teachers' practices, we must also look into ways of reforming current teacher education programs so as to set teachers on an "inbound trajectory" to nurture them as active practitioners of the history discipline and not as passive outsiders of the field (Sears, 2014, p. 13). In other words, teachers must feel they belong to history communities of practice not so much to become professional historians themselves, but rather to understand and gradually develop expertise in the practice of constructing and deconstructing historical narratives – an essential competence of historical consciousness.

While there is no magical formula for reforming teacher education, as so much is dependent upon professional context and culture, we propose in conclusion a strategy that better mobilizes research findings in history education. We understand that preparing teachers to teach narrative competence cannot be accomplished by simply telling them what to do, but by encouraging them to take part in meaningful and relevant practices of making history. As such, our approach should be understood as a roadmap to set them on an inbound trajectory, moving them from the periphery to the core of history. First, our research suggests that *teachers should adopt a constructivist view of history learning* and teach in reference students' own narrative ideas of the collective past. Language, gender, culture, and identity seriously affect how students engage with the past and create narrative representations of it. As a matter of fact, the more students develop identification with events and predecessors, the more likely are they to engage with this past. In our most recent study, for example, students who strongly identified as Québécois or French Canadians were also those who produced the most engaged, militant stories of the collective past. They often related directly to predecessors and used the first-person plural narrative voice to establish a sense of collective identification and historical continuity ("we lost the Seven Years' War," "they took our rights away"). Educators should strategically use this personal connection to explore students' identity, prior knowledge and historical beliefs; in order words, what they accept, ignore, or (dis)miss in their understandings of the past. As Keith Barton and Alan McCully contend, students "need more opportunity to engage with their own perspectives" (Barton & McCully, 2012, p. 398).

If we want students to progress in their historical thinking, we need to develop progression models and related learning strategies that allow teachers to see where their individual students stand with regard to their (mis)conceptions about the past and about the nature of history so as to gradually transform these into more powerful and usable ones. "A correct understanding of the epistemological underpinnings of history," as Jeffery Nokes and Susan De La Paz note, is a prerequisite for argumentative historical narrative (Nokes & De La Paz, 2017, p. 560). For instance, those who look for evidence as hard facts and trust narratives to reveal the true representation of the past have a hard time accepting that history is based on disciplinary criteria and evaluative principles for crafting historical interpretations. These naive beliefs create one of the greatest challenges for engaging students in thinking historically. Cultivating narrative competence thus requires some important understandings of learners' beliefs about historical knowledge and narratives. To be truly usable for orienting their lives in modern, complex societies, the stories students employ should not only be coherent and useful, they also ought to be historically rigorous and well-founded in evidence.

Second, and in relation to the first element, *teachers should move students beyond their previous historical understandings by opening up their minds to alternatives perspectives* they would not have considered otherwise. A key principle of schooling in modern, liberal democracy is precisely to cultivate open-mindedness and perspective-taking. As societies become increasingly complex and culturally diversified, with people asserting radically different views and stories of the collective past, it is imperative to teach students multiperspectivity and empathetic connectivity. By encountering the experiences of different groups and individuals in society (French vs. English Canadians, Aboriginals, marginalized groups, etc.) students can learn to appreciate and care for varied perspectives on past and present realities and not exclusively the dominant ones of their own community. In this sense, as Maria Grever and Robbert-Jan Adriaansen argue, multiperspectivity is "as much about learning about the past as it is about learning about the limits of [one's] own assumptions and prejudices" (Grever & Adriaansen, 2019, p. 824).

Our own studies have shown that even within the context of the classroom, students are far from a monolithic group when it comes to history. For the same collective past, they often generate as narrators very different historical narratives, placing emphasis on distinctive periods, events, agents, and perspectives. Discrepancies, both among stories and with students' own prior understanding, can potentially lead them "to seek additional evidence that would confirm or contradict particular accounts" (Barton & McCully, 2012, p. 398). In this way, perspective-taking or empathy might motivate students to engage in more sophisticated narrations that are founded in evidence (as opposed to mythistories) and informed by alternative perspectives; in other words, narrative representations that are "polythetic" in nature (Shemilt, 2000).

Finally, *teachers should reveal and teach the construction of historical narratives.* In many ways, students learn history through narratives (e.g., classroom lectures, textbooks, movies, and oral stories). But rarely are these problematized and deconstructed in any disciplinary way in class (Lévesque, 2016). If we want students to think historically and develop more critical approaches to the collective past, teachers should explain the structure, the story space, the grammar, the rhetorical mode of expression, and the purpose of historical narratives in society. Stories may well be a primary mode of thought but they are not the past, and hence set limits on what we consider significant enough to tell and remember. Many narratives are told in society which may contradict, compete with, or complement one another, and this means that students should be equipped with the intellectual tools to deal with the constructed nature of these representations of the past.

Argumentative narrative writing is a complex and difficult activity that is still poorly integrated into Canadian history education. In fact, narrative writing is still seen as a literacy skill (taught in language arts), not a feature of history learning. Equally problematic, when narrative is used in school, it is often naively presented in opposition to argumentation (the typical five-paragraph essay) as if narration was exclusively a literary genre and not a form of historical knowledge production. Once history is understood and taught as a form of knowledge that contributes to the construction of narrative competence then it becomes possible to envisage various strategies that will help students to create more valid representations of the past that will, in turn, shape their historical consciousness. This does not mean that students will become "mini-historians," as some critics have wrongly claimed, but it does mean that they will gradually learn how narrative interpretations are created and used in society.

In conclusion, we might say that simplified representations of the collective past breed exclusion and mutual ignorance; they do not equip 21st century people with the tools to orient themselves in highly diverse political communities made up of contesting memories and living stories. To be truly usable, students' historical narratives must be more complex and open to different voices, ethical choices, and perspectives. In order to play a more productive role in this process, we believe that history education needs to prepare teachers to be more conscious of and proactive in providing students with what we call engagement in narrative competence.

References

Barton, K.C./McCully, A.W.: "Trying to 'see things differently': Northern Ireland students' struggle to understand alternative historical perspectives". *Theory and Research in Social Education* 40 (4), 2012, pp. 371–408.

Calder, L.: "The stories we tell". *OAH Magazine of History* 27 (3), 2013, pp. 5–8.

Cardin, J.-F.: "La place du récit dans l'apprentissage de l'histoire". *À l'école de clio: Histoire et didactique de l'histoire* 1, pp. 1–1, retrieved 1.11.2020, from https://ecoleclio.hypotheses.org/275.

Chapman, A.: *Developing Students' Understanding of Historical Interpretations*. Edxecel/Pearson: Oxford, 2016.

Duquette, C.: "Relating historical consciousness to historical thinking through assessment". In Ercikan, K./Seixas, P. (eds.): *New Directions in Assessing Historical Thinking.* Routledge: New York, 2015, pp. 51–63.

Eliasson, P./Alvén, F./Yngvéus, C.A./Rosenlund, D.: "Historical consciousness and historical thinking reflected in large-scale assessment in Sweden". In Ercikan, K./Seixas, P. (eds.): *New Directions in Assessing Historical Thinking.* Routledge: New York, 2015, pp. 171–182.

Ercikan, K./Seixas, P. (eds.): *New Directions in Assessing Historical Thinking.* Routledge: New York, 2015.

Éthier, M.-A.: "Contre la pensée narrative téléologique". *Le Devoir,* 27.2. 2014, pp. 1–13, retrieved 1.11.2020, from https://www.ledevoir.com/opinion/idees/401205/lar.

Ferry, V.: "Le paradoxe de la preuve en histoire. Une approche rhétorique de l'écriture d'Arlette Farge". *Mots. Les langages du politique* 95 (Mars), 2011, pp. 114–137.

Genette, G.: *Narrative Discourse: An Essay in Method.* Cornell University Press: Ithaca, NY, 1980.

Gibson, L.: *Constructing students' Historical Reference Frameworks in Canadian History Using Visual Source-Based Timelines.* Paper presented at the annual meeting of the American Education Research Conference, San Antonio, TX, April 2017.

Granatstein, J. L.: *Who Killed Canadian History?* Harper Collins: Toronto, 1998.

Grever, M./Adriaansen, R.J.: "Historical consciousness: The enigma of different paradigms". *Journal of Curriculum Studies* 51 (6), 2019, pp. 814–830, retrieved 1.11.2020, from DOI: 10.1080/00220272.2019.1652937.

Kölbl, C./Konrad, L.: "Historical consciousness in Germany: Concept, implementation, assessment". In Ercikan, K./Seixas, P. (eds.): *New Directions in Assessing Historical Thinking.* Routledge: New York, 2015, pp. 17–28.

Körber, A.: "Historical consciousness, historical competencies – and beyond? Some conceptual development within German history didactics", 2015, 56 p., retrieved 1.11.2020, from S. -URN: urn:nbn:de:0111-pedocs-108118. https://www.pedocs.de/volltexte/2015/10811/pdf/Koerber_2015_Development_German_History_Didactics.pdf.

Körber, A./Meyer-Hamme, J. (2015): "Historical thinking, competencies, and their measurements: Challenges and approaches". In Ercikan, K./Seixas, P. (eds.): *New Directions in Assessing Historical Thinking.* Routledge: New York, 2015, pp. 89–101.

Lantheaume, F./Létourneau, J. (eds.): *Le récit du commun: L'histoire nationale racontée par les élèves.* Presses universitaires de Lyon: Lyon, 2016.

Létourneau, J.: *Je me souviens? Le passé du Québec dans la conscience de sa jeunesse*. Fides: Anjou, QC, 2014.

Létourneau, J.: "Remembering our past: An examination of the historical memory of young Québécois". In Sandwell, R. (ed.): *To the Past: History Education, Public Memory, and Citizenship in Canada*. University of Toronto Press: Toronto, 2006, pp. 70–87.

Lévesque, S.: "Going beyond 'narratives' vs. 'competencies': A model for understanding history education". *Public History Weekly* 4, 2016, retrieved 1.11.2020, from DOI: dx.doi.org/10.1515/phw-2016-5918.

Lévesque, S./Croteau J.-P.: *Beyond History for Historical Consciousness: Students, Narrative, and Memory*. University of Toronto Press: Toronto, 2020.

Lévesque, S./Croteau, J.-P./Gani, R.: "Conscience historique des jeunes francophones d'Ottawa: Sentiment d'appartenance franco-ontarienne et récit du passé". *Revue du Nouvel-Ontario* 40, 2015, pp. 177–229.

Martineau, R.: *Fondements et pratiques de l'enseignement de l'histoire à l'école: Traité de didactique*. Presses de l'Université du Québec: Québec, 2010.

Munslow, A.: *Narrative and History*. Palgrave Macmillan: London, 2020.

Nokes, J./De La Paz, S.: "Writing and argumentation in history education". In Metzger, Scott Alan/McArthur Harris, Lauren (eds.): *Wiley International Handbook of History Teaching and Learning*. Wiley-Blackwell: Hoboken, NJ, 2017, pp. 551–578.

Osborne, K.: " 'Our history syllabus has us gasping': History in Canadian schools: Past, present, and future". *Canadian Historical Review* 81 (3), 2000, pp. 404–435.

Peck, C.L.: " 'It's not like [I'm] Chinese and Canadian. I am in between': Ethnicity and students' conceptions of historical significance". *Theory and Research in Social Education* 38 (4), 2010, pp. 574–617.

Propp, V.: *Morphology of the Folktale* (2nd Edition) (trans. Louis Wagner). University of Texas Press: Austin, 1968.

Reich, G.A./Buffington, M./Muth, W.R.: "(Dis) Union at 150: Collective memories of Secession". *Theory and Research in Social Education* 43 (4), 2015, pp. 499–527.

Ricoeur, P.: *Temps et récit*. 3 vols. Éditions du Seuil: Paris (1983–1985).

Rüsen, J.: *History: Narration, Interpretation, Orientation*. Berghahn Books: New York, 2005.

Sandwell, R.: " 'We were allowed to disagree, because we couldn't agree on anything': Seventeen voices in Canadian debates over history education".

In Taylor, T./Guyver, R. (eds.): *History Wars and the Classroom: Global perspectives.* Information Age Publishing: Charlotte, NC, 2012, pp. 51–76.

Sears, A.: "Moving from the periphery to the core: The possibilities for professional learning communities in history teacher education". In Sandwell, Ruth/Amy von Heyking (eds.): *Becoming a History Teacher: Sustaining Practices in Historical Thinking and Knowing.* University of Toronto Press: Toronto, 2014, pp. 11–29.

Seixas, P.: "A model of historical thinking". *Educational Philosophy and Theory 49* (6), 2015, pp. 1–13, retrieved 1.1.2020, from DOI: 10.1080/00131857.2015.1101363.

Seixas, P.: "Teaching rival histories". In Åström Elmersjö, Henrik/Clark, Anna/Vinterek, Monika (eds.): *Search of Narrative Plausibility. In International Perspectives on Teaching Rival Histories.* Palgrave MacMillan: London, 2017, pp. 253–268.

Shemilt, D.: "The caliph's coin: The currency of narrative frameworks in history teaching". In Stearns, P.N./Seixas, P./Wineburg, S. (eds.): *Knowing, Teaching, and Learning History: National and international perspectives.* New York University Press: New York, 2000, pp. 83–101.

Staub, J. (ed.): *Narration, Identity, and Historical Consciousness.* Berghahn Books: New York, 2005.

Waldis, M./Hodel, J./Thünemann, H./Zülsdorf-Kersting, M./Ziegler, B.: "Material-based and open-ended writing tasks for assessing narrative competence among students". In Ercikan, K./Seixas, P. (eds.): *New Directions in Assessing Historical Thinking.* Routledge: New York, 2015, pp. 117–131.

Wertsch, J.V.: "Collective memory and narrative templates". *Social Research* 75 (1), 2008, pp. 133–156.

White, H.: "The value of narrativity in the representation of reality". *Critical Inquiry* 7 (1), 1980, pp. 5–27.

Wineburg, Sam: "Making historical sense". In Stearns, Peter/Seixas, Peter/Wineburg, Sam (eds.): *Knowing, Teaching, and Learning History: National and International Perspectives.* New York University Press: New York, 2000, pp. 306–325.

Wineburg, Sam/Mosborg, Susan/Porat, Dan/Duncan, Ariel: "Forrest Gump and the future of teaching the past". *Phi Delta Kappan* 89 (3), 2007, pp. 168–177.

Zanazanian, P.: "Historical consciousness and being Québécois: Exploring young English speaking students' interactions with Quebec's master historical narrative". *Canadian Ethnic Studies* 47 (2), 2015, pp. 113–135.

Fredrik Alvén

Chapter 4 Please let it be a history test with a content I can handle: How girls and boys perform in the national test in history in Sweden[1]

Abstract: In Sweden a national test in history since 2013 in the spring the last term for the compulsory school is performed. During the period 2013 to 2019 the boys have been performing all the better. In the year of 2019, the boys even had the same results as the girls at the national test. This does not apply to other national tests, where the girls outperform the boys. Research describes the boys' poor performance in school with rather predestining hypotheses about a deconstructive macho culture among the boys, biological differences between the genders or discrimination towards the boys in the school. This study assumed that the answer to the boys' better results at the history test over time must be found in the test itself and the boys' understanding of history, a subject matter education approach, that is. With statistical methods it is shown that a certain type of history and forms of the tasks suites boys better and vice versa for the girls. These results involve difficult questions to history as a subject to be assessed and graded.

Keywords: History teaching, Assessment, National test, Gender differences, Fairness.

Research explaining the boys' poor results in the school

Both in Sweden and in the Western world overall boys have been performing worse and worse in the school during the new century (Clark et al., 2008; OECD, 2015). This development worries authorities in Sweden, while it can lead to a severe gender gap on the labor market and, in the future, a gender inequality overall (SOU, 2009: 64). In Sweden the girls have had better grades for a long

1 This study is the result of the research project entitled "Fair assessment in the national test in history," subsidized by the Crafoord Foundation, Sweden; Also is result of research projects granted by "Spanish Ministry for Science, Innovation and Universities. Secretary of State for Universities, Research, Development and Innovation," grant number PGC2018-094491-B-C33; and "Seneca Foundation. Regional Agency for Science and Technology," grant number 20874/PI/18.

time in most subjects and the last decade they also have outperformed the boys in math and natural sciences (Zimmerman, 2018). More girls are getting the highest grades and more boys are not even approved a grade (Skolverket SIRI 2020.04.24)). In Sweden gender seems to explain school performance better than belonging to a minority group or a social class without privileges (Skolverket, 2006). The gap between boys and girls are also widening faster in Sweden than in other OECD countries. The ones hardest hit are the low-performing boys (OECD, 2015). As a result, we today see more women at the universities and especially at those with the highest status (Zimmerman, 2018; UKÄ, 20.04.24).

The research trying to explain these gender differences in school performance is vast and complex. Overall, they cluster around three categories: social and cultural factors, biological factors and, finally, school factors.

The social and cultural research describing the gap between boys and girls in school performance points to different sets of behaviors, attitudes and values. These different norms between boys and girls are understood in cultural contexts with norms of masculinity and femininity (Epstein et al., 1998; Weaver-Hightower, 2003). The boys' underachievement is explained as a result of a macho culture that forces them to a behavior that doesn't fit to school achievement. This includes disruptive behavior, anti-school attitudes and a longing for masculine attributes (Francis, 1999). These norms operate collectively, and the boys are ridiculed by their friends if they work hard and show that they want to perform well in school. On the contrary, they are requested to behave challenging to the school norms (Warrington et al., 2000). That's the way the boys are gaining their masculine identity and appreciation among their friends (Jackson, 2003). This also seems to hinder the boys to ask for help in school (Ryan et al., 2009). Meanwhile girls have more positive attitudes towards school (OECD, 2009). The girls also seem to value hard work and self-discipline higher than boys (Costa et al., 2001). Values that are related to school success.

The biological research explains the differences between boys and girls with physical differences. The brain organization, hormones and genetics would then be to the boys' disadvantage in the school work (Cahill, 2006; Gibb et al., 2008). The boys seem to mature later and are hopelessly behind the girls already at the school start. Some research even interprets the boys' worse school performance as biologically determined (Gurian, 2001; Sax, 2005). In Sweden some researches mean that the boys' way of thinking don't fit in today's school and point to the fact that boys today are oftener labelled with some kind of learning disability (Levander, 1990).

Research that finds hinder for the boys in the school context itself points out that today's school is better suited to girls than to boys. Some researchers describe the school as feminized (Delamont, 1999). This perspective includes explanations about a lack of male role models in the school, more gentle discipline than before, denial of competition and even a feminization in the curricula. Other research shows that the teachers don't have the same high expectations on boys as they have on girls. The boys are supposed to be louder and messier than the girls. The teachers therefore mostly reprimand the boys while they instead teach the girls (Fredriksson & Taube, 2012; Zimmermann, 2018). In Sweden some research even points out that the boys are getting worse grades for the same results or products as the girls, including when national tests are graded (Hinnerich et al., 2011; Hinnerich & Vlachos, 2013; Berg & Palmgren, 2019).

As it comes to history in the compulsory school in Sweden the girls get better final grades than the boys. This confirms the contemporary research and development overall. However, the results for the national test in history, written during the last term of the compulsory school and supposed to weigh heavily to support the final grade, shows interesting results over the time they have been performed, 2013 an onwards, that is. The statistical data base SiRiS at The National Agency for Education in Sweden has been used to construct figure 1. Each grade renders points in the data base: A 20, B 17,5, C 15, D 12,5, E 10 and F (not approved) 0. The value for each year is the mean for the final grades and test results for the boys respective the girls. The girls have much higher final grades than the boys the whole period. For both sexes the final grades are higher than the test grades. This is congruent with other national tests in Sweden. More interesting is that the boys are performing all the better on the national tests in history over time. There is a clear improvement from the year 2015, and in 2019 the boys perform on the same level as the girls. From 2016 the boys' gap between the final grades and the test grades also is narrower than the girls' gap. Something has made the boys performing better at the national test in history. This improvement, over such a short period, is unreasonable to explain with former researchers' hypothesis about why girls and boys perform differently in the school. Instead this study seeks explanations in the test itself using theories emanating from the history didactic field, a subject matter approach, that is.

Figure. Means for the final grades boys and girls and means for the test results for boys and girls 2013–2019. Source: SIRIS data base, The National Agency for Education.

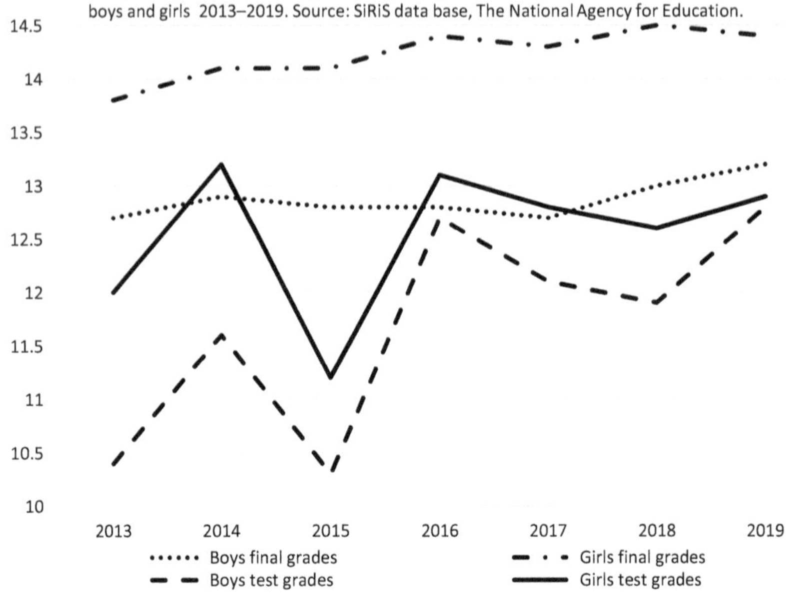

A history didactic approach

There is not much contemporary research about how girls respective boys handle and perceive history in school (Kitson et al., 2011). That is strange while Thompson as early as 1962 stated:

> *General notions exist that there are significant differences according to sex, boys being particularly interested in certain topics, for example, warfare and battles, these having much less appeal to girls, who presumably are thought to be more readily involved in the gentler aspects of the human past.*
>
> (Thompson, 1962)

The quote, however, makes it fair to think that different forms and aspects of history in the school either appeals to boys or to girls. Further on, this can affect their grades and performances on tests. For some decades the history didactic field at the same time has described history as something that effects our entire existence and our understanding of the world around us. This is tightly connected to our identity. To change your perception of history is not easily made, that is (Rüsen, 2005). How we interpret the past and think about the future affects our perception of and actions in the contemporary. In northern Europe the concept of historical consciousness has been used to describe this human mental process, and the concept has in Sweden been explicitly outlined in the curricula since 1994. Either the subject of history should give the students a historical consciousness or develop their already existing historical consciousness (Utbildningsdepartementet, 1994; Skolverket, 2000; Skolverket, 2011). In the theoretical concept of historical consciousness there are some important consumptions, that must affect the way we understand and teach history in school. The concept emanates from the hermeneutical philosophy and states that we in many ways are history. We cannot look at the past in an objective way, but are always biased from the culture we look upon the past from, that is (Karlsson, 2004; Gadamer, 2006). But at the same time there are possibilities to emancipate from the historical cultural fabric we live in. During the 1970s and the 1980s German researchers as Klaus Bergman and Anette Kuhn discussed this possibility for the subject of history (Kuhn, 1979; Bergman, 1985). Jeismann (1979) combined the two perspectives and meant that we both are and do history. If we understand history in that way, that we mostly perceive history as a heritage, but can interpret some things in new ways, this must be true to the students as well. A lot of research also consistently have shown that students from minority groups do not agree upon or understand the history they meet in school. Instead they carry a strong perception of history formed in their families (Rosenzweig & Thelen, 1998; Wertsch, 2000). Of course, this must influence how they manage the subject in school. The fact that historical narratives and the selection from the past to build them also are affected by cultural defined moral statues and political understanding make it even harder for some students to agree upon the schools' master narratives (Gaddis, 2002; Wertsch, 2002; Rüsen, 2004; Alvén, 2017). Why could this not also apply different to boys and girls?

The subject of history has a long tradition of building the nation state and nationalistic feelings among its citizens with captivating stories about former heroes (Barton & Levstik, 2004; Foster, 2011). Today these stories in the Western world largely are changed to stories that strengthen democratic values, peace and internationalism (Pingel, 2010; Carretero, 2017). But still the story is used as a tool to foster the citizens. That is not strange while storytelling both is a convincing method and an effective way to learn about moral and ethical values (White, 1973; Coffin, 2006; Kessler & Wong-MingJi, 2009). Karlsson writes that history that is included in the curriculum always has been a tool to legitimize a perspective of the world the ones' with power want (Karlsson,

1999). Some researchers even have given up the idea of a neutral and value-laden history education, and instead search the values and moral that could be constructive in the history education today (Ammert & Sharp, 2016; Edling & Mooney Simmie, 2020). Actually, in many ways the introduction of the concept of historical consciousness to history education was a desire to morally deal with a dark past, with the Holocaust as the most significant example, to be able to form a better future (Karlsson, 2009; Wilschut, 2012). Not least, research with this concept as a starting point at the same time has shown that it is not easy to retrain peoples' relationship to history.

While history has been understood as a subject anchored in a certain culture, with distinctive moral and ethical values and using the story as a convincing tool the student is supposed to interpret and do history herself. This approach started in Great Britain during the 1970s. Lee and Ashby write that the disciplinary concepts of history came in to focus in this shift (Lee & Ashby, 2000). A new way of teaching history – the New History – was introduced by the Schools Council History Project (Berger, 2012). Influential researchers such as Ashby, Blow, Lee and Shemilt favored a history teaching grounded in the discipline of history (Shemilt, 2000, 2009; Lee, 2004; Lee & Howson, 2009; Ashby & Edwards, 2010; Blow et al., 2015). Much of their research is focused on second-order historical disciplinary concepts (Lee & Ashby, 2000). Concepts as source criticism, historical causation, historical significance and historical empathy, are supposed to help students understand the epistemology of history. Lee summarizes the change:

> *... it is absurd to say that schoolchildren know any history if they have no understanding of how historical knowledge is attained... The ability to recall accounts without any understanding of the problems involved in constructing them or the criteria involved in evaluating them has nothing historical about it.*
>
> (Lee, 1991, p. 48–49)

After this epistemological turn, much research about history teaching has focused on the students' own construction of history, that is, how they can learn to think as historians and construct their own historical narratives (Laville, 2004; Semmet, 2012).

Other contemporary processes have also challenged a national master narrative in the classroom, for instance globalization and migration (Rüsen, 2011). These processes mean new ways of interpreting and understanding the past in the classroom. According to many researchers, the answer to this challenge in the classroom is to look upon history as a subject that draws its strength from the ability to many different perspectives (Seixas, 2009; Grever, 2012; Nordgren & Johansson, 2015).

In the wake of this the curricula in some countries demand teachers to develop the students' competences to make their own historical narratives, often via the concept of historical consciousness (Eliasson et al., 2015; Kölbl & Konrad, 2015). VanSledright (2014) understand this to mean that the students are supposed to *do history*. Reasonably, this means that history can no longer be taught as a national master narrative that students are supposed to recall in tests and essays. Instead, students should learn how to make their own historical narratives. Consequently, the students' own historical narratives must be assessed. Some researchers also have tried to find ways to assess the students' historical thinking skills (Seixas & Morton, 2013; Ercikan & Seixas, 2015). However, they have mostly focused on the historical thinking dimension, the scientific approach to history, that is (Ercikan & Seixas, 2011; Eliasson et al., 2015). The moral and ethical dimensions are not developed in these models. At the same time, ethical dimensions seem to be taken into account as the teachers assess the students' narratives (Alvén, 2019).

Meanwhile research indicates that how we understand and use history is strongly dependent on our identity (Ahonen, 2012; Grever, 2012). Historical concepts as historical culture, historical memory and historical consciousness affects our ways of understanding history (Rüsen, 2007). On the whole, it is reasonable to assume that, some students will construct historical narratives which violates those presented by their schools. Stobart argues that:

> *There is no cultural neutrality in assessment or in the selection of what is to be assessed. This applies as much to mathematics as it does to history, and attempts to portray any assessment as acultural are a mistake.*
>
> (Stobart, 2005, p. 282)

He also writes that assessment often strive to standardize knowledge, and this encourages assimilation. Altogether there seem to be several things challenging a reliable, fair and neutral assessment of the students' historical understanding today if we are supposed to take the concept of historical consciousness seriously.

As said, many researchers have shown that the ethnical dimension means that many minority groups react and protest against the school's history teaching (Rosenzweig & Thelen, 1998; Wertsch, 2002; Barton & McCully, 2005). But almost nothing is done to understand if boys and girls understand history differently. Some research, however, indicate that they do (Andersson Hult,

2016; Sandberg, 2019; Alvén, 2020). If so, assessment of history competences should take this into account.

Method

The empirical material in the study are the national tests themselves in history between the years 2013–2019, and statistical data from samples each year which makes it possible to see the students' gender and grading for each item and their totally grade for the test. In the test the students do not collect points, but proof of a quality for each item. In Sweden there is knowledge requirements in the curriculum to help to mark the students. The knowledge requirements are built up in three steps with different qualities, which corresponds to the grades E, C and A. Below is an example for the competence of using sources. The bolded words show the progression between the knowledge requirements.

- Pupils can use historical source material to draw **simple and to some extent** informed reasoning about people's living conditions, and apply **simple and to some extent** informed reasoning about the credibility and relevance of sources.
- Pupils can use historical source material to draw **developed and relatively well** informed conclusions about people's living conditions, and apply **developed and relatively well** informed reasoning about the credibility and relevance of their sources.
- Pupils can use historical source material to draw **well developed and well** informed conclusions about people's living conditions, and apply **well developed and well** informed reasoning about the credibility and relevance of various sources.

In the statistical material the proof of quality for each item is changed to 0 points if not approved and 1 point for the first level and so on. So, a student never can have more than three points on an unique item, and that corresponds to an answer on the highest level, grade A.

Each year Malmö University, where the test is constructed, collects this data on a platform. The sample contains the results for every student born on a certain date. Approximately 25 000 students take the test in history each year. The sample is each year about 1500 to 2000 students' results, so about 6–8 % of the population. The sample is also randomized and must be understood as representative.

The total points for each year have been compared between the girls and boys with an independent t-test, where the mean is compared. The result has been

Tab. 1: Number of collected results and the distribution between boys and girls 2013–2019

Test year	*2013*	*2014*	*2015*	*2016*	*2017*	*2018*	*2019*
N totally	*1699*	*1760*	*2010*	*1577*	*1764*	*1794*	*1831*
N Boys	*861*	*883*	*1039*	*820*	*912*	*882*	*913*
N Girls	*838*	*887*	*971*	*757*	*852*	*912*	918

interpreted as significative if the p-value has been <0,05. The results have also been calculated with Cohen's d to calculate the effect size. The t-test, however, only compare the boys' and girls' average total scores on the tests. If differences in achievement vary among the items and within a group, then gender gaps computed from the overall scores is not a good tool, while it won't tell us were the boys respective the girls success or fail.

So called DIF analysis (differential item functioning) is used to identify those test items for which different groups of students perform differently. It encompasses a set of approaches for comparing performance of groups on individual items while at the same time considering the students' potential to score well on the test. In that way, DIF analysis is more useful than comparing total scores for identifying differences between groups taking a test (Martinková et al, 2017). DIF analysis can, in other words, identify differences that are not revealed when comparing total scores. But DIF-analysis can be interpreted in different ways. In this study the categories that ETS (Educational Testing Service) use are practiced. Items with a severe difference is categorized as C or CC (Zwick, 2012).

After identifying the items with major differences, they were categorized after a model constructed by the researchers Ammert and Eliasson (2019). They have evaluated the tests' content and form, between 2013–2017, with their model. The items' content is categorized as political/ideological, military or social history and sometimes as knowledge about epochs. The items' level means that macro should be understood as history with consequences for several countries, meso for regions and micro for local places. Analyze about trigger is what forces history to become, is it actors or structures? The format is either multiple-choice questions or constructed responses. Ammert and Eliasson actually use a third category, a hybrid between the two formats. That one is not used in this study. Instead such an item is interpreted from what form that gives the most points, either the multiple choice form or the constructed response form.

Tab. 2: Model to categorize the items

Content	Level	Trigger	Format	Ability
Political/ ideological history	Macro	Actor	Multiple Choice	Reference
Military history	Meso	Structure	Constructed response	Source use
Epochs	Micro	Actor/ Structure		Use of history
Social history				Use of Concepts
Economical history				

The syllabus in history contains four abilities that the history education should develop among the students.

These are the ability to:

- use a historical frame of reference that incorporates different interpretations of time periods, events, notable figures, cultural meetings and development trends,
- critically examine, interpret and evaluate sources as a basis for creating historical knowledge,
- reflect over their own and other's use of history in different contexts and from different perspectives, and
- use historical concepts to analyze how historical knowledge is organized, created and used.

The different categories can sometimes be mixed, and this will be shown in the results.

Results

Table three consolidates and explain figure one. From the year 2015 and on the boys' results continuously approaches the girls' results. 2019 there is no differences between the test results. The t-test is significative all the way to year 2018, but Cohen's d is decreasing over time.

To be able to tell what happens to test items each item was analyzed with DIF-analyses. Tab. 4 shows every item that has benefited the boys in each test.

Tab. 3: Total points, mean for boys and girls, t-test significance and Cohen's d 2013–2019

Test year	2013	2014	2015	2016	2017	2018	2019
Total points	50	44	60	56	60	60	60
Mean boys	19	20	26	27	28	28	30
Mean girls	23	23	28	28	30	30	30
t-test significance	Yes	Yes	Yes	Yes	Yes	Yes	No
Cohen's d	0,396	0,322	0,181	0.094	0,161	0,15	0

There are obvious similarities between these items. 70 % have a content of political history or military history. If you add economical history and knowledge about epochs the number changes to 81 %. As much as 74 % of the items take place on a macro level. The trigger is structural processes in 67 % of the items. As it comes to the formats 85 % is multiple-choice questions, and from the four abilities in the curriculum 81 % represents knowledge of references to history.

Tab. 5 shows the items that favored the girls. This table's content is quite different from Tab. 4. In all of the items the content is social or political/social history. The items with only social history constitute 68 % of the items. The level is either meso or micro in 82 % of the items. In 77 % of the items there is an actor, and in 50 % the actor is crucial for the historical course of events. The form is totally dominated by constructed responses, 95 %. The abilities are, however, evenly distributed, and one cannot say that any ability seems to favor the girls.

Overall a multiple-choice item with a political or military content on a level of macro with a structural perspective that asks for reference knowledge seem to favor the boys. The girls are favored by a constructed response item with a social content on a micro or meso level and with an explicit actor. To the girls it does not seem to matter which kind of ability it is that will be assessed.

The last part of the analysis was a qualitative reading and understanding of all the items in the history national tests between 2013–2019. The analyze model in Tab. 2 was used, when categorizing all the items. Thereafter each subcategory's share in per cent in each test was calculated. In that way each test could be analyzed from each category if it contained a majority of items that either favored the girls or the boys. Contents that were supposed to favor the boys were military, political and economic history and knowledge about epochs. For the girls it was a content with either pure social history or a political history with obvious social consequences. The macro level was supposed to favor the

Tab. 4: Tests and items with a DIF that favor the boys

Test 2013	**Content**	**Level**	**Trigger**	**Format**	**Ability**
Item A 1	Political	Macro	Structure	Multiple Ch.	Reference
Item A 2	Military	Macro	Structure	C. Response	Reference
Item A 7	Political/ ideological	Macro	Actor	C. Response	Use of hist.
Item A 10	Epochs	Macro	Structure	Multiple Ch.	Reference
Item B 9	Political/ ideological	Meso	Actor	Multiple Ch.	Concepts
Test 2014					
Item 3	Political/ Social	Meso	Actor	C. Response	Use of hist.
Item 5	Political	Macro	Structure	Multiple Ch.	Reference
Item 6	Economical	Macro	Structure	Multiple Ch.	Reference
Item 8	Social	Micro	Actor	Multiple Ch.	Reference
Item 12	Military	Macro	Structure	Multiple Ch.	Concepts
Item 14	Social	Micro	Actor	Multiple Ch.	Sources
Item 17	Political/ Social	Meso	Structure	Multiple Ch.	Reference
Test 2015					
Item 1	Mixed	Meso	Structure	Multiple Ch.	Reference
Item 9	Military	Macro	Structure	Multiple Ch.	Reference
Item 10	Epochs	Macro	Structure/ Actor	Multiple Ch.	Reference
Test 2016					
Item 1	Mixed	Macro	Structure	Multiple Ch.	Reference
Item 4	Political/ Social	Macro	Structure	Multiple Ch.	Reference
Item 8	Military	Macro	Structure	Multiple Ch.	Reference
Item 15	Military	Macro	Structure	Multiple Ch.	Reference
Item 16	Military	Macro	Structure	Multiple Ch.	Reference
Test 2017					
Item 4	Political/ Military	Macro	Structure	Multiple Ch.	Reference
Item 16	Military	Macro	Structure	Multiple Ch.	Reference
Test 2018					
Item 11	Military	Macro	Structure	Multiple Ch.	Reference
Item 20	Military	Macro	Structure	Multiple Ch.	Reference
Test 2019					
Item 3	Mixed	Meso	Structure	Multiple Ch.	Reference

Tab. 4: Continued

Test 2013	Content	Level	Trigger	Format	Ability
Item 11	Politcal/ Military	Macro	Structure/ Actor	Multiple Ch.	Reference
Item 16	Military	Macro	Structure/ Actor	C. Response	Reference

Tab. 5. Tests and items with a DIF that favor the girls

Test 2013	Content	Level	Trigger	Format	Ability
Item A 9	Political/Social	Meso	Actor	C. Response	Use of hist.
Item B 5	Social	Micro	Actor	C. Response	Sources
Test 2014					
Item 16	Social	Micro	Structure	Multiple ch.	Reference
Test 2015					
Item 8a	Social	Micro	Actor	C. Response	Reference
Item 18	Social	Micro	Actor	C. Response	Reference
Item 20a	Social	Meso	Actor/Structure	C. Response	Reference
Item 20b	Social	Meso	Actor/Structure	C. Response	Concepts
Test 2016					
Item 11	Social	Micro	Actor	C. Response	Sources
Item 18	Political/Social	Macro	Structure	C. Response	Reference
Item 21	Social	Micro	Actor	C. Response	Concepts
Test 2017					
Item 5	Political/Social	Meso	Actor	C. Response	Sources
Item 20	Social	Meso	Structure	C. Response	Reference
Item 23	Social	Macro	Actor	C. Response	Reference
Test 2018					
Item 5	Social	Meso	Actor/Structure	C. Response	Reference
Item 18	Political/Social	Macro	Actor/Structure	C. Response	Use of hist.
Item 19	Social	Micro	Actor	C. Response	Use of hist.
Item 20	Social	Micro	Actor	C. Response	Use of hist.
Item 21	Political/Social	Meso	Actor/Structure	C. Response	Use of hist.
Test 2019					
Item 2	Political/Social	Micro	Actor	C. Response	Sources
Item 8	Social	Meso	Actor	C. Response	Use of hist.
Item 9	Social	Meso	Structure	C. Response	Reference
Item 19	Political/Social	Macro	Structure	C. Response	Reference

boys, while meso and micro levels were supposed to favor the girls. Items with a structural explanation were interpreted as favoring the boys, while items with an actor were supposed to favor the girls. The multiple-choice questions were analyzed as favoring the boys and constructed responses as favoring the girls. The ability to use reference knowledge was supposed to favor the boys while no special ability was understood as favoring the girls.

Even Tabs. 6 and 7 support a historical didactic explanation of why the boys have been performing all the better on the history tests in Sweden over time. In several of the categories there seem to be a radical shift between the years 2014 and 2015. In the test 2014, 21 % of the content favored the boys, while the number in 2015 was 38 % . For the girls the number decreased from 75 % 2014 to 55 % 2015. In the test 2014 only 16 % were on a level of macro while it was 35

Tab. 6: The test benefits for the boys in per cent, Content, level, trigger, form and abilities

Test	Military, Political, Economic, Epochs	Macro	Structural	Multiple Choice Questions	Reference
2013	22 %	29 %	47 %	16 %	34 %
2014	21 %	16 %	55 %	14 %	30 %
2015	38 %	35 %	53 %	53 %	38 %
2016	44 %	45 %	61 %	55 %	46 %
2017	29 %	43 %	58 %	50 %	47 %
2018	50 %	47 %	65 %	50 %	47 %
2019	37 %	43 %	50 %	48 %	60 %

Tab. 7: The test benefits for the girls in per cent, Content, level, trigger, form and abilities

Test	Social Social/Political	Meso Micro	Actor	Constructed Responses	Ability
2013	70 %	71 %	53 %	84 %	-
2014	75 %	84 %	45 %	86 %	-
2015	55 %	65 %	47 %	47 %	-
2016	45 %	55 %	39 %	45 %	-
2017	57 %	57 %	42 %	50 %	-
2018	33 %	53 %	35 %	50 %	-
2019	33 %	57 %	50 %	52 %	-

% in the test 2015. In the test 2018, 47 % were on a macro level. On the contrary, structural history has dominated the whole period, 47 % in the test 2013 and 65 % in the test 2018. This should favor the boys. The form of the items has changed drastically. Only 14 % in the test 2013 and 53 % in the test 2015 were multiple-choice questions. The tested abilities have also changed over time. 34 % were reference knowledge in the test 2013 and 60 % in the test 2019.

Nonetheless, the trend is clear. Over time the national tests in history has moved from a content and a form that favor girls to a content and a form that more favor boys.

Discussion

The fast and distinct improvement for the boys and their results on the national test in history is hard to explain from established research about the boys' and the girls' differences in school performance. In six years, it is hard to believe that something biological could have happened to the boys' brains so that they would perform better. Also, cultural differences that could turn the boys from a macho and anti-school attitude would take a longer time than six years, and in many ways the gender roles still exist among the students in school. Instead this study has focused at the subject of history and subject matter education theories to explain the change in results. After all, the pedagogical and sociological established gender research still plays an important role to explain a macro perspective on differences between the sexes and how they manage school. Nonetheless, sometimes subject matter education can explain important differences emanating from the subject itself and the students' perception of it. Such research can give new explanations and understanding about different groups performing different in school.

It seems like Thompson was right already 1962 when he said that boys and girls understand and perceive history in different ways. That is important knowledge as it comes to teach history, and not least assess historical knowledge. To assess in a fair way a lot of things must be ensured. First and foremost, there must be a clear vision of what historical knowledge is and by extension what is supposed to be assessed (Ercikan & Seixas, 2015). That demands a good subject knowledge as well as an unanimous understanding among the assessors what the core of the subject is. The fact that history as a subject in the school has become more complex, where the students' own interpretations of the past and their narratives are supposed to be assessed, during the latest decades makes such an unanimous understanding even harder. Several researchers have tried to elaborate models and concepts that the history teaching could lean on

(Stearns et al., 2000; VanSledright & Limon, 2006). These models differ, however, and there is, among else, a distinct difference between an Anglo-Saxon tradition that focus on historical methods and a northern European tradition that starts in the concept of historical consciousness. The former starting in historical methods, while the latter one is more interested in philosophical perspectives on history and the existential role it plays for the student. There are attempts that have tried to unite the two traditions, but these are not totally convincing, while it is hard to, on the one hand, search for historical truths and, on the other hand, affirm different historical perspectives and identities (Shemilt, 2000; Lee, 2004; Seixas, 2004). Sharpened, one could say that the Anglo-Saxon tradition answers the didactic question *How to teach*, and the northern Europe tradition the didactic question *Why to teach*, and the answers are today not really compatible.

The uncertainty of teaching history one also can find among the teachers in Sweden (Nygren, 2009; Berg, 2014). Some history teachers want to give the students factual knowledge and a cultural heritage, and meanwhile others want to give them a critical thinking approach. This means that the teachers are looking for different qualities as they assess the students' historical knowledge (Alvén, 2017; Alvén, 2019). At the same time the students, including boys and girls it seems, also have different perceptions of history, what they find important and interesting, that is (Andersson Hult 2016; Strandberg, 2019; Alvén, 2020).

The results in this study therefore ask difficult questions to the researchers in history didactic, to the test constructors of tests in history, the history teachers and the National Agency for Education in Sweden.

As it comes to the researchers, they must continue to explore and try to understand what consequences it has when a subject with deep roots in a nationalistic identity and meaningful metanarratives tries to change to a subject where the students are supposed to interpret the past themselves and construct their own narratives. This leads to a difficult tension between history teaching as bringing a cultural heritage to the becoming citizens and history as a subject for critical thinking, historical methods and as a possibility to affirm the students' own identities. In Sweden the nationalistic mission to the subject of history was clear before the second world war (Zander, 2001). After the second world war, with its' obvious proof of what destructive nationalism can lead to, history was relegated. Instead civic education should form the new citizens to become democratic, following the human rights (Englund, 1986; Larsson, 2011). Today teaching of history does not strive for nationalism but demands new values as democracy and pluralism. The methods are, however, still the same, using convincing moral stories about good and bad (Barton & Levstik, 2008; Alvén, 2019). At the same time convincing research shows that students belonging to different minorities often react negative to the school's history teaching. These students can feel excluded by the metanarrative in the school (Rosenzweig & Thelen, 1998; Wertsch, 2002; Nordgren, 2006; Epstein & Schiller, 2009). It is not unreasonably that these students can understand and interpret history in different ways than what is expected from their history teacher. This is an urgent question when history teaching more and more demands the students to do their own history. That is why history didactic must look upon what it means when subgroups of students perceive history in different forms – what is the consequences for the subject of history and what is the consequences for the students? Among such different groups are boys and girls.

With the knowledge that historical content and item forms is perceived in different ways between sub groups of students, among else girls and boys, the test constructors must take this into account as they formulate tests in history. What is a good balance between different sorts of content and item forms? We can obviously not think that a test of a student's source skill is independent of what the historical source contextualizes, and what form the item has. This makes the question of validity very difficult. Should different groups of students show their abilities from different sorts of historical content? Is it a methodological knowledge or historical factual knowledge that is at the core of the assessment? This question is directly addressed to the tension that exists in the subject of history itself – history as knowledge about a metanarrative orienting the students in the contemporary or history as methodological skills. The national test in Sweden in the beginning of the studied period mostly asked for methodological skills and later on more asked for factual knowledge. At the same time girls seem to handle the first kind of knowledge better and boys the latter one. Basically, the test constructors could decide if they want boys or girls to perform best, and this gives them a difficult morally responsible. What principles can they lean on as they construct the test?

In the same way history teachers must decide what they attend to understand as core knowledge in history and by extension what to assess. They have a metanarrative in the core content in the curriculum they can teach and test using multiple-choice questions. Source criticism, to analyze use of history and handle important methodological concepts are, however, harder to teach as facts and to test using multiple-choice questions. If these abilities form the basis for the subject, history must both be taught in another way as well as assessed in another way. And what about a reliable grading on a national level if the teachers can understand the mission in diverse ways? And which teaching prepare the students best for the national test in history?

The National Agency for Education, on the other hand, must clarify how to understand and teach history in school. Now they give the direction that different parts of the core content not must be taught equally much, and that the teacher can decide what to focus on. At the same time the knowledge requirements, used to assess the students' knowledge, contain some core content, and this makes a paradox for the teachers. Is the core content or the abilities the most important?

Today it is a bit provocative to test factual knowledge while research about assessment the latest decades have strived for assessment of abilities, competences and authentic methodological knowledge. Meanwhile, do we want becoming citizens that may be out of knowledge about the Holocaust, Gulag or the Trans-Atlantic slave trade? To teach the students a metanarrative and then test their factual knowledge with multiple-choice questions would, at the same time, probably be reliable because of the transparency of what to assess, but would it not be at the expense of what we mean that the history subject is supposed to give the students – a competence to independently navigate in a complex global society?

References

Ahonen, Sirka. *Coming to Terms with a Dark Past: How Post-Conflict Societies Deal with History*. Peter Lang: Frankfurt am Main 2012.

Alvén, Fredrik. Pojkarna, jämställdheten och framtiden. Skola och Samhälle, 2020, https://www.skolaochsamhalle.se/flode/skola/fredrik-alven-pojkarna-jamstalldheten-och-framtiden/, retrieved 2020-08-03.

Alvén, Fredrik. "Bias in teachers' assessments of students' historical narratives". *History Education Research Journal*, 16(2), 2019, pp. 306–321.

Alvén, Fredrik. *Tänka rätt och tycka lämpligt*. Malmö Universitet: Malmö 2017.

Ammert, Niklas/Eliasson, Per. *Externa historieprov. En studie av validitetsfrågor*. Malmö Universitet: Malmö 2019.

Ammert, Niklas/Sharp, Heather. "Working with the cold war: Types of knowledge in Swedish and Australian history textbook activities". *Journal of Educational Media, Memory, and Society*, 8, 2016, pp. 58–82.

Andersson Hult, Lars. Historia i bagaget. En historiedidaktisk studie om varför historiemedvetande uttrycks i olika former. Umeå universitet: Umeå 2016.

Ashby, Rosalyn/Edwards, Cristopher. "Challenges facing the disciplinary tradition: Reflections on the history curriculum in England". In Nakou,

I. & Barca, I. (eds.), *Contemporary Public Debates Over History Education*. IAP: Charlotte, NC 2010.

Barton, Keith C./Levstik, Linda S. *History*. In Arthur, J., Hahn, C., & Davies, I. (eds.), *Handbook of Education for Citizenship and Democracy*. Sage: London 2008.

Barton, Keith.C./Levstik, Linda. S. *Teaching History for the Common Good*. Lawrence Erlbaum Associates: Mahwah, NJ 2004.

Barton, Keith C./McCully, Alan W. "History, identity, and the school curriculum in Northern Ireland: An empirical study of secondary students' ideas and perspectives". *Journal of Curriculum Studies*, 37(1), 2005, pp. 85–116.

Berg, Mikael. *Historielärares ämnesförståelse: centrala begrepp i historielärares förståelse av skolämnet historia*. Karlstads universitet: Karlstad 2014.

Berg, Petter/Palmgren, Ola/Tyrefors, Björn. *Gender Grading Bias in Junior High School Mathematics*. Research Institute of Industrial Economics: Stockholm 2019.

Berger, Stefan. "De-nationalizing history teaching and nationalizing it differently! Some reflections on how to defuse the negative potential of national(ist) history teaching". In Carretero, M., Asensio, M., & Rodríguez-Moneo, M. (eds.), *History Education and the Construction of National Identities*. Information Age Publishing: Charlotte, NC 2012.

Bergmann, Klaus. "Emanzipation". In Bergman, K., Kuhn, A., Rüsen, J., & Schneider, G. (eds.), *Handbuch der Geschichtsdidaktik*. Kallmeyer'sche Verlagsbuchhandlung: Düsseldorf 1985.

Blow, Frances/Rogers, Rick/Shemilt, Denis/Smith, Claire. "Only connect: How students form connections within and between historical narratives". In Chapman, A. & Wilschut, A. (eds.), *Joined-Up History: New Directions in History Education Research*. Information Age Publishing: Charlotte, NC 2015.

Cahill, Larry. "Why sex matter for neuroscience". *Nature Reviews. Neuroscince*, 7(6), 2006, pp. 477–484.

Carretero, Mario. "Teaching history master narratives: Fostering Imagi-Nations". In Carretero, M., Berger, S., & Grever, M. (eds.), *Palgrave Handbook of Research in Historical Culture and Education*. Palgrave Macmillan: London 2017.

Clark, Mary Ann/Lee, Sang Min/Goodman, William/Yacco, Summer. "Examining male underachievment in public education: Action research at a district level". *NASPP Bulletin*, 92(2), 2008, pp. 111–132.

Coffin, Caroline. *Historical Discourse: The language of time, cause and evaluation*. Continuum: London 2006.

Costa, Paul T./Terraciano, Antonio/McCrae, Robert. R. "Gender differences in personality traits across cultures: Robust and surprising findings". *Journal of Personality and Social Psychology*, 81, 2001, pp. 322–331.

Delamont, Sara. "Gender and the discourse of derision". *Research papers in Education*, 14(1), 1999, pp. 3–21.

Edling, Silvia/ Mooney Simmie, Geraldine. *Democracy and Teacher Education. Dilemmas, Challenges, and Possibilities*. Routledge: London, NY 2020.

Eliasson, Per/Alvén, Fredrik/Axelsson Yngvéus, Cecilia/Rosenlund, David. "Historical consciousness and historical thinking reflected in large-scale assessment in Sweden". In Ercikan, K. & Seixas, P. (eds.), *New Directions in Assessing Historical Thinking*. Routledge: New York 2015.

Englund, Tomas. *Curriculum as a Political Problem: Changing Educational Conceptions, with Special Reference to Citizenship Education*. Lunds Universitet: Uppsala 1986.

Epstein, Debbie/Elwood, Janette/Hey, Valerie/Maw, Janet. "Schoolboy frictions: Feminism and failing boys". In Epstein, J., Elwood, J., Hey, V., & Maw, J. (eds.), *Failing Boys? Issues in Gender and Achievement*. Open University Press: Buckingham 1998.

Epstein, Terrie/Shiller, Jessica. "Race, gender, and the teaching and learning of national history". In Parker W. C. (ed.), *Social Studies Today: Research and Practice*. Routledge: New York 2009.

Ercikan, Kadriye/Seixas, Peter. (eds.). *New Directions in Assessing Historical Thinking*. Routledge: New York 2015.

Ercikan, Kadriye/Seixas, Peter. "Assessment of higher order thinking: The case of historical thinking". In Schraw, G. & Robinson, D.R. (eds.), *Assessment of Higher Order Thinking Skills*. Information Age Publishing: Charlotte, NC 2011, pp. 245–261.

Foster, Stuart. "Dominant traditions in international textbook research and revision". *Education Inquiry*, 2(1), 2011, pp. 5–20.

Francis, Becky. "Lads, lasses and (new) labour: 14–16–year–old students' responses to the laddish behavior and boys underachievement debate". *British Journal of Sociology of Education*, 20(3), 1999, pp. 355–371.

Fredriksson, Ulf/Taube, Karin. *Läsning, läsvanor och läsundersökningar*. Studentlitteratur: Lund 2012.

Gadamer, Hans-Georg. *Truth and Method*. Continuum Impacts: London & New York 2006.

Gaddis, John Lewis. *The Landscape of History: How Historians Map the Past*. Oxford University Press: Oxford 2002.

Gibb, Sheree J./Fergusson David, M./Horwood, John. "Gender differences in educational achievement to age 25". *Australian Journal of Education*, 52(1), 2008, pp. 63–80.

Grever, Maria. "Dilemmas of common and plural history: Reflections on history education and heritage in a globalizing world". In Carretero, M., Asensio, M., & Rodríguez-Moneo, M. (eds.), *History Education and the Construction of National Identities*. Information Age Publishing: Charlotte, NC 2012.

Gurian, Michael. *Boys and Girls Learn Differently! A Guide for Teachers and Parents*. Jossey-Bass: San Francisco, CA 2001.

Hinnerich, Björn/Höglin, Erik/Johannesson, Magnus. "Are boys discriminated in Swedish high schools?". *Economics of Education Review*, 30, 2011, pp. 682–690. https://www.sciencedirect.com/science/article/abs/pii/S0272775711000227?via%3Dihub, retrieved 2020-04-24.

Hinnerich, Björn/Vlachos, Jonas. *Lika för Alla? Omrättning av Nationella Prov i Grundskolan och Gymnasieskolan under tre år.* Skolinspektionen Resultatbilaga, Dnr U2009/4877/G: Stockholm 2013.

Jackson, Carolyn. "Motives for Ladishness at school: Fear of failure and fear of the feminine". *British Educational Research Journal*, 29, 2003, pp. 583–598.

Jeismann, Karl Ernst. "Geschichtsbewußtsein". In Bergman, K./Fröhlich, K., & Kuhn, A. (eds.), *Handbuch der Geschichtsdidaktik*. Kallmeyer: Düsseldorf 1979.

Karlsson, Klas-Göran. "Historia, Historiedidaktik och historiekultur – teori och perspektiv". In Karlsson, K.-G. & Zander, U. (eds.), *Historien är närvarande. Historiedidaktik som teori och tillämpning*. Studentlitteratur: Lund 2014.

Karlsson, Klas-Göran. "Historiedidaktik: begrepp, teori och analys". In Karlsson, K.-G. & Zandler, U. (eds.), *Historien är nu. En introduktion till historiedidaktik*. Studentlitteratur: Lund 2009.

Karlsson, Klas-Göran. *Historia som vapen*. Natur och Kultur: Stockholm 1999.

Kessler, Eric/Wong-MingJi, Diana. (eds.). *Cultural Mythology and Global Leadership*. Edward Elgar Publishing: Cheltenham 2009.

Kitson, Alison/Husbands, Chris/Steward, Susan. *Teaching and Learning History 11–18, Understanding the Past*. Open University Press: Berkshire 2011.

Kölbl, Carlos/Konrad, Lisa. "Historical consciousness in Germany: Concept, implementation, assessment". In Ercikan, K. & Seixas, P. (eds.), *New Directions in Assessing Historical Thinking*. Routledge: New York 2015.

Kuhn, Anette. *Einführung in der Didaktik der Geschichte*. Kösel-Verlag: München 1980.

Larsson, Hans-Albin. *Mot bättre vetande: en svensk skolhistoria*. SNS förlag: Stockholm 2011.

Laville, Christian. "Historical consciousness and historical education: What to expect from the first for the second". In Seixas, P. (ed.), *Theorizing Historical Consciousness*. University of Toronto Press: Toronto 2004.

Lee, Peter. "Understanding history". In Seixas, P. (ed.), *Theorizing Historical Consciousness*. Toronto: University of Toronto Press 2004.

Lee, Peter. "Historical knowledge and the national curriculum". In Aldrich, R. (ed.), *History in the National Curriculum*. Kogan Page: London 1991.

Lee, Peter/Ashby, Rosalyn. "Progression in historical understanding among students ages 7–14". In Stearns, P.N., Seixas, P., & Wineburg, S. (eds.), *Knowing, Teaching, and Learning History: National and International Perspectives*. New York University Press: New York 2000.

Lee, Peter/Howson, Jonathan. "Two out of five did not know that Henry VIII had six wives: History education, historical literacy, and historical consciousness". In Symcox, L. & Wilschut, A. (eds.), *National History Standards: The Problem of the Canon and the Future of Teaching History*. Information Age Publishing: Charlotte, NC 2009.

Levander, Sten. "Tar skolan hänsyn till könsskillnader?" *Nämnaren 2*, 1990. http://ncm.gu.se/pdf/namnaren/1117_90_2.pdf, retrieved 2020-04-24.

Martinková, Patricia/Drabinová, Adéla/Liaw, Yang- Ling/Sanders, Elizabeth/McFarland, Jenny/Price, Rebecca. "Checking equity: Why differential item functioning analysis should be a routine part of developing conceptual assessments". *CBE-Life Sciences Education*, *16*(2), 2017.

Nordgren, Kenneth. *Vems är historien?: historia som medvetande, kultur och handling i det mångkulturella Sverige*. Karlstads universitet: Karlstad 2006.

Nordgren, Kenneth/Johansson, Marie. "Intercultural historical learning: A conceptual framework". *Journal of Curriculum Studies*, 47(1), 2015, pp. 1–25.

Nygren, Thomas. *Erfarna lärares historiedidaktiska insikter och undervisningsstrategier*. Uppsala universitet: Uppsala 2009.

OECD. "What lies behind gender inequality in education?". *PISA in Focus*, No. 49, OECD Publishing: Paris, 2015, https://doi.org/10.1787/5js4xffhhc30-en, retrieved 2020-04-24.

OECD. *Equally Prepared for Life? How 15-Year-Old Boys and Girls Perform in School: Programme for International Student Assessment*. OECD: Paris 2009.

Pingel, Falk. *UNESCO Guidebook on Textbook Research and Textbook Revision*. United Nations Educational, Scientific and Cultural Organization: Paris 2010.

Rosenzweig, Roy/Thelen, David. *The Presence of the Past: Popular Uses of History in American Life*. Columbia University Press: New York 1998.

Rüsen, Jörn. "Forming historical consciousness: Towards a humanistic history didactics". In Nordgren, K., Eliasson, P., & Rönnqvist, C. (eds.), *The Processes of History Teaching*. Karlstads universitet: Karlstad 2011.

Rüsen, Jörn. "Memory, history and the quest for the future". In Cajani, L. & Ross, A. (eds.), *History Teaching, Identities, Citizenship*. Trentham Books: Trentham 2007.

Rüsen, Jörn. *History: Narration, Interpretation, Orientation*. Berghahn Books: New York 2005.

Rüsen, Jörn. "Historical consciousness: Narrative structure, moral function, and ontogenetic development". In Seixas, P. (ed.), *Theorizing Historical Consciousness*. University of Toronto Press: Toronto 2004.

Ryan, Allison/Shim, Serena/Lampinkins-Thando/Shawn, Kiefer, Sarah/Thompson Geneene. "Do gender differences in help avoidance vary by ethnicity? An examination of African American and European American students during early adolescence". *Developmental Psychology*, 45(4), 2009, pp. 1152–1163.

Sandberg, Karin. "Manligt och kvinnligt i skolämnet historia. Vad har det för konsekvenser för elevernas historiekultur?" *Nordidactica 2019:1*, 2019.

Sax, Leonard. *Why Gender Matters. What Parents and Teachers Need to Know about the Emerging Science of Sex Differences*. Doubleday: New York 2005.

Seixas, Peter. "A modest proposal for change in Canadian history education". *Teaching History*, 137, 2009, pp. 26–32.

Seixas, Peter. *Theorizing Historical Consciousness*. University of Toronto Press: Toronto 2004.

Seixas, Peter/Morton, Tom. *The Big Six Historical Thinking Concepts*. Nelson Education: Toronto 2013.

Semmet, Sylvia. "Controversiality and consciousness: Contemporary history education in Germany". In Taylor, T. & Guyver, R. (eds.), *History Wars and the Classroom: Global Perspectives*. Information Age Publishing: Charlotte, NC 2012.

Shemilt, Denis. "Drinking an ocean and pissing a cupful: How adolescents make sense of history". In Symcox, L. & Wilschut, A. (eds.), *National History Standards: The Problem of the Canon and the Future of Teaching History*. Information Age Publishing: Charlotte, NC 2009.

Shemilt, Denis. "The caliph's coin: The currency of narrative frameworks in history teaching". In Stearns, P.N., Seixas, P., & Wineburg, S. (eds.), *Knowing,*

Teaching, and Learning History: National and International Perspectives. New York University Press: New York 2000.

Skolverket (National Agency for Education). *Database SIRI* 2020, https://www.skolverket.se/skolutveckling/statistik/sok-statistik-om-forskola-skola-och-vuxenutbildning?sok=SokC, retrieved 2020-04-24.

Skolverket (National Agency for Education). *Läroplan för grundskolan, förskoleklassen och fritidshemmet 2011*. Skolverket: Stockholm 2011.

Skolverket (National Agency for Education). *Könsskillnader i måluppfyllelse och utbildningsval*. Skolverket: Stockholm 2006.

Skolverket (National Agency for Education). *Grundskolan: kursplaner och betygskriterier*. Skolverket: Stockholm 2000.

SOU (2009:64). *Flickor och pojkar i skolan – hur jämställt är det?* Delbetänkande av Deja-delegationen för jämställdhet i skolan. Fritze: Stockholm 2009.

Stearns, Peter/Seixas, Peter/Wineburg, Sam. *Knowing, Teaching, and Learning History: National and International Perspectives*. New York University Press: New York 2000.

Stobart, Gordon. "Fairness in multicultural assessment systems". *Assessment in Education: Principles, Policy and Practice*, 12(3), 2005, pp. 275–287.

Thompson, D. "Some psychological aspects of history teaching". In Burston, W.H. & Green, C.W. (eds.), *Handbook for History Teachers*. Methuen: London 1962.

UKÄ (Swedish Higher Education Authoriy). *Statistiska meddelanden*, UM 46, SM 1841, 2018. https://www.uka.se/download/18.7edd426c16658763f4b9e26/1542725338190/2018-11-21-statistiskt-meddelande-sokande-och-antagna-till-hogskoleutbildning-grundniva-och-avancerad-niva-hostterminen-2018.pdf, retrieved 2020-04-24.

Utbildningsdepartementet. *Kursplaner för grundskolan*. Stockholm 1994.

VanSledright, Bruce A. *Assessing Historical Thinking and Understanding: Innovative Designs for New Standards*. Routledge: New York 2014.

VanSledright, Bruce A./Limon, Margareta. "Learning and teaching social studies: A review of cognitive research in history and geography". In Alexander, P.A. & Winne, P.H. (eds.), *Handbook of Educational Psychology*. Erlbaum: Mahwah, NJ 2006.

Warrington, Molly/Younger, Mike/Williams, John. "Student attitudes, image and the gender gap". *British Educational Journal*, 26(3), 2000, pp. 393–407.

Weaver-Hightower, Marcus. "The boy turn in research on gender and education". *Review of Educational Research*, 73(4), 2003, pp. 471–498.

Wertsch, James V. *Voices of Collective Remembering*. Cambridge University Press: Cambridge 2002.

White, Hayden. *Metahistory: The Historical Imagination in Nineteenth-Century Europe*. Johns Hopkins University Press: Baltimore, MD 1973.

Wilschut, Arie. *Images of Time: The Role of HiCo of Time in Teaching History*. Information Age Publishing: Charlotte, NC 2012.

Zander, Ulf. *Fornstora dagar, moderna tider: bruk av och debatter om svensk historia från sekelskifte till sekelskifte*. Lunds Universitet: Lund 2001.

Zieky, Michael. *A DIF Primer*. Educational Testing Service: Princeton, NJ 2003. www.ets.org/s/praxis/pdf/dif_primer.pdf, retrieved 2020-08-05.

Zimmerman, Fredrik. *Det tillåtande och det begränsande: en studie om pojkars syn på studier och ungdomars normer kring maskulinitet*. Göteborgs universitet: Göteborg 2018.

Zwick, Rebecca. *A Review of ETS Differential Item Functioning Assessment Procedures: Flagging Rules, Minimum Sample Size Requirements, and Criterion Refinement*. ETS RR 12-08: Princeton NJ 2012.

Part II Researches about opinions and perceptions of teachers in initial training and in practising

David Parra Monserrat, Xosé Manuel Souto González, and Nancy Palacios Mena

Chapter 5 The representation of geographical and historical education among teachers-in-training: An Ibero-American perspective[1]

Abstract: Due to the importance of rethinking and reframing the epistemology, methodology and objectives that have traditionally shaped the disciplinary codes of geography and history, this chapter tackles the representation of said school subjects. This research has been carried out through the study of the memories and conceptions of 200 students (teachers-in-training) from the Bachelor's Degree in Primary Education Teaching and the Master's degree in Secondary Education Teaching (or equivalents in Spain, Colombia, Chile, Brazil and Portugal). Tradition and individual memory, both based on life experiences and personal memories, reshape and mix with school's collective memory. This research aims to prove that the influence of this phenomenon has created a consensual common sense that hinders change, naturalizes codes and guarantees the continuity of certain narratives, practices and uses that are not useful from the perspective of a citizenship education. To this end, a tool that evaluates the degree of influence of school experiences in the participants, the impact of pedagogic and disciplinary education and the effect of curricular system and framework where the configuration and reproduction of said representations takes place is used.

Keywords: Geography Education, History Education, Social representations, Teacher training, School memory

Introduction

It is not simple to explain how we understand and teach social problems in schools, as it depends on a number of factors, some of which are related to personal decisions and others to the normative framework and the academic context (here we are focusing on initial teacher training). Furthermore, personal decisions are loaded with memories, feelings, expectations, and desires which

---

1 This study is the result of the research Project entitled "The social representations of educational content in the development of teaching competences" (PGC2018-094491-B-C32), funded by the Spanish Ministry of Science, Innovation and Universities and co-funded by the ERDF.

have been shaped, often unconsciously, by the influence of the dominant culture. That is why it is vital to study how students preparing to be teachers perceive the academic discipline they will teach, given that these representations have an impact both on the discourse and actions of teachers and pupils, as well as on the objectives attributed to academic knowledge.

Addressing this subject of study (the representation that future teachers of these subjects have of geography and history) and doing so, moreover, from an Ibero-American perspective (the sample includes students from Brazil, Colombia, Chile, Spain and Portugal) has theoretical and methodological implications. On the one hand, we must of course refer to the context of social representations, a clearly interdisciplinary theoretical framework which coincides with the so-called behavioral shift in human and social sciences (Jodelet, 1985; Abric, 2001; Moscovici, 2011; Sammutt et al., 2015). Second, given that we are dealing with a comparative study, we must use analytical tools and categories which allow us to carry out this comparison, taking into account the particularities of the various territories/contexts studied in this research, but also the greater or lesser extent of disciplinary didactic training received during undergraduate and/or postgraduate study.

Thanks to the research we have carried out to date, we are able to specify the problem we wish to address (Souto, Martínez, & Parra, 2014; Souto, 2016; Parra & Fuertes, 2019). As such, we can formulate it as the persistence of routines, narratives and uses in geography and history teaching which send us back to the standardized models institutionalizing the school system. As a result, a large proportion of pupils show apathy toward, if not completely rejecting, these subjects. In fact, numerous studies identify social sciences as the least relevant subjects in the school curriculum for secondary school pupils. This topic has been studied from various perspectives (psychological, sociological, epistemological) both for history and geography teaching (González, 2002; Alonso-Tapia et al., 2008). This phenomenon can be explained in relation to two main factors: on the one hand, an epistemological foundation for history education, which consisted of an overview of general histories thus presenting a teleological timeline from the origins of nations to the present (Maestro, 2005) and, on the other hand, a description of different territories (regional and national) for geography education, which combined factors concerning the physical environment and human activity (Lambert & Morgan, 2010). Both factors generated an explanation of political legitimacy, since these subjects presented an immutable territory, on which we developed our history, which contributed to naturalizing and perpetuating the great narratives of the Nation or Modernity (Souto, 2010). Furthermore, another vital aspect is the impact of social agents who took part in the creation of a disciplinary code, which demonstrated a summary of prestigious academic historical knowledge with geography presented as a harmonious scenario between resources and the population (Cuesta, 1997; Luis & Romero, 2007). These two explanations have provided us with a critical approximation of teachers' programs, resources, and actions. However, it is not sufficient to challenge a model which has been repeated one academic year after another (Tutiaux-Guillon, 2008; Cuban, 2016; Harris & Burn, 2016; Markuszewska, Tanskanen, & Vila, 2018). It is necessary to explore the mechanisms which impede or paralyze innovative activities in the teaching of geography and history, and understand those which activate the practicality of teachers who tend to take a stand on more than one occasion against the criteria which they claim to ideologically accept.

To this end, it is essential to mention *conventional wisdom* that has been formulated about the teaching profession, which comes with certain expectations and behavior or, in other words, the development of *practicality*, routines and habits which are very difficult to change (Fernández & Gurevich, 2018). For this reason, we opted for a mixed approach (quantitative and qualitative) in order to be able to uncover the intentions concealed by apparent behavior and verbal statements, because we need to understand how certain decisions are objectified as if they were natural (e.g. following a syllabus) and how teaching behavior is anchored in cultural academic expectations, which tend to be criticized in daily conversations.

As a result, the main objective of this research is to identify the impact of school recollections and academic contexts, as well as the influence of disciplinary training, on the representation of geography and history among future primary and secondary school teachers. In this vein, this study will try to answer five main questions: (1) What impact do school recollections have on the representation of geography and history among future teachers? (2) What impact do the educational settings experienced by these teachers have on upholding or breaking down these ideas? (3) Does greater or lesser epistemological disciplinary training have an impact on these representations? (4) Are there significant differences between students training to become primary school teachers and those training to become secondary school teachers? (5) How is disciplinary content related with the everyday concerns of society, in particular with the citizens we know as pupils?

Using these questions, we aim to fuel a discussion ranging from the personal expectations and experiences of teachers in training to the definition of the conceptual framework of disciplinary standards which are perpetuated in school activities. As such, this study is part of a wider field and program of research, which links initial training with continuous training, in line with suggestions to make academic research compatible with school innovation.

Method

The methodology cannot be reduced to a few methods of empirical verification: it starts with selecting the problem that is the object of study and continues with formulating the hypothesis. It implies an intrinsic relationship between the subject and the object of knowledge, which is why it is necessary to have a method and procedures which are used to achieve the objectives we outlined previously.

Given that the objectives involve identifying the impact of school recollections, the institutional framework of teacher training and epistemological knowledge on the expectations of future geography and history teachers in primary and secondary school, it is necessary to carry out a series of intellectual activities contrasted with data originating from the designed tools.

Research design, context, and participants

This study used a sample of 200 students from 5 Ibero-American countries (Brazil, Chile, Colombia, Spain and Portugal) and 11 universities (University of São Paulo, Federal University of São João del Rei, University of La Serena, Francisco José de Caldas District University, University of Tolima, Complutense University of Madrid, University of Granada, University of Valencia, University of Coimbra, University of Lisbon and University of Porto) (see Tab. 1). The students participated during the 2019/20 academic year and the sample was non-probabilistic (available or convenient sample), given that we deliberately chose future social science, history and geography teachers, both for primary

Tab. 1: Participant characteristics

Country	Degree in Primary Education or equivalent	Master's Degree in Secondary Education or equivalent	Total
Brazil	11	13	24
Chile	17	18	35
Colombia	17	8	25
Spain	38	41	79
Portugal	10	27	37
Total	93	107	200

and secondary education; it is a casual sample as we chose participants to whom we had access (Flick, 2009). Furthermore, when selecting the sample group, we took into account the particularities of initial training in each country. In Brazil and Portugal, history and geography are taught as two separate subjects in secondary school teacher training, while in Chile, Colombia and Spain the subjects are taught together in conjunction with other social sciences.

Tool

In order to gather data, we designed a mixed questionnaire categorized in three sections and made up of open and closed questions. The questionnaire was verified by three international experts. We provided them with a validation guide which explained the objectives of each question and the analytical category the response may pertain to, as shown in Tab. 2, so that they could evaluate the pertinence, clarity and relevance of the questions.

Although the questionnaire provided us with a wealth of easy-to-process information in comparative terms, we are aware of the limitations of this type of tool, both from a technical perspective and in terms of further detail needed to understand the phenomena it claims to address (Escofet el al., 2016). That is why we believe that the data gathered and analyzed in this chapter should serve, above all, to present the problems we wish to explore, in particular the associated routines and practicality which are "naturally" formulated in the initial training process and which we can only outline in the following pages.

Tab. 2: Example of the validation guide for the questionnaire

QUESTIONS	OBJECTIVES	CATEGORY
2.- On a scale from 1 to 5 (1: extremely poor; 5: excellent), how do you evaluate your understanding of history and geography? Geography 1 2 3 4 5 History 1 2 3 4 5	To understand their own perceptions of their learning and education of these subjects to evaluate if their previous qualifications have any impact.	Relationship with the subject
3.- How would you define these disciplines? History: Geography:	To identify the epistemological disciplinary background of the participants in order to evaluate the link between their recollections and future teaching practice.	Epistemological disciplinary conception

In this sense, other methodological strategies could go into more detail, such as discussion groups or profound analysis of case studies, in order to delve into the complex web of topics such as participants' expectations of their future teaching practice. As a result, it is clear that we need to complement quantitative methods with qualitative techniques for critical methodology, which involve modifying questionnaires and debates in relation to the needs of the subject and context.

As mentioned previously, the questionnaire included 13 questions categorized in three sections: a) General considerations about the subjects of geography and history in the educational sphere; b) School recollections of these subjects; 3) Predisposition to practice in their future teaching work (Tab. 3).

As we can observe in Tab. 3, the first section focused, on the one hand, on the nature and uses of geography and history and, on the other, on emotions, preferences or affective aspects which interfere with the way in which we represent a subject. Given the importance of school experiences and recollections in the development of representations of school disciplines, the questionnaire continued with questions related to content, methodological strategies, resources used and the objectives identified in the classes they remembered (McCrum, 2013). Finally, the third section focused on aspects related to the future professional practice of the participants: specifically their idea of innovation, how they see the role of the teacher and the types of skills that should be promoted in the teaching/learning of geography and history. All of these aspects have an impact on their representation of the discipline and their own personal trajectory.

We analyzed the responses using a qualitative approach, with an open codifying procedure which led to the development of a number of categories. Moreover, these categories were submitted for verification by research professors at the universities which participated in this study.

Results and discussion

Perception and recollections of geography and history lessons

As previously mentioned, social representations and the derived personal decisions or actions are deeply linked with memories, emotions, expectations and desires which shape and fuel them (Castorina & Barreiro, 2012; Palacios, 2018a; Domingos & Diniz, 2019). That is why we believe it is pertinent to start with the participants' (future teachers) emotional relationship with the subjects they are planning to teach in the future. Neuro-educational studies also reveal

Tab. 3: Questionnaire about recollections and representations of school history and geography

Section I. General considerations about geography and history
1. Do you enjoy history and geography? (1: not at all, 2: a little, 3: indifferent, 4: somewhat, 5: a lot) Why?
2. How do you evaluate your understanding of history and geography? (1: extremely poor, 5: excellent)
3. How would you define these disciplines? (geography/history)
4. Please write, in order of relevance, three words you associate with history and geography.
5. Why are these subjects useful? What do you think their purpose should be in the education of primary and secondary school pupils? (geography/history)
Section II. Recollections of school experiences with history and geography
6. In general, what do you remember about history and geography classes during secondary school and baccalaureate? (content, methodologies, resources, teacher's role/attitude. . .)
7. At university, has there been any change in the type of content, methodologies or resources used in history and geography classes? If yes, what are the most significant differences?
8. At any point, did you think that the history and geography lessons you were given were ideologized or politicized? Do you think it is good for teachers to take a stance in class? Why?
Section III. Considerations about future teaching practice in history and geography
9. As a future teacher, what should the foundations of educational innovation in the didactics of history and geography be?
10. Do you think it is better to have one teacher specialized in history and geography or should there be two teachers, one for each subject? Why?
11. Do you agree (YES) or disagree (NO) with the following phrases? — The teacher should aim, above all, to understand the great social problems of the modern world so that they can be the subject of analysis in school. — The teacher should follow the instructions of official journals to plan their subject matter. — The teacher should adjust their syllabus so that the content of the textbook can be studied throughout the duration of the school year. — The subject syllabus should take into account the local environment where the school is located. — What pupils, their families and neighbors think should be part of the educational content included in the subject syllabus. — The most important part of creating a syllabus is understanding the subject you teach.

Tab. 3: Continued

12. In your opinion, what are the most important skills that a geography and history teacher must have? Put the following skills in order of importance (1st, 2nd, 3rd, 4th, 5th, 6th).
— Their ability to connect with pupils' families.
— Their ability to select the problems to be studied in the educational content.
— Their aptitude to create a sequence of didactic activities over a set period of time.
— Their character to maintain order in the classroom and prevent an uncomfortable environment
— Their contribution to the school's educational project
— Their mastery of their subject
13. Do you agree (YES) or disagree (NO) with the following phrases?
— Objective exams are the most important part of assessment
— Assessments should include a report of the pupil's participation delivered to the Assessment Board
— Assessments are an evaluation tool to improve the pupil's learning
— The main objective of assessments is to inform families of the control they have over the pupil's learning.

that, as with other fields of knowledge, emotions have an impact on our way of reasoning. Meanwhile, research in sociology of education demonstrates that information saturation, in the search for homogenous concepts, simplifies and guides thinking about social problems.

In this vein, the surveys revealed that the participants in this study generally have a relatively positive perception of the subjects we are looking at, given that the overall results, as shown in Tab. 4 (the last two columns), show a score of over 4 points (on a scale from 1 to 5) both for history and geography, regardless of educational background. If we look at the results per country, we note certain differences but under no circumstances is the score lower than 3.5 points. As a result, we can confirm that a large majority of the sample participants quite or greatly enjoy the disciplines that are the subject of this study (we must put this into context in terms of the formulation of the question and the type of student, as it is difficult not to enjoy the path we have chosen, as well as the convenient selection of the sample, which implies the willingness of the participants). Beyond the quantitative data, the qualitative arguments provided to support their ratings are relevant. Accordingly, both for history and geography, the arguments in favor concentrate on the usefulness to "understand the present," "understand relationships between human beings and the environment" or "not to repeat the mistakes of the past." Furthermore, in some cases,

Tab. 4: Do you enjoy history and geography? (1: not at all, 2: a little, 3: indifferent, 4: somewhat, 5: a lot)

	Brazil		**Chile**		**Colombia**		**Spain**		**Portugal**		**Total**	
	DPE	**MSE**	**DPE**	**MSE**	**DPE**	**MSE**	**DPE**	**MSE**	**DPE**	**MSE**	**DPE**	**MSE**
History	4.8	4.6	4.6	4.8	4.5	4.5	3.8	4.8	3.5	3.9	4.2	4.5
Geography	4.2	4.5	4.0	3.6	4.1	4.8	3.6	4.5	4.7	4.9	4.1	4.5

DPE: Degree in Primary Education/MSE: Master's Degree in Secondary Education

the opportunity to contribute to the education of critical and engaged citizens with intercultural or democratic values was highlighted:

Br_10: "*É a minha disciplina de formação, é tenho a convicção que o conhecimento histórico da Humanidade pode levar a mudanças sociais e políticas que tanta necessitamos*".

Ch_34: "*I like history because it addresses the development of natural cultures in Latin America and some historical processes which are relevant for the development of modern society, such as the dictatorships experienced in Latin America during the 20th century.*"

Co_4: "*I feel a deep affinity for the recognition of urban space, spatial and territorial recognition and geographical knowledge in general – I believe these aspects contribute to active citizens with regard to reflecting about the development and construction of space.*"

Participants who do not enjoy the subject consistently bring up aspects linked to teaching methods (memorizing, reproducing lessons) and the lack of pupil participation, which often leads them to describe the subject as "boring":

Es_10: "*I don't like it much because the subject involves a lot of dates and the only way to learn is by memorizing.*"

Pt_24: "*Por outro lado história não me cativa tanto precisamente pelas aulas expositvas e pouco dinamicas que tinha*"

In both cases, most participants talk about their own school experience and the influence of their teachers in order to evaluate their perception of the subject positively or negatively:

Ch_13: "*It's because at school and university the teaching of geography tends to involve memorizing and rarely leaves room to apply knowledge based on geographical problems.*"

Ch_14: "*It was one of the most relevant subjects in terms of content during early and middle education. There were also optional modules in which we could participate.*"

Es_6: "*It's thanks to the teachers who used to teach me this subject and the one I have now, because they gave us a broader and more interesting vision of what geography is.*"

As a result, the responses give us a good impression with regard to the objectives of geography and history education, while negative comments concentrate on methodological aspects. In other words, most of the participants enjoy the "why" and the "purpose," but they reject, as a result of their recollections, concrete aspects of "how." This all means that it is fundamental to go into more detail in the school memories of participants, with some individual recollections which, when combined with the overall recollection of school, have led to certain consensuses and *a practicality* which will condition their future professional practice.

As a consequence, in relation to their school recollections, we noted that the types of content presented by both future primary school and secondary school teachers are remarkably similar. There were also no great differences between countries of origin. In the field of geography, participants focused on two traditional aspects of the discipline: physical (hydrology, geomorphology, climatology, etc.) and human (demography, development, land use, political organization, economy, etc.) with a special mention for national geography in the case of the three Latin American countries. Moreover, it should be noted that in Chile numerous participants highlighted the scarce discussion of geographical content during secondary school. With regard to history, there is a clear prevalence of factual content (mainly of a political nature) or important socio-political and economic processes which make up the central core of main school narratives. In this vein, various references to national histories stood out (which reveals the persistence of certain overriding narratives and identity-based uses of historical education), in addition to references to European history. On the latter point, it is striking that, in the three Latin American countries, there are many more references to European history than, for example, the history of pre-Colombian America or the proverbial "native or indigenous peoples":

Br_1: "*Predominava o estudo da história brasileira e da européia.*"

Ch_8: "*What I remember most is the history of my own country (Chile), independence and how our nation was built. With regard to global history, there was always an emphasis on the world wars, always addressed from a Euro-centric point of view."*

Co_9: "*In history classes, content referring to "Occidental History", with the European man as the focal point of history, was predominant.*"

However, some Brazilian participants acknowledge that the Workers' Party (PT) government made changes aimed at widening the view and presenting a more inclusive and not so Euro-centric history during their mandate. This trend has stagnated (and even shrunk) in recent years.

Br_18: "*Hoje, como pesquisador em Educação, entendo que muitos conteúdos ficaram de fora de nossas aulas, Não estuamos por exemplo, as histórias da Asia e da África e pouco vimos da história dos povos nativos das Américas. Nos anos seguintes, já durante a minha graduação em História, foram implementadas leis para assegurar a existência dos conteúdos citados, nos currículos da educação básica e também nos cursos de formação inicial de professores das áreas de linguagens e ciências humanas*".

As regards teaching strategies and resources, we also did not note any significant differences between the recollections of different participants (Graph 1). Use of the textbook is predominant in all the groups analyzed, both for history and geography lessons (always with a value over three on a Likert scale from 1 to 4). It is mentioned both in relation to following lessons (reading, highlighting, etc.) and completing exercises (generally mechanic exercises of reproduction requiring a low cognitive level). Lectures or note-taking also appear with higher values, as well as notes provided by teachers or maps, especially in geography classes. Some categories which were mentioned less frequently included the use of audio-visual materials, technological resources, or field trips, although there were some exceptions (especially in recollections of history education in Chile and geography education in Portugal). All of this confirms other research carried out in the specific context of Spain (Ramiro, 2008; Martínez, 2014).

Despite these mostly traditional recollections, many participants present, as we have seen, a positive image of these lessons and their teachers, who the future secondary school teachers also consider to be part of the reason why they chose this course of study at university. As various studies have highlighted (Martínez, Souto, & Beltrán, 2006; Barton, 2008), the involvement, attitude and behavior of teachers plays a crucial role in school memories; but also in identification with the disciplinary code, which inevitably conditions how we understand and evaluate teaching practice (Cuesta, 1997; Grant, 2014; Cuban, 2016). In this vein, we must emphasize the clear relationship between this concept

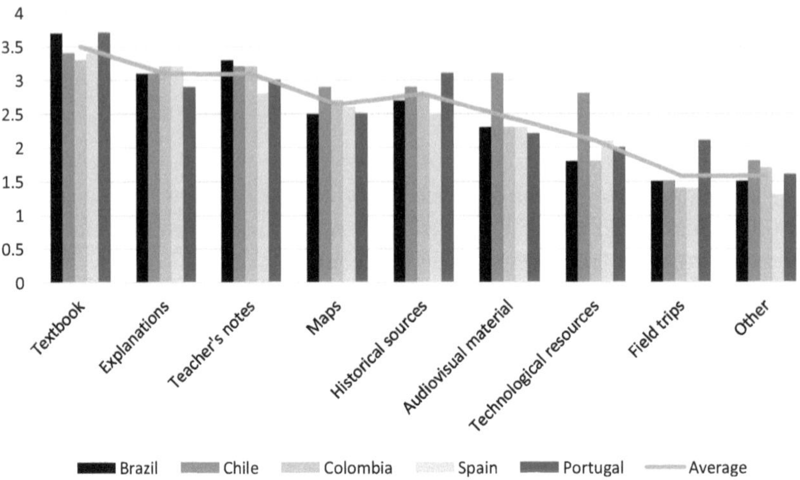

Graph 1: Recollections of resources used in history lessons

and social representations, given that, like the above, the disciplinary code has an impact on the discourse, actions and routines that teachers and pupils implement on a daily basis in the school environment. In this way, they end up creating a *habitus*, a *consensus on conventional wisdom*, a *practicality* which

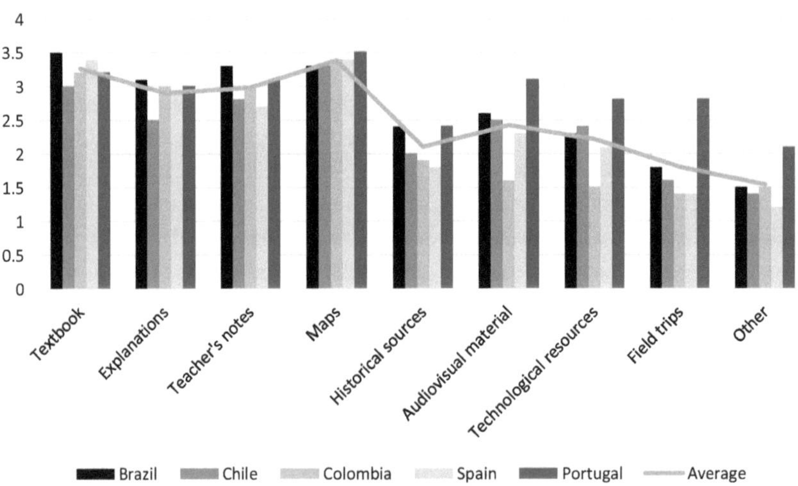

Graph 2: Recollections of resources used in geography lessons

naturalizes the teaching/learning of these subjects and makes change difficult (Castorina, 2017; ; Palacios, 2018b; Campo, García, & Souto, 2019; Domingos, 2019; Diniz, 2019; Parra, 2019).

Consequently, it is relevant to note that we observe more similarities than differences among the recollections of geography and history lessons, despite the different samples and traditions. Once again, this points to the strength of certain transnational disciplinary codes which perpetuate narratives, practices, and objectives on both sides of the Atlantic.

Representations and uses of geography and history

Moving beyond perceptions and recollections of history and geography, the questionnaire explored the conception and uses that future teachers attribute to these subjects.

The results regarding the definition of history show differences in terms of country of origin and the higher or lower degree of disciplinary training. As such, on the one hand we notice that more traditional characterizations (D1, D2 and D3) continue to be used by both future primary and secondary school teachers (Tab. 5), although they tend to appear more often among future primary school teachers. In this vein, the higher percentage of historicist responses (D2) stands out, focused on the narration/description of relevant facts in the past. However, we notice that the greater the disciplinary didactic training, the greater the number of more complex or more epistemologically rich answers (D4 and, especially, D5) related with the contributions of social history, *Annales* or, even, the cultural shift. Furthermore, our attention is called to the scarce use of response D5 among future primary school teachers, but also among future secondary school teachers in Portugal. This can probably be accounted for by the fact that teacher training in Portugal differentiates history and geography as two different subjects and most of the Portuguese participants have disciplinary training in geography, not history.

As concerns geography (Tab. 6), the main responses given by undergraduate students in all countries fell back on the definition of science "which describes planet Earth, its physical phenomena and characteristics" (over 35 % in all cases). The considerable support shown by Spanish students (both undergraduate and postgraduate) for this definition is particularly striking, which could be related once again with the scarce geography education of the survey participants (both future primary school teachers and secondary school teachers, most of whom have qualifications in history and art history). The

Tab. 5: Definitions of history

	Brazil		Chile		Colombia		Spain		Portugal	
	DPE	MSE	DPE	MSE	DPE	MSE	DPE	MSE	DPE	MSE
D1	0	0	0	0	1 (5.9 %)	1 (12.5 %)	5 (13.2 %)	0	1 (10 %)	2 (7.4 %)
D2	4 (36.4 %)	1 (7.7 %)	6 (35.3 %)	6 (33.3 %)	10 (58.8 %)	0	17 (44.7 %)	11 (27.5 %)	4 (40 %)	6 (22.2 %)
D3	2 (18.2 %)	1 (7.7 %)	1 (5.9 %)	2 (11.1 %)	2 (11.8 %)	1 (12.5 %)	8 (21.1 %)	1 (2.5 %)	1 (10 %)	6 (22.2 %)
D4	5 (45.5 %)	8 (61.5 %)	9 (52.9 %)	8 (44.4 %)	2 (11.8 %)	2 (25 %)	11 (28.9 %)	13 (32.5 %)	3 (30 %)	6 (22.2 %)
D5	0	3 (23.1 %)	1 (5.9 %)	3 (16.7 %)	2 (11.8 %)	4 (50 %)	0	15 (37.5 %)	0	3 (11.1 %)

D1) Important events which have taken place since the beginning of time.
D2) The science that studies and describes/narrates relevant events from the past.
D3) The science which presents and explains the reality of the past.
D4) The science that explains political, social, and cultural changes and processes throughout the years.
D5) Analysis/interpretation of past phenomena by historians using sources.
DPE: Degree in Primary Education/MSE: Master's Degree in Secondary Education

other responses are varied, although they tend to highlight the study of the relationship between human beings and the environment (D4) or the explanation of social, political and economic relations and the changes produced as a result (D5). Although this involves going beyond a descriptive or eminently physical geography, on very few occasions is it presented from a perspective of radical geography which problematizes these relationships. This all points toward a predominance of more classic perspectives of geography and, in general, visions that scarcely problematize or are open to more recent epistemological approaches related with the cultural or behavioral shift (Straforini, 2018). Despite this, as we have already seen for history, we notice differences between the various groups, given that the more rich and complex perspectives (D4, D5 and, especially, D6) are greater among postgraduate rather than undergraduate students. Furthermore, although the sample size is small, we notice a considerable difference with Brazil and Portugal, as well as undergraduate students in Chile, with regard to social commitment to a more political geography related with the idea of change; this conjecture is endorsed by other studies (Souto & García, 2019) which compare certain results regarding the definition of school geography in Spain with those given by students in Peru, Colombia and Brazil.

However, they are no more than initial interpretations which allow us to go into further depth in the future. To do this, we need to identify the connections of key words which define school geography (maps, environment, or landscape) with the use that is made of these concepts in the interpretation of political territories and personal spaces.

The ways in which disciplines are conceived have a clear incidence on the uses and objectives that participants attribute to geographical and historical education, as shown below.

With regard to the objectives attributed to history (Tab. 7), "classic" answers (O1, O2, O3, O6) are found alongside other more cognitive procedural (O7) and other distinctly critical answers (O8 and O9). The response based on the use of the discipline to "understand the present" is notably remarked in all groups. Behind this idea, there may be a certain finalism or teleologism, which is why, in general, the historical perspective supported by these teachers in training is more oriented toward tracking and identifying the origin of their world, their values or their identity in the past, rather than acknowledging its cultural dimension and dissociating it. In this way, we distinguish these responses from those that discuss "understanding the present to improve or transform the future" which are much fewer. Once again, the most relevant finding is that the most critical objectives (O7, O8 and O9) tend to be mentioned, as a whole, more often by Master's students who have greater epistemological disciplinary training than Degree students. In the same way, some of the more classic responses are almost not mentioned at all by students who have more education in history and geography.

With reference to the uses of geography, we also detect the presence of more or less "classic" objectives (O1, O2, O3) combined with others that focus on problematizing the local environment and the active role of humans on its transformation (O5 and O6). Once again, we note that, in general, the greater the disciplinary training, the greater the inclination toward this type of objective.

Even more meaning was attributed to all of these responses when we asked participants, in the last section of the questionnaire, to indicate what the foundations should be of innovation in geographical and historical education.

If we analyze the data in line with the participants' different educational backgrounds, we notice in all cases the clear presence of the predominant methodology (understood as the incorporation of certain "groundbreaking" resources) and educational psychology as a preferred option when defining educational innovation. For most participants, innovating in the instruction of geography and history involves adopting active methodologies which give the pupil a more active role and makes them the focus of learning. Similarly, many

Tab. 6: Definitions of geography

	Brazil		Chile		Colombia		Spain		Portugal	
	DPE	MSE	DPE	MSE	DPE	MSE	DPE	MSE	DPE	MSE
D1	1 (9.1 %)	0	0	0	2 (11.8 %)	0	0	0	0	0
D2	4 (36.4 %)	1 (7.7 %)	6 (35.3 %)	7 (38.9 %)	7 (38.9 %)	2 (25 %)	24 (63.2 %)	21 (51.2 %)	4 (40 %)	5 (18.5 %)
D3	1 (9.1 %)	0	0	1 (5.6 %)	1 (5.6 %)	0	8 (21.1 %)	0	1 (10 %)	1 (3.7 %)
D4	1 (9.1 %)	3 (23.1 %)	5 (29.4 %)	6 (33.3 %)	4 (23.5 %)	2 (25 %)	2 (5.3 %)	18 (43.9 %)	1 (10 %)	8 (29.6 %)
D5	2 (18.2 %)	6 (46.2 %)	5 (29.4 %)	3 (16.7 %)	3 (16.7 %)	2 (25 %)	5 (13.2 %)	2 (4.9 %)	2 (20 %)	7 (25.9 %)
D6	1 (9.1 %)	3 (23.1 %)	0	1 (5.6 %)	0	2 (25 %)	0	0	0	0

D1) Science that teaches us how to find and locate geographical points
D2) Science which describes the planet Earth, its physical phenomena, and characteristics
D3) Understanding of the territory where physical and/or social phenomena occur
D4) Science which establishes the relationship between human beings and the environment
D5) Science which explains social, political, and economic relations and the changes brought about in the environment
D6) Understanding/problematizing physical and social spaces as subjective spaces which shape identity
DPE: Degree in Primary Education/MSE: Master's Degree in Secondary Education

participants highlighted the introduction of new strategies or resources such as ICT, gamification, or cooperative work, which leads to more "participative," "interactive" and "fun" learning. Making use of active pedagogical styles often turns into ingenuous activism which is more focused on motivating and entertaining than learning; however, it must be noted that some responses justify the introduction of these practices and resources by referring to constructivist approaches and the need to ensure meaningful learning.

A second block of responses focuses on the development of abilities and, very specifically, historical, and geographical thinking skills. Moreover, an even smaller group underlines the low number of responses which link innovation to re-evaluating/problematizing the content, epistemological approaches, or objectives. Once again, in these types of responses we notice a significant difference between the participants, given that Master's students tend to refer to these types of questions more often than participants training to become primary school teachers.

Tab. 7: Objectives of history

	Brazil		**Chile**		**Colombia**		**Spain**		**Portugal**	
	DPE	**MSE**	**DPE**	**MSE**	**DPE**	**MSE**	**DPE**	**MSE**	**DPE**	**MSE**
O1	0	0	0	0	1 (5.9 %)	0	5 (13.2 %)	3 (7.3 %)	2 (20 %)	0
O2	1 (9.1 %)	0	2 (11.8 %)	2 (11.1 %)	4 (23.5 %)	0	6 (15.8 %)	5 (12.2 %)	1 (10 %)	3 (11.1 %)
O3	2 (18.2 %)	0	2 (11.8 %)	0	2 (11.8 %)	0	8 (21.1 %)	0	1 (10 %)	3 (11.1 %)
O4	3 (27.3 %)	4 (30.8 %)	3 (17.7 %)	7 (38.9 %)	5 (29.4 %)	2 (25 %)	10 (26.3 %)	19 (46.3 %)	2 (20 %)	5 (18.5 %)
O5	0	0	1 (5.9 %)	1 (5.6 %)	0	0	0	8 (19.5 %)	2 (20 %)	2 (7.4 %)
O6	2 (18.2 %)	1 (7.7 %)	0	2 (11.1 %)	2 (11.8 %)	4 (50 %)	7 (18.4 %)	4 (9.8 %)	0	3 (11.1 %)
O7	1 (9.1 %)	2 (15.4 %)	3 (17.7 %)	2 (11.1 %)	0	0	2 (5.3 %)	5 (12.2 %)	1 (10 %)	1 (3.7 %)
O8	5 (45.5 %)	5 (38.5 %)	5 (29.4 %)	6 (33.3 %)	4 (23.5 %)	1 (12.5 %)	1 (2.6 %)	13 (31.7 %)	1 (10 %)	5 (18.5 %)
O9	0	1 (7.7 %)	0	0	0	2 (25 %)	0	2 (4.9 %)	0	0

O1) Acquire general knowledge
O2) Know our roots and our country's history
O3) Know the reality of the past
O4) Understand the present
O5) Understand the present to improve/transform the future
O6) Learn from the past (be aware of the mistakes)
O7) Think historically and develop historical knowledge
O8) Develop critical citizen behaviors
O9) Denaturalize our identity and values
DPE: Degree in Primary Education/MSE: Master's Degree in Secondary Education

Conclusions

Despite the small sample size, this research presents results which point in the same direction as other studies and invite us to delve deeper into a set of key questions. We present these again below for reference, using the questions we asked ourselves for this research:

– Differences between the various participating groups' school recollections were not significant in general terms. School recollections recount the importance of the disciplinary code for geography and history and the numerous

Tab. 8: Objectives of geography

	Brazil		Chile		Colombia		Spain		Portugal	
	DPE	MSE	DPE	MSE	DPE	MSE	DPE	MSE	DPE	MSE
O1	1 (9.1 %)	2 (15.4 %)	1 (5.9 %)	2 (11.1 %)	2 (11.8 %)	0	3 (7.9 %)	0	2 (20 %)	1 (3.7 %)
O2	2 (18.2 %)	0	5 (29.4 %)	2 (11.1 %)	3 (17.7 %)	0	8 (21.1 %)	1 (2.4 %)	2 (20 %)	4 (14.8 %)
O3	4 (36.4 %)	1 (7.7 %)	5 (29.4 %)	6 (33.3 %)	3 (17.7 %)	3 (37.5 %)	11 (28.9 %)	34 (82.9 %)	3 (30 %)	6 (22.2 %)
O4	0	1 (7.7 %)	3 (17.7 %)	0	0	0	8 (21.1 %)	5 (12.2 %)	0	0
O5	5 (45.5 %)	8 (61.5 %)	1 (5.9 %)	2 (11.1 %)	2 (11.8 %)	4 (50 %)	0	0	1 (10 %)	6 (22.2 %)
O6	1 (9.1 %)	2 (15.4 %)	2 (11.8 %)	6 (33.3 %)	3 (17.7 %)	2 (25 %)	4 (10.5 %)	6 (14.6 %)	2 (20 %)	7 (25.9 %)

O1) Read maps, learn how to find and locate geographical points
O2) Know the historical evolution of the planet
O3) Understand the territory we live in
O4) Understand nature, prevent natural disasters and look after the environment
O5) Problematize the surroundings and develop critical citizen behaviors that convert the pupil into an active agent of change
O6) Understand the impact of human activities on the environment and promote sustainable development
DPE: Degree in Primary Education/MSE: Master's Degree in Secondary Education

continuities which continue to be covered in the everyday school life of the various countries analyzed. The students (future teachers) are able to criticize the methodologies that characterized their classes, but they consider a range of factors when evaluating their recollections, including the role of the teacher and the usefulness of the subjects.

- We observe that school recollections act as a sense of practicality which anchors conventional wisdom in teaching geography and history. In other words, they remember concepts and procedures (climate, maps) which are set by a traditional conception of their use in schools, thus making it difficult to innovate and improve learning if they don't question the methodological foundations of their teaching practice.
- Education has an impact on the representations of geography and history and their uses. It all indicates that better disciplinary training allows students to acquire and develop didactic knowledge of the content which is much richer

and more complex in epistemological terms and defends socio-educational objectives which are entrenched in critical pedagogy. Nevertheless, the sample should be extended and explored in further depth using other sorts of strategies and tools in order to draw more compelling conclusions.
- Finally, the research also indicates that initial teacher training provides future teachers with the tools to present alternatives from a methodological and even educational psychological perspective. However, without a solid disciplinary foundation, it is difficult to break down certain consensuses regarding the development of materials, their public purposes, or their educational practices.

References

Abric, Jean-Claude: "A abordagem estrutural das representações sociais". In: Moreira Antonia S.P./Oliveira, Denize C. (eds.): *Estudos interdisciplinares de representações sociais*. AB: Goiânia. 2001, pp. 27–38.

Alonso-Tapia, Jesús/Abad, Lourdes/Sánchez, Sergio: "Estrategias para la evaluación de la comprensión del mundo social en el marco de la enseñanza de la Geografía Humana durante la Enseñanza Secundaria". *Journal for the Study of Education and Development, Infancia y Aprendizaje* 31(1), 2008, pp. 25–44.

Barton, Keith C.: "Research on students' ideas about history". In: Levstik, Linda S./Tyson, Cynthia A. (eds.): *Handbook of Research in Social Studies Education*. Routledge: New York/London. 2008, pp. 239–258.

Campo, Benito/García-Monteagudo, Diego/Souto, Xosé M.: "Tradiciones escolares en la educación geográfica. Impugnar sus rutinas para favorecer la innovación". In: Parra, David/Fuertes, Carlos (eds.): *Reinterpretar la tradición, transformar las prácticas: Ciencias Sociales para una educación crítica*. Tirant lo Blanch: Valencia. 2019, pp. 45–72.

Castorina, José A.: "Las representaciones sociales y los procesos de enseñanza-aprendizaje de conocimientos sociales". *Psicologia da Educaçao* 44, 2017, pp. 1–13, retrieved on 7.15.2020, from DOI: http://dx.doi.org/10.5935/2175-3520.20170001.

Castorina, José A./Barreiro, Alicia: "Los usos de las representaciones sociales en la investigación educativa". *Educación, Lenguaje y Sociedad* 9(9), 2012, pp. 15–40.

Cuban, Larry: *Teaching History Then and Now: A Story of Stability and Change in Schools*. Harvard Education Press: Cambridge. 2016.

Cuesta, Raimundo: *Sociogénesis de una disciplina escolar: la Historia*. Pomares-Corredor: Salamanca. 1997.

Diniz, André A.: "Ensinar, cuidar e sofrer: habitus e representação social do ser professor do ensino fundamental". In: Barbosa, Maria F. (ed.): *Bourdieu e Moscovici: fronteiras, interfaces e aproximações*. Editora Universitária Leopoldianum: São Paulo. 2019, pp. 47–68.

Domingos, Moisés: "Habitus docente, sens pratique e representações sociais". In: Barbosa, Maria F. (ed.): *Bourdieu e Moscovici: fronteiras, interfaces e aproximações*. Editora Universitária Leopoldianum: São Paulo. 2019, pp. 19–46.

Domingos, Moisés/Diniz, André A.: "Representações Sociais, Sens Pratique, Poder Simbólico e o Processo de Construção do Ser Docente". *Arxius de Sociologia* 41, 2019, pp. 43–64.

Escofet, Anna/Folgueiras, Pilar/Luna Ester et al.: "Elaboración y validación de un cuestionario para la valoración de proyectos de aprendizaje-servicio". *Revista Mexicana de Investigación Educativa* 21(70), 2016, pp. 929–949.

Fernández, María Victoria/Gurevich, Raquel: "Educación geográfica e investigación educativa en la formación de profesores". In: García de la Vega, Alfonso (ed.): *Reflexiones sobre educación geográfica. Revisión disciplinar e innovación educativa*. Universidad Autónoma de Madrid and Edições Lagim: Madrid. 2018, pp. 27–44.

Flick, Uwe: *An Introduction to Qualitative Research*. Sage Publication: London. 2009.

González, María Carmen: *La enseñanza de la Historia en el nivel medio. Situación, tendencias e innovaciones*. Anaya: Madrid. 2002.

Grant, S. G.: *History Lessons: Teaching, Learning and Testing in US High School Classrooms*. Routledge: London. 2014.

Harris, Richard/Burn, Katharine: "English history teachers' views on what substantive content young people should be taught". *Journal of Curriculum Studies* 48(4), 2016, pp. 518–546, retrieved on 7.15.2020, from DOI: https://doi.org/10.1080/00220272.2015.1122091.

Jodelet, Denise: "La representación social, fenómenos, concepto, teorías". In Moscovici, Serge (ed.): *Psicología social II, pensamiento y vida social*. Paidós: Barcelona. 1985, pp. 469–494.

Lambert, David/Morgan, John: *Teaching Geography: A Conceptual Approach*. Open University Press: Maidenhead. 2010.

Luis, Alberto/Romero, Jesús: *Escuela para todos: conocimiento académico y geografía escolar en España (1830–1953)*. Universidad de Cantabria: Santander. 2007.

Maestro, Pilar: "La idea de España en la historiografía escolar del siglo XIX". In: Morales, Antonio/Esteban, Mariano (eds.): *¿Alma de España? Castilla* *en las interpretaciones del pasado español*. Marcial Pons: Madrid. 2005, pp. 179–181.

Markuszewska, Iwona/Tanskanen, Minna/Vila, Josep: "New ways to learn geography - challenges of the 21st century". *Quaestiones Geographicae* 37(1), 2018, pp. 37–45, retrieved on 7.15.2020, from DOI: https://doi.org/10.2478/quageo-2018-0004.

Martínez, Nicolás/Souto, Xosé M./Beltrán, José: "Los profesores de Historia y la enseñanza de la Historia en España. Una investigación a partir de los recuerdos de los alumnos". *Enseñanza de las Ciencias Sociales* 5, 2006, pp. 55–71.

McCrum, Elizabeth: "History teachers' thinking about the nature of their subject". *Teaching and Teacher Education* 35, 2013, pp. 73–80, retrieved on 7.15.2020, from DOI: https://doi.org/10.1016/j.tate.2013.05.004.

Moscovici, Serge: *A invenção da sociedade. Sociologia e Psicologia*. Editora Vozes Ltda.: Petropolis, Rio Janeiro. 2011.

Palacios, Nancy: *La experiencia social de la educación: un estudio de tres instituciones educativas de secundaria de Colombia*. Ediciones Uniandes: Bogotá. 2018a.

Palacios, Nancy: "Perceptions of secondary school students of the social science class: A study in three Colombian institutions". *International Journal of Instruction* 11(4), 2018b, pp. 353–374.

Parra, David: "Representación de la historia escolar y crítica del consenso del sentido común". In: Parra, David/Fuertes, Carlos (eds.): *Reinterpretar la tradición, transformar las prácticas. Ciencias Sociales para una educación crítica*. Tirant Humanidades: Valencia. 2019, pp. 73–98.

Parra, David/Fuertes, Carlos (eds.): *Reinterpretar la tradición, transformar las prácticas. Ciencias Sociales para una educación crítica*. Tirant Humanidades: Valencia. 2019.

Ramiro, Enric: "¿Alguien se prepara las clases de Geografía? Comentarios y resultados sobre distintas variables en la preparación de las actividades geográficas y sus recursos en la Enseñanza Obligatoria". In: Marrón, Mª Jesús/Rosado, Mª Dolores/Rueda, Carmen (eds.): *Enseñar geografía: la cultura geográfica en la era de la globalización*. Grupo de Didáctica de la AGE y Universidad de Jaén: Jaén. 2008, pp. 161–176.

Sammut, Gordon/Andreouli, Eleni/Gaskell, George et al.: "Social representations: A revolutionary paradigm?" In: Sammut, Gordon/Andreouli, Eleni/Gaskell, George et al. (eds.): *Cambridge Handbook of Social Representations Cambridge*. Cambridge University Press: Cambridge. 2015, pp. 3–11.

Souto, Xosé M.: "¿Qué escuelas de Geografía para educar en ciudadanía?". *Didáctica de las Ciencias Experimentales y Sociales* 24, 2010, pp. 25–44.

Souto, Xosé M.: "La investigación cualitativa y la innovación didáctica en geografía. El trabajo cualitativo como recurso didáctico en geografía". Actas XI Congreso Nacional Didáctica de la Geografía. In: Alanis, Leonardo et al. (eds.): *Nativos digitales y geografía en el siglo XXI: Educación geográfica y sistemas de aprendizaje*. AGE, Univ. Sevilla: Sevilla. 2016, pp. 80–101.

Souto, Xosé M./García, Diego: "Conocer las rutinas para innovar en la geografía escolar". *Revista de Geografía Norte Grande* 74, 2019, pp. 207–228.

Souto, Xosé M./Martínez, Nicolás/Parra, David: "La representación poliédrica de la memoria escolar. Un proyecto de investigación sobre las clases de Historia de España". In: Martínez, Nicolás (ed.): *La Historia de España en los recuerdos escolares. Análisis, interpretación y poder de cambio de los testimonios de profesores y alumnos*. Nau Llibres: Valencia. 2014, pp. 29–52.

Straforini, Rafael: "O ensino de Geografia como prática espacial de significação". *Estudos Avancados* 32(93), 2018, pp. 175–195, retrieved on 7.15.2020, from DOI: https://doi.org/10.5935/0103-4014.20180037.

Tutiaux-Guillon, Nicole: "Interpréter la stabilité d'une discipline scolaire: l'histoire-géographie dans le secondaire français". In: Audigier, François/Tutiaux-Guillon, Nicole (eds.): *Compétences et contenus. Les curriculums en question*. De Boeck : Brussels. 2008, pp. 117–146.

María de la Encarnación Cambil, Esther Jiménez, Mª Montserrat Pastor and Ander Delgado

Chapter 6 An analysis of the perceptions of trainee Secondary Education teachers on current historiographical trends[1]

Abstract: This chapter contains research on the perceptions of trainee Secondary Education teachers, following the completion of their teaching practice, on traditional and current historiographical approaches. The objective is to analyse which historiographical trends currently have more presence in the teaching and learning of history, how they influence teaching methodology and the approach to activities carried out in the classroom. The chapter also examines the difficulties and advantages of their use and the challenge that methodological innovation implies from the most current historiographical approaches, such as the history of the present, the history of everyday life, gender perspective, etc., and how, based on these aspects, historical thinking skills in secondary education can be improved. To this end, a questionnaire validated by experts and evaluated using a Likert-type assessment scale (1–5) was prepared. The sample consisted of 646 students from 23 Spanish universities studying a master's degree in Secondary and Baccalaureate Education, specialising in geography and history, which enables them to become history teachers.

Keywords: History Education, Teacher Education, Questionnaire, Historiography, Secondary Education

Introduction

We present a study carried out among students of a master's degree in Secondary and Baccalaureate Education in the speciality of geography and history, with the aim of analysing their perception regarding the historiographical trends which are currently most prevalent in the teaching and learning of history, their influence on teaching methodology and on the planning of activities implemented

1 This research is part of the coordinated project COMPROP (PGC2018-094491-B-C31: *Emotional dimension and controversial issues in teacher training*; and PGC2018-094491-B-C3: *Methodological concepts and active learning methods to improve the teaching skills of teachers*) with funding from the Spanish State Research Agency and European Regional Development Fund.

in the classroom, along with the difficulties and benefits related with their use. It is of great interest to examine the challenge implied by methodological innovation from the most current historiographical approaches such as, the history of the present, the history of mentalities or everyday life, gender perspectives, oral history, microhistory, and eco-humanist history, which is so popular today, and how they can contribute towards improving historical thinking skills in compulsory secondary education (ESO in Spanish) (12–16 years of age) and baccalaureate (16–18 years of age). The sample consists of 646 students from 23 Spanish universities who studied a history degree. Data was collected via an *ad hoc* questionnaire which was validated by experts. The results were evaluated via a closed Likert-type scale (1–5).

Justification

In the context of education, the presence of history in the curriculum is of great importance due to its great educational power, as mentioned by Prats and Santacana (2011): "history is a subject which makes it possible to reflect on the reasons which lie behind historical events and phenomena; it enables people to understand the complexity of social events and phenomena from both the past and the present; it is essential in the formation of the individual and in determining their position in society, allowing them to develop their own criteria and to hold a critical perspective of the present."

There are many factors which have an influence on students' perceptions of history as a science, among which are epistemology, the theoretical rationale which originates from the discipline itself and general scientific theory. For the teaching and learning of history, it is necessary to master knowledge generated in this scientific area and to hold a perspective of the dominant historiographical trends at each moment which influence the methodology and learning activities and which are also present in textbooks in the classroom. Many authors, for example Valls (1999, 2002, 2018), Saiz and Valls (2014) and Gómez Contreras (2019), have based their research on the analysis of textbooks. These studies, via the analysis of the historiographical trends which guide the contents of textbooks, have made it possible to glimpse the different conceptions of history which have been created over the course of time in the classroom. Therefore, the analysis of the perceptions of teacher training students of the predominant historiographical trends of today in the teaching of history is one way of advancing and improving the work of training the teachers of tomorrow from the necessary reflection of what is done and how it is done.

Bearing in mind the fact that the consideration of history as a science depends on its relationship with its object by way of knowledge (Miralles, Molina/Ortuño 2011), many definitions have been made of history and many are the reflections regarding its social usefulness. Lucien Febvre notably defined history as "the science of men, but of men in time" and Pierre Vilar considered that "history is the study of the mechanisms which link the dynamics of structures to the succession of events." Another definition, that of E. H. Carr, considered it as "a continuous process of interaction between the historian and the events, an unending dialogue between the present and the past" (cited in Gómez Carrasco/Rodríguez 2017, p. 268). However, these definitions contradict the reality of the classroom both in terms of the methodology employed for the teaching and learning of history and the perception which secondary pupils have of history, considering it as knowledge based on memorising dates, information and relevant characters, all of which they consider to be far removed from their reality.

At present, there are many research projects which aim to make progress in the teaching and learning of history by overcoming a linear narrative and favouring the knowledge of different approaches which serve to demonstrate why there is a dichotomy existing over time in societies between constant change and continuity. For this reason, along with an epistemological foundation, it is important to introduce historiography into the world of didactics in order to enable trainee teachers to acquire knowledge which will allow them to understand the present and the future of our societies and which, at the same time, will serve as a tool to understand the key concepts of the subject from a historiographical perspective. These aspects are necessary for the teaching of history in the classroom as they facilitate knowledge of how the science of history has changed over time, demonstrating that there is no one vision of events, thereby favouring and promoting the development of a critical spirit.

Innovation in the teaching of history must be oriented towards encouraging the development of intellectual and social skills. The objective is for students to understand the complexity of historical knowledge, for them to be able to identify causal relations, to construct contextualised explanations based on evidence and to take cultural diversity into account.

These capacities form part of the values which the citizens of a democratic society must possess. They can be applied in their daily lives and are essential for the preservation of memory understood as the present of the past (Ricoeur 1999), recognising that the events and historical changes of the past have been understood and interpreted from different points of view by different societies, cultural spaces and ideologies (Jelin 2002). These events and changes lack sense and must be given an interpretation which must be carried out from the present, establishing causalities and interdependencies and bearing in mind and favouring our own individual and collective cultural and social identity.

History as a social science, aside from any differences, opposition and even disagreements which the study of the different trends of historical thinking may manifest at first glance, presents a great level of openness and capacity for dialogue between the different paradigms upon which its historiography has been constituted and which have influenced the field of education (Cambil/Tudela 2016). In this regard, positions and controversies which arose in the first third of the 20th century with the appearance of so-called "new history" have progressively been overcome. This "new history" opposed positivism, a historiographical paradigm inherited from the 19th century and developed around the figure of Auguste Comte, who considered political history and the biographies of great characters to be true historical knowledge (Le Goff/Revel/Chartier 1978). In the field of education, positivism, from a psychological basis of the subject, led to a behaviouristic line which attempted to teach pupils valid, trustworthy and applicable knowledge. Interest was focused on delimiting objectives and accommodating them to behaviour with the aim of achieving the desired learning, based on the idea that a good teaching process would generate a good product. Thus originated the tendency of this historiographical school of building taxonomies or classifications, objectives to be achieved and reliable systems of evaluation. Teachers became the subject of knowledge and of how things were to be done and the only judges with competence to evaluate students. The methodology applied left little space for creativity and improvisation given that the objectives, teaching programmes and proposals were laid out beforehand, presupposing homogeneity among the students, with no room for diversity, and that they would respond positively to the effectiveness of the system (Cambil/Tudela 2016).

"New history" would position itself against positivism based on the analysis of structures from two different conceptions: Marxism and the eco-demographics of the Annales school. Marxist historiography reacted against positivism by placing emphasis on social and economic history of a structural nature. It reached its prime in the 1950s, based on the "British Marxist school," the main organ of dissemination of which was the journal *Past and Present*. The response of historians such as E. P. Thompson and Pierre Vilar to the abuses of the theory, along with the great debates and discussions among professional historians, revitalised this historiographical school and lent it a more social dimension (Miralles/Molina/Ortuño 2011). One of its greatest contributions was the critical function of history, which would be taken up by many Germany authors, leaders of the so-called critical social history, characterised by stressing the necessity of history applied to the service of society (Gómez Carrasco/Rodríguez 2017). In the field of education, Marxist historiography reinterpreted by critical theory denounced and combatted facts hidden beneath an apparently neutral school reality and its character as an agency of the reproduction of value systems and visions of the world at the conscious or unconscious service of interests, objectives and powers. It stressed the need for students to become aware of their own system of values and of the keys and influences behind their position in the world in such a way as to be able to establish a critical and autonomous reflection. History must attempt to highlight the underlying intention in significant and urgent social events, such as dignity, human rights, inequality, poverty, discrimination and the domination of human beings, the great world conflicts, in order to bring about a critical and clear reflection among students from different points of view which allow for possible alternatives and proposals.

The *Annales* school, founded byMarc Bloch and Lucien Febvre in 1928, and its tool of dissemination the journal *Annales d'histoire économique et sociale*, broadened the object of study of history, highlighting its scientific basis and focusing on social structures and on the human relationship with these structures. Its founders rejected the history of events in order to reinterpret history as a social science. They called for an analysis which would pay more attention to the role of social and economic phenomena when determining the evolution of historical events. In the field of education, the French *Annales* school had a great influence on changes in the official programmes of history teaching during the last quarter of the 20th century, as has been mentioned by authors such as Paniagua (1996), Miralles (2005), and Gómez Carrasco and Miralles (2016). This influence is present in the division made in both the curriculum and in school textbooks between political history, social history, economic history, the history of thinking and culture, and the history of art in search of the ideal of total history, from an essentially structural perspective.

These two historiographical trends continued until the end of the 1970s, with research focused on society, the economy and mankind, albeit within great structures. In Europe, from the end of the 1970s to the present day, the object of history has changed, leading to a diversification of themes focused on culture, mentalities and the exploration of new horizons versus structuralism. All of this is linked to the controversy generated around the historical narrative and the development of post-modern thought, which reflects on the relativity of knowledge and, particularly, the implications which different historical events have had, and still have, for the prevailing inheritance in the contemporary world, such as the fall of the Berlin Wall and the end of the Cold War and the concentration of wealth in the new neoliberal economic phase. The consequence of this has been that historians have accused the historiographical schools of the middle of the 20th century of being deterministic and of "depersonalising" history. In this way, the so-called "new history" came, in the latter decades of the 20th century, to be known as "old new history" (Burke 2001). The new historiographical trends which arose from this crisis aimed to explore new themes and analytical methods in an attempt to escape from the historical discourse of the "prison of the structural" and the habit of building structures and social groups without venturing into them. This new conception has discovered innovative ways of explaining change, mobility and the historical process in general, such as representations, symbols, ways of perceiving the world and the social reality as a cultural product (Gómez Carrasco/Rodríguez 2017). Transformations have given way to postures found around the reconstruction of the past, theory and the relationship with other disciplines and social sciences. Along these lines, in the 1990s, a great debate arose between supporters and detractors of the new historiographical practices emerging from the confrontation between structures and individuals, a situation which could have led to the error of abandoning the study of social reality and replacing it with symbols (Casanova 1997). In the field of education, this way of approaching history and the problems of our time lead to great possibilities for learning, promoting debate and opening up a practically unlimited space of contexts, ideas and interpretations.

The different historiographical trends have, without doubt, had an influence on the teaching of history. However, their evolution over the last few decades has not been reflected in the necessary didactic transposition for the teaching and learning of this subject. Innovation and research along these lines are becoming ever more frequent, with the results being implemented in the classroom in methodology and in textbooks. However, according to Miralles, Molina ans Ortuño (2011), the reality is that, instead of requiring pupils to memorise contents and political information, they are now forced to memorise historical interpretations with a greater socio-economic burden, which are ultimately explanations far removed from the pupils' reality. For the reasons presented above, this research is considered necessary in order to know and be able to analyse how trainee teachers perceive the historiographical trends which are most present in the teaching of history, with the aim of advancing and attaining the quality of education required by the society of the 21st century.

Methodology

Objectives

The objective of this research is to analyse the reality of trainee teachers when showing their preference for one type of more or less recent historiography over another and for the historical contents which they consider important when teaching history in secondary and baccalaureate education. It is, therefore, an approach to historical themes from the perspective of those who are most involved, the teachers giving classes in schools. In order to achieve this general objective, the following specific aims were proposed:

(a) To use a tool for measurement to enable us to study the evaluations and opinions of trainee secondary education teachers regarding the historical contents derived from different historiographies.
(b) To cross these evaluations with data such as the sex and university of origin of the participants in order to draw conclusions.
(c) To respond to the results obtained both in the analysis and in the conclusions, based on personal motivations, contents learned throughout school and university education, attachment to textbooks and deference to one historiographical trend or another which is more useful or motivating for the teacher.

Participants

The data of the population studied in this research was obtained via a prior sociodemographic analysis which has allowed for a better fit of the results taking into account variables such as the university of origin and sex. The questionnaires were completed by trainee teachers studying a master's degree in Secondary and Baccalaureate Education in the speciality of geography and history. Efforts were made to obtain a large sample with a wide territorial range. Thus, a total of 646 students from 23 Spanish universities participated in this study (Fig. 1), with a minimum participation of 1.3 % (9 students) from the University of Oviedo and from Jaume I University and a maximum participation of 10.8 % with a total of 70 students from the University of Valencia. The percentage of participation from the remaining universities oscillated between 1.3 % and 10.8 %.

The age of the participants ranged from 18 to 34 years of age (Tab. 1). 51.7 % of those surveyed were between 21 and 23 years of age, having recently completed their degrees in history, geography or the history of art. 37.9 % were between 25

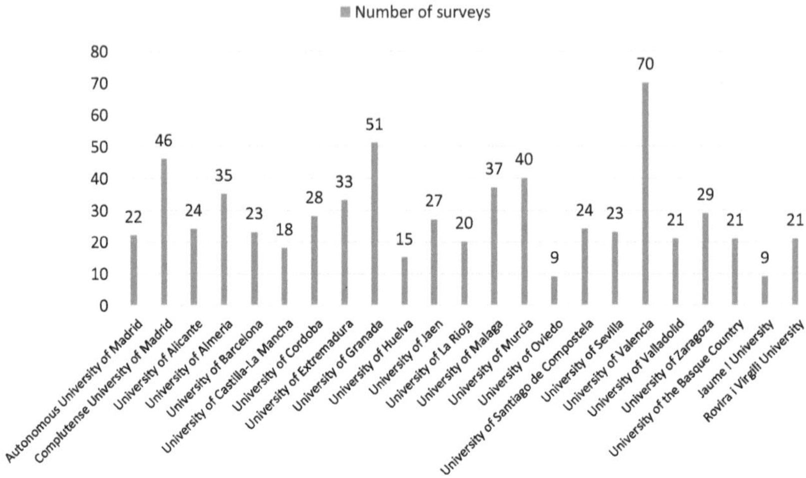

Fig. 1: Bar graph of the number of surveys per university of the participants. Authors' own work

and 34 years of age. Thus, it can be calculated that 89.6 % of participants who completed the questionnaires were between 21 and 34 years of age.

The parameter of gender was also taken into account, with slightly more men participating in the study than women (Tab. 1). 55.4 % of the participant were men (n=358) compared with 43.3 % of women n=280).

The age of the participants is directly related with the studies completed and the academic training they have received (Tab. 1). 97 % of those surveyed (627 out of a total of 646) had completed their undergraduate studies in one of the three branches of the social sciences: history, geography and the history of art. Only 1.1 % (7 participants) had completed studies higher than a master's degree or an MAS. Furthermore, 0.8 % (5 participants) had finalised their doctoral studies in one of the three human sciences.

In order to obtain even more information regarding academic training, the participants were asked whether they had done any kind of teaching innovation course (Tab. 1). 89.3 % responded that they had not whereas 9.9 % replied that they had indeed received some training in this field of education. The type of teaching innovation courses received were particularly related with new active methodologies, the improvement of educational technology, the elaboration of teaching resources, and special needs education. To a lesser degree, some participants had taken courses related to the school management and values education.

Tab. 1: Social and educational profile of the participants. Authors' own work

CHARACTERISTICS	Nº	%
Age	646	100
18–24	334	51.7
25–34	245	37.9
35 or older	49	7.6
No answer	18	2.8
Sex	646	100
Male	358	55.4
Female	280	43.3
No answer	8	1.3
Academic level	646	100
Doctorate	5	0.8
Master's degree/MAS	7	1.1
Degree	627	97
Diploma course	1	0.2
No answer	6	0.9
Teaching innovation course	646	100
Yes	64	9.9
No	577	89.3
No answer	5	0.8

Tool

The tool employed was a quantitative non-experimental questionnaire drawn up *ad hoc* by the coordinators of the research project and validated by experts. The first part collected the data of personal variables (age, sex, academic training), with anonymity always being maintained, while in the second part different questions were asked regarding historical contents, issues relating to competences, etc. In this study, we focus on the data collected relating to the following question: *In your opinion, which topics do you believe are most relevant for teaching in history classes?* In order to respond to this question, a total of 19 items were presented with history topics linked to different historiographical trends. These items were measurable via a closed Likert-type evaluation scale (1–5), in which each participant was required to score each value (with 1 being extremely irrelevant and 5 extremely relevant). The historical topics evaluated are shown in Tab. 2. In all of the evaluations, the participants relied on their own personal opinion, their training experience and the importance they attributed to some historical issues compared to others when teaching secondary and baccalaureate pupils.

Tab. 2. Items of the survey relating to topics considered to be more or less relevant for teaching in secondary and baccalaureate history classes. Authors' own work

Item 21	Main historical events which make it possible to understand the origin of the nation
Item 22	Main political characters and military leaders
Item 23	Economic issues: the development of agriculture, trade and industry
Item 24	The great processes of humanity (the Neolithic Revolution, the Industrial Revolution, etc.).
Item 25	The history of transversal issues which may motivate pupils: the history of housing, the history of food, the history of transport, etc.
Item 26	The history, culture and heritage of your local area and region
Item 27	The history, culture and heritage of our country
Item 28	The history, culture and heritage of other, more distant, countries
Item 29	The history of interesting characters: conquerors, inventors, artists, etc.
Item 30	The everyday life of ordinary people
Item 31	The history of women: writers, scientists, etc.
Item 32	Genocide and dictatorships
Item 33	The disappeared, mass graves and political and social repression
Item 34	The development of democracy and political participation
Item 35	The development of human rights, social demands and struggles for equality
Item 36	Migratory and religious conflicts
Item 37	Feminism, sexism, gender identities
Item 38	Environmental issues, landscape and the exploitation of natural resources
Item 39	Identities, social representations and cultural practices

Procedure

The questionnaires were applied by teachers in the area of the teaching of the social sciences in the master's degree in Secondary and Baccalaureate Education from the 23 Spanish universities. First of all, there was a presentation phase of the study for the participants to inform them of the purpose and the objectives, which provoked a high degree of motivation resulting in a high level of participation among the students. They were reminded to consider their own opinion regarding the history topics to be taught and the historiographies which they knew and which of them they would like to teach in their workplace, thinking at all times of the wellbeing of their future secondary and baccalaureate pupils and of the attractiveness of the topics reflected in the survey.

Tab. 3: Items of the survey which respond to the question: Which topics do you believe to be most relevant for teaching in history classes? Authors' own work

	N	Min	Max	Mean	Median	%1	%2	%3	%4	%5	Sd
Item 21	644.00	1	5	3.42	4.00	7.45	14.60	25.90	32.30	19.70	1.17
Item 22	645.00	1	5	3.06	3.00	8.99	20.00	36.60	25.30	9.15	1.08
Item 23	643.00	1	5	3.96	4.00	0.47	4.04	22.60	44.60	28.30	0.84
Item 24	644.00	1	5	4.43	5.00	0.16	0.78	7.76	38.00	53.30	0.68
Item 25	643.00	1	5	4.01	4.00	2.02	5.60	20.70	32.80	38.90	1.00
Item 26	645.00	1	5	4.29	4.00	0.62	2.48	13.50	34.00	49.50	0.84
Item 27	643.00	1	5	4.27	4.00	0.47	2.18	12.80	39.30	45.30	0.80
Item 28	643.00	1	5	3.93	4.00	0.78	4.82	24.10	41.50	28.80	0.89
Item 29	645.00	1	5	3.61	4.00	2.48	12.70	27.30	36.70	20.80	1.03
Item 30	643.00	1	5	4.16	4.00	0.93	3.89	16.20	36.50	42.50	0.90
Item 31	644.00	1	5	4.37	5.00	1.24	1.09	13.00	29.00	55.60	0.84
Item 32	642.00	1	5	4.33	4.00	0.31	1.87	11.10	38.30	48.40	0.77
Item 33	644.00	1	5	4.28	4.00	1.24	1.86	12.30	36.80	47.80	0.84
Item 34	645.00	1	5	4.47	5.00	0.62	0.62	7.91	33.00	57.80	0.72
Item 35	643.00	1	5	4.59	5.00	0.62	0.93	5.13	25.50	67.80	0.69
Item 36	644.00	1	5	4.31	4.00	0.16	1.40	11.30	41.10	46.00	0.74
Item 37	643.00	1	5	4.33	5.00	2.95	3.27	9.64	26.30	57.90	0.98
Item 38	644.00	1	5	4.26	4.50	0.78	3.88	14.00	31.40	50.00	0.89
Item 39	644.00	1	5	4.32	4.00	0.93	1.86	10.40	37.70	49.10	0.81
Total items	643.74	1	5	4.13	4.24	1.75	4.62	15.91	34.74	42.98	0.87

Finally, the responses were collected and unified from all of the universities in order to move on to the analytical phase.

Analysis, results and discussion

Tab. 3 shows the descriptive statistics of *Block II.1. Opinions and conceptions on the teaching of history*, in which the trainee teachers evaluated the topics which they considered to be more relevant for teaching in secondary and baccalaureate history classes. This first stage of the analysis focused on the items which were best or worst evaluated by the trainee teachers, representing this information in graphs to order the data. The second analysis focused on the evaluations according to sex.

Analysis I: Best and worst valued items according to new and old historiographical trends

The most attractive topics for the participants can be observed in Tab. 3 whilst Graph 1 shows the items valued from best to worst. The best valued item was number 35, corresponding to "The development of human rights, social demands and struggles for equality," which received a mean score of 4.59 out of 5 (Graph 1) and was judged by the participants to be the most relevant topic for teaching in secondary and baccalaureate classes. Indeed, 67.8 % of the participants awarded this item with the highest score (Tab. 3). The second-best valued item was number 34, corresponding to the topic of civic education ("The development of democracy and political participation"), which received a mean score of 4.47 (Graph 1) and was considered to be extremely relevant for teaching by 57.8 % of the participants. Items 35 and 36 both respond to the educational effectiveness of civic education, which has gained acceptance over the last two decades in the educational community in the teaching of the social sciences. From civic education, social competences are taught, the democratic values of respect and tolerance are instilled, and answers are sought to social problems, thereby improving coexistence. In third place was item 24 regarding "The great processes of humanity (the Neolithic Revolution, the Industrial Revolution, etc.)," obtaining a mean score of 4.43 (Graph 1) and being considered as extremely important for teaching by 53.3 % of the participants (Tab. 3).

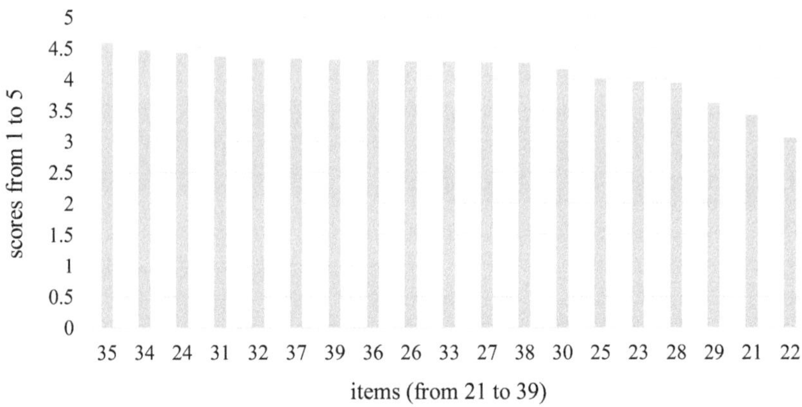

Graph 1: Items ordered according to the mean from best to worst valued. Authors' own work

Tab. 4: Items of the questionnaire, their mean scores and relationship with historiographical trends. Authors' own work

Item	Mean	Topic	Historiographical trends
35	4.59	The development of human rights	Critical theory/History of the present/ Oral history
34	4.47	The development of democracy	History of the present: citizen participation
24	4.43	The great processes of humanity	Critical social history applied to society
31	4.37	The history of women	Gender history
32	4.33	Genocide and dictatorships	History of the present/Historical memory
37	4.33	Teaching feminism	Gender history
39	4.32	Identities, cultural practices	History of the present: Identities/Total history
36	4.31	Migratory and religious conflicts	History of the present/Historical memory/Oral history
26	4.29	The history and culture of local heritage	Local history/Microhistory/Annales
33	4.28	The disappeared, political repression	History of the present/Historical memory/Oral history
27	4.27	The history, culture and heritage of the country	History of Culture/History of heritage/Annales
38	4.26	Environment, landscape, resources	Eco-humanist history/Environmental history
30	4.16	The everyday life of ordinary people	History of Mentalities/History of everyday life
25	4.01	The history of housing, transport, etc.	History of Mentalities/History of everyday life
23	3.96	Economic, trade, agricultural issues	Marxist historiography/Economic history
28	3.93	The history, culture and heritage of distant countries	History of Culture/History of heritage/Annales
29	3.61	The history of interesting characters	Positivist historiography/Historical biography
21	3.42	Historical events of the origin of the nation	Positivist historiography: Political history
22	3.06	Political and military characters	Positivist historiography: Political and military history

The worst valued historical topics with a score of 1 (extremely irrelevant) were, first of all, item 22 (Tab. 3), concerning the "main political characters and military leaders," with a mean score of 3.06 (Graph 1) and a high percentage of participants who would not teach it (8.9 % in total). Secondly, item 21 ("Main historical events which make it possible to understand the origin of the nation") also obtained a low mean score of 3.42 and was declared not very relevant by 7.4 % of the participants. Item 29 ("The history of interesting characters: conquerors, inventors, artists, etc.") was also valued badly, receiving a mean score of 3.61 out of 5. Therefore, it can be observed that the leading topics of the historiography which perpetuates or inherits the ideals of positivism (biographies and heroic deeds which give identity to the nation) are not motivating or relevant in the opinion of future teachers. Curiously, an effective renewal of the historical contents of textbooks has still not arrived, and these still maintain an eminently positivist framework. However, the arrival of new historiographical trends has been observed since the middle of the 20th century.

By way of conclusion, we present Tab. 4, which shows the items with their topics, the mean score of each one and their relationship with historiographical trends:

The question that should be asked is to what these preferences of the master's degree students are due. As mentioned above, the explanation is fundamentally motivated by the students' experience of education, both that received in the subjects of social sciences and history as past pupils of secondary and baccalaureate education (which gave them their first contact with historical contents and themes) and in the education received in their undergraduate history studies,[2] which gave them a deeper knowledge of the work of the historian and the extremely intimate relationship which exists between the content taught and its epistemology and the historiographical trend. Thus, it is no coincidence that the new historiographical trends which arose in the 1980s are more highly valued (critical theory, the history of the present/historical memory, gender history, the history of culture, etc.), than the themes established in 19th-century tenets (the positivist trend: political and military history, historical biography, etc.

This formative stage presents a common element, which has a great influence in showing this evolution of trends when carrying out historical work: the capacity of textbooks to influence the learner. According to Díaz-Salazar (2016)

---

2 It should be remembered that 627 of the total of 646 students had graduated in history, geography and the history of art, with the first option being the most numerous among the participants.

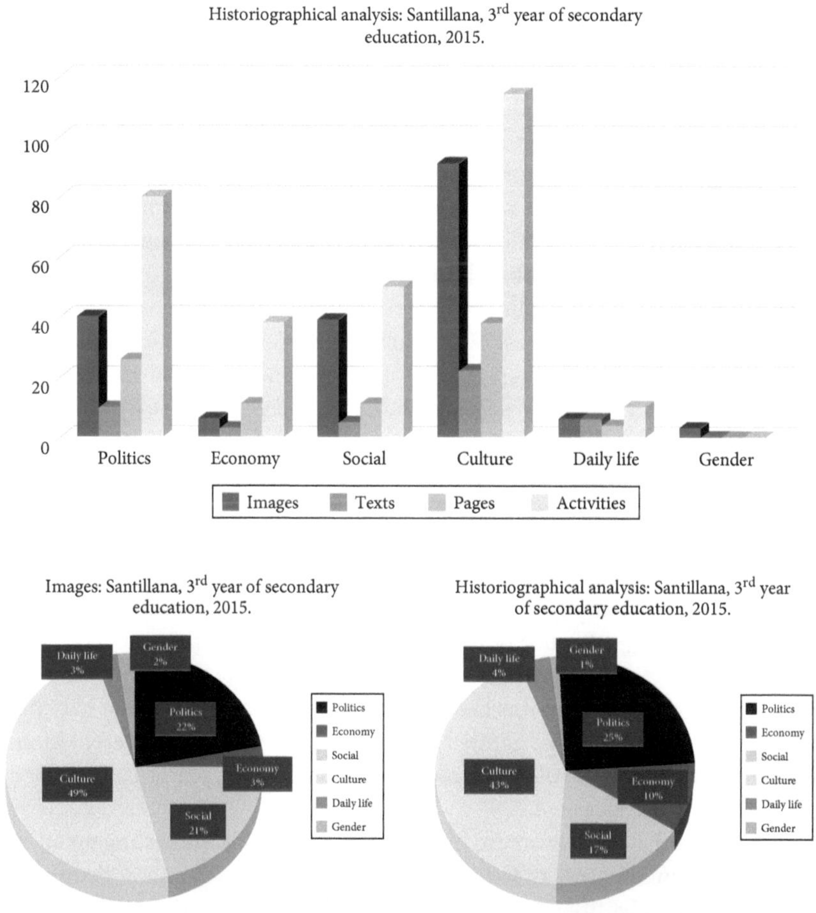

Fig. 2: Historiographical analysis: Santillana, 3rd year of secondary education, 2015. Gómez Contreras, J.M. p. 271

"the use of textbooks as a support mechanism for the presentation of the subject is habitual" (p. 207). This fact would not be relevant if the methodology used by the teacher in general terms were different to the traditional expositional model. However, Díaz-Salazar's research concluded that the significance of the textbook in the present day has its own character as a study reference, with a clear role as a planning guide and verifier of knowledge in the subject of history.

All of this is reflected in the vision of historical contents of the trainee teachers participating in this study, and is, therefore, reflected in this research.

Gómez Contreras (2019, p. 13) points out the fact that there are two fundamental premises. The first is that the textbook is the main tool for the study and learning of social sciences and history in secondary and baccalaureate education. The second is that historiographical trends are unequally distributed in the discourse established in textbooks and are extremely significant in foreseeing the historical conceptions which are being created in our classrooms. For example, in the textbook published by Santillana for the 3rd year of secondary education in 2015, edited by Teresa Grence, the culture variable stands out with 43 %, whereas the rest of the variables are left in the background. The second variable (politics) occupies 25 % of the content, the social variable 17 %, and the economy 10 %. Everyday life, with 4 %, and gender history possess an insignificant importance.

The predominance of the cultural variable is clear in both graphs relating to historiographical analysis and to images. This is a good example of what textbooks offer to pupils.

Thus, the different historiographical trends form the backbone of the textbooks used by our pupils during their secondary and baccalaureate education, with a strong influence of the history of culture and political, economic and social aspects, which are still addressed without due reflection and criticism, although they are approached in a different way the to positivist narrative with the inclusion of a large number of images and activities of different kinds. It is evident that there is still much room for improvement, particularly as far as gender history is concerned.

In this regard, it is necessary to highlight one extremely interesting detail. As a result of the momentum of feminism in Spanish society, item 37 ("Feminism, sexism, gender identities") obtained one of the highest scores if the high number of trainee teachers who considered it an extremely relevant topic is taken into account (57.9 % gave it a score of 5). However, it also received one of the highest negative scores as 2.95 % of the participants gave it a score of 1 ("of little relevance" for teaching) (Tab. 3). It may be that this percentage of participants who do not consider it a necessary topic in social science classes hold this opinion due to the fact that they judge it has already been integrated into society, meaning that it is not necessary to reinforce the point in the classroom. Alternatively, it may be that they consider it as an ideology which should not be present in schools.

What can be stated is that we are faced with a diversity of historiographical trends which have one thing in common: they consider history as a social science which takes its own methodological resources from the diversity of sciences which make up the social sciences (Alcaraz/Pastor 2012, p. 137).

Analysis II: Best and worst valued items according to sex

If the evaluations of the history topics are studied according to sex, some extremely striking results can be observed. To begin with, as far as the mean scores of each item are concerned, it can be stated that there is a good degree of parity with the means oscillating between 3.06 and 4.77 on the value scale of 1 to 5. Thus, at first sight, all of the topics for teaching in history, independently of the historiographical trend to which they belong, are valued between "neither very relevant nor of little relevance" and "extremely relevant," in such a way that the evaluations are generally positive both for men and for women. However, attention must be paid to some particular items in order to distinguish the differences more clearly. Furthermore, it must be taken into account that there was a total of 358 men among the total of 646 participants, compared with a lower number of women (280).

It should be highlighted that, of all the 19 items (from number 21 to number 39 of the survey), only four were considered to be more relevant by the male participants than the female respondents. This data is shown in Graph 2 below.

These four items were item 24 "The great processes of humanity (the Neolithic Revolution, the Industrial Revolution, etc.)," evaluated by 54.6 % of the male participants as the most relevant, compared to 52.1 % of female respondents; item 23 "Economic issues: the development of agriculture, trade and industry," which was considered important by 29.5 % of male participants, compared with 26.8 % of women; item 21 "Main historical events which make it possible to understand the origin of the nation," which was relevant for 21 % of men, compared with 18.2 % of women; and item 22 "Main political characters and military leaders, which, although it was not very highly valued, is significant in that 9.78 % of men and 8.57 % of women considered it relevant. The remaining items were valued higher by the female participants, even though there were less of them.

Graph 3 shows the values of the items awarded the highest scores by the female trainee teachers.

For the female participants, the most relevant item was item 31, the teaching of "the history of women: writers, scientists, etc.," which was valued as being extremely relevant by 80.3 % of the female participants, compared with 57.7 % of men. Item 35 "The development of human rights, social demands and struggles for equality" and item 37 "Feminism, sexism, gender identities" were

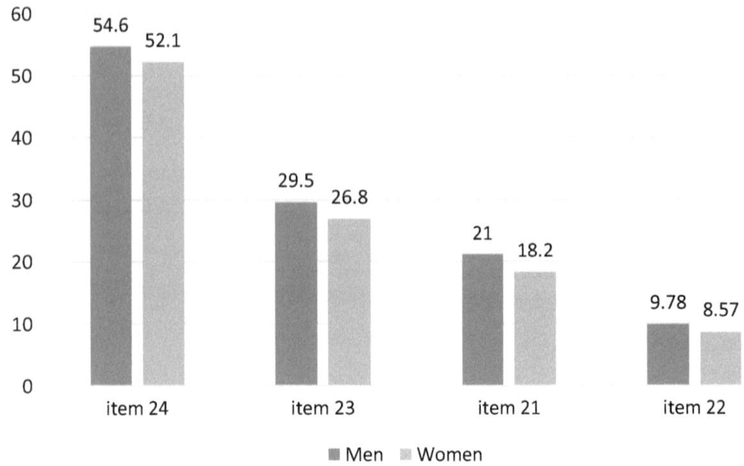

Graph 2: Items valued better by men than by women. Authors' own work.

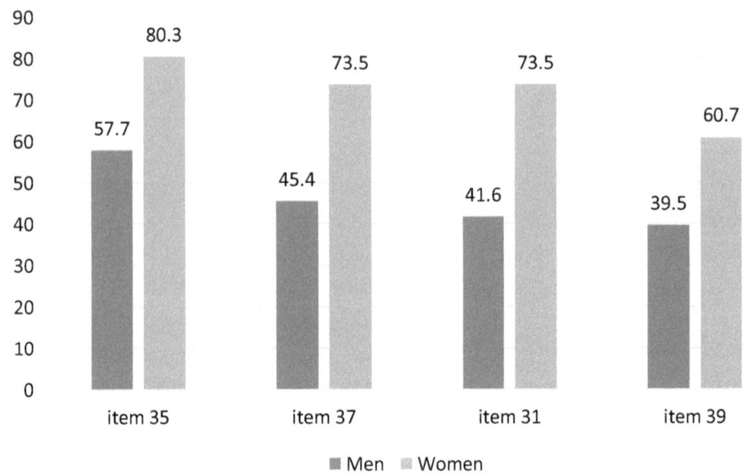

Graph 3: Items valued better by female participants. Authors' own work.

also valued as being extremely important for teaching at school by 73.5 % of the female trainee teachers. However, these two items were considered as important by a considerably lower number of men (45.5 % of men for item 35 and 41.6 % for item 37). Finally, item 39 "Identities, social representations and cultural practices," relating to topics from which cultural diversity can be promoted, was considered fundamental by 60.7 % of women, compared with only 39.5 % of men.

Conclusions

It is stating the obvious that in order to teach any subject, first of all, it is necessary to have knowledge of it. In the case of history, this does not exclusively mean possessing an "encyclopaedic" knowledge of the most significant events and characters of different historical periods. Rather, it is also necessary to have an appropriate epistemological knowledge of the subject, to know how historians approach historical problems, how they select and organise information and build their explanations, to understand the different historiographical trends which have built up historical knowledge, etc. In other words, it is necessary to have sufficient knowledge of how history is made, not only of "what happened" in the past.

Our research, the results of which are analysed in this chapter, is an attempt to examine in a comprehensive manner the opinions and evaluations of future secondary education history teachers in Spain regarding epistemological and historiographical knowledge. Thanks to the large number of people surveyed and their geographical dispersion, the results obtained provide a comprehensive view of historiographical preferences and of the historical contents which seem most relevant to future teachers for inclusion in the school curriculum.

The analysis of the data obtained from the questionnaire and the discussion thereof throughout this chapter have revealed a series of relevant aspects which should be highlighted in these conclusions. First of all, the interest of the participants in issues which make reference to or are connected with life in society and the idea of citizenship is clear. The interest in global human rights and democracy and the main historical processes of humanity point towards a preference for dealing with social problems in the classroom with the aim of achieving a better understanding of them and promoting the ideal of coexistence in society. A glance at Tab. 4 perfectly reflects this viewpoint among the respondents. However, issues which could be considered as being closer to a positivistic view of history, focusing on the most important events and characters linked with the idea of memorisation in the teaching of history were the worst valued aspects in this survey. The explanation for this situation may lie in the fact that the majority of those surveyed were graduates in history, geography or the history of art, hence this inclination in the choice of these issues due to their deeper understanding of this knowledge area which makes it possible for them to apply historical method in the treatment of these issues. In addition, the participants' own experience as students of a system excessively based on the memorisation of data and the use of textbooks may lead them to reject this type of methodological approach.

Last of all, as far as the participants' responses are concerned, no great differences were observed in the evaluations of the contents chosen according to the sex of the respondents. There was only a greater incidence among women in issues directly related with gender perspectives, which received a higher score among women than men. Issues relating to the history of women, feminism, equality and gender relations attracted the attention of the female participants in a significant way. This circumstance reflects the points of interest of future female teachers and the social problems which exist in this area in Spanish society as a whole.

References

Alcaraz, Amparo/Pastor, María Montserrat: "Tendencias de la Historia como objeto de enseñanza a través de la Historiografía". *Didácticas Específicas*, nº 6 junio 2012, pp. 114–139.

Burke, Peter: *Historia de los acontecimientos y renacimiento de la narración.* Alianza: Madrid 2001.

Cambil, María de la Encarnación/Tudela, Antonio: "Epistemología de las Ciencias Sociales: Características, concepto y ámbitos del conocimiento social". In: Liceras, Ángel/Romero, Guadalupe (eds.): *Didáctica de las Ciencias Sociales. Fundamentos, contextos y propuestas.* Pirámide: Madrid 2016, pp. 23–45.

Casanova, Julián: *La historia social y los historiadores. Cenicienta o princesa.* Crítica: Barcelona 2003.

Chartier, Roger: *El mundo como representación. Historia Cultural: entre práctica y representación.* Gedisa: Madrid 1995.

Díaz-Salazar, Rafael: *Educación y cambio ecosocial: del yo interior al activismo ciudadano.* PPC: Madrid 2016.

Gómez Carrasco, Cosme J./Miralles, Pedro: "¿Pensar históricamente o memorizar el pasado? La evaluación de los contenidos históricos en la educación obligatoria en España". *Revista de Estudios Sociales* 52, 2015, pp. 52–68.

Gómez Carrasco, Cosme J./Miralles, Pedro: "Historical Skills in Compulsory Education: Assesment, Inquiry Based Strategies and Argumentation". *New Approaches in Educational Research* 5(2), 2016, pp. 130–136.

Gómez Carrasco, Cosme J./Miralles, Pedro: *Los espejos de Clío. Usos y abusos de la Historia en el ámbito escolar.* Sílex: Madrid 2017.

Gómez Carrasco, Cosme J./Ortuño, Jorge/Molina, Sebastián: "Aprender a pensar históricamente. Retos para la historia del siglo XXI". *Tempo e Argumento* 6(11), 2014, pp. 1–25.

Gómez Carrasco, Cosme J./Rodríguez, Raimundo A.: "La historia como materia formativa. Reflexiones epistemológicas e historiográficas". *Revista de Historiografía* 27, 2017, pp. 265–286.

Gómez Carrasco, Cosme J./Rodríguez, Raimundo A./Mirete, Ana Belén: "Percepción de la enseñanza de la historia y concepciones epistemológicas. Una investigación con futuros maestros". *Revista Complutense de Educación* 29(1), 2018, pp. 237–250.

Gómez Contreras, José Manuel: *La enseñanza de la Historia de los textos escolares (2000-2015): Historiografía, Metodología y Formación de Identidades.* Unpublished doctoral thesis, Autonomous University of Madrid 2019.

Jelin, Elizabeth: *Los trabajos de la Memoria.* Siglo XXI: Madrid 2002.

Le Goff, Jacques/Revel, Jacques/Chartier, Roger (eds.): *La nouvelle histoire.* Retz: Paris 1978.

Miralles, Pedro: "Las tendencias historiográficas recientes y la enseñanza de la Historia en Bachillerato". *Revista de Historiografía* 2, 2005, pp. 158–166.

Miralles, Pedro/Molina, Sebastián/Ortuño, Jorge: *La importancia de la historiografía en la enseñanza de la historia.* Geu: Murcia, 2011.

Paniagua, Francisco J.: "Dejad a los políticos en la cuneta. La historia social busca su propio espacio". *Íber. Didáctica de las Ciencias Sociales, Geografía e historia* 12, 1997, pp. 25–36.

Prats, Joaquín/Santacana, Joan: "¿Por qué y para qué enseñar historia?" In: *Enseñanza y aprendizaje de la Historia en la Educación Básica.* Secretaría de Educación Pública: Cuauhtémoc, Mexico City 2011, pp. 18–68. Accessed 15th September 2017. http://www.ub.edu/histodidactica/images/documentos/pdf/ensenanza_aprendizaje_historia_educacion_basica.pdf.

Ricoeur, Paul: *La lectura del tiempo pasado. Memoria y olvido.* Arrecife: Madrid 1999.

Saiz, Jorge/Valls, Rafael: "Libros de texto de historia y nivel cognitivo: modelo conceptual e hipótesis sobre actuales manuales españoles". In: Prats, Joaquín/Barca, Isabel/López Facal, Ramón (eds.): *Historia e Identidades culturales.* Centro de Investigação em Educação: Braga 2014, pp. 477–488.

Valls, Rafael: "La Historia en sus manuales escolares: entre textos, contextos y competencias". In: Monteagudo, José/Escribano, Ainoa/Gómez Carrasco, Cosme J. (eds.): *Educación histórica y competencias transversales: Narrativas, TIC y competencia lingüística*. Murcia: Editum 2018, pp. 15–34.

Vansledright, Bruce/Reddy, Kimberly: "Changing Epistemic Beliefs? An Exploratory Study of Cognition Among Prospective History Teachers". *Revista Tempo e Argumento* 6(11), 2014, pp. 28–68.

Vilar, Pierre: *Pensar históricamente. Reflexiones y recuerdos*. Planeta: Barcelona 1997.

Juan Carlos Colomer, Jorge Sáiz, and Antonio J. Morales

Chapter 7 History and geographical thinking skills: Between disciplinary knowledge and knowledge for teaching in Teacher Education[1]

Abstract: This chapter aims to analyze trainee teachers' historical and geographical knowledge. This is based on previous studies that have defined the History and geographical thinking skills in a sample of students of a bachelor's degree in Primary Education Teaching and a master's degree in Secondary Education Teaching (Universitat de València). These researches mentioned in this chapter tried to evaluate the extent to which the aforementioned skills are acquired, and their professional abilities to put them into practice in different educational stages. To do this, the authors resorted to questionnaires combining both closed questions and practical exercises regarding editing sources and the elaboration of narratives and pictorial representations. This was done with the purpose of assessing trainee teachers' educational knowledge, and the impact that disciplinary and pedagogical training have had in the development of their professional competences. For that matter, it is especially relevant to ensure that trainee teachers not only have acquired these skills at a theoretical level, but that they are also capable of putting them into practice when necessary.

Keywords: Disciplinary knowledge, History Education, Geography Education, Teacher Education

Introduction

The overall training and education of a future social sciences teacher is a complex task at every educational stage. In all cases, future teachers do not just require general educational knowledge, an understanding of teaching centers, life in the classroom and the psychological evolution of pupils. They also need to learn basic disciplinary content in both history and geography as well as specialized professional teaching skills: how to tackle the instructional use of these subjects, which is known as Pedagogical Content Knowledge (Shulman, 1986).

1 This study is the result of the research Project entitled "Las representaciones sociales de los contenidos escolares en el desarrollo de las competencias docentes" (PGC2018-094491-B-C32), funded by the Ministry of Science, Innovation and Universities of Spain and co-financed with EU ERDF funds.

In other words, this is the method of teaching (history and geography) content and the perception and understanding that teachers bring to this content with the objective of explaining it to their pupils. As such, it is an understanding which links the perception of disciplines (history and geography) with the objectives of their teaching (Souto &García-Monteagudo, 2019).

That being said, the disciplinary content for geography and history, like with any other taught subject, is not derived solely from substantive or first-rate knowledge, understood as geographical and historical data and phenomena. It also stems from methodological or second-rate knowledge, understood as cognitive skills, expertise, or abilities to represent the socio-spatial reality of geography and history as scientific subjects (Seixas & Morton, 2013). As a result, we will start with a basic assumption for this chapter: the historical and geographical Pedagogical Content Knowledge that future teachers require must include the portrayal of both subjects as a proactive learning technique which involves understanding and knowing how to use certain historical and geographical thinking skills. From this perspective, we will try to answer the following specific questions in the subsequent pages: What do these historical and geographical skills involve and what do they bring to teacher training? How can we determine if future teachers learn them and use them? On this matter, what skill levels do primary and secondary school teachers in training demonstrate? To answer these questions, we will use the results of teaching research carried out under the framework of the University of Valencia to make relevant suggestions for improvement.

Theoretical basis

Historical and geographical thinking as a basic resource to train future primary and secondary school social science teachers

Teacher training does not just involve the development of teaching or pedagogical skills which are needed to transpose academic content. In this vein, educational research has shown that, for different school subjects, it is difficult to develop consistent teaching knowledge without a solid understanding of the subject material (Hashweh, 2005; Abell et al., 2009; König, 2013). This type of knowledge must absolutely be linked to historical and geographical education in the school environment. It's clear that it cannot be the same as the knowledge derived from the disciplinary academic environment, which is learned in a specific context: academic culture and the reality of school disciplines are firmly entrenched in codes perceived as specific traditions and routines (Cuesta, 1997,

1998; Viñao, 2002). However, that does not mean that we should not strive for teachers in training to learn cognitive abilities in the fields of geography and history as scientific disciplines based on essential subject matter. From the angle of social science didactics, these skills are considered necessary as the foundation of teacher training and the path to improving and enriching teaching practice, as they provide a dimension of active learning (Gómez, Ortuño, & Miralles, 2018), removed from the traditional practices of memorizing and repeating. We believe that these skills must be integrated into teacher training in order to develop fundamental Pedagogical Content Knowledge as a professional teaching skill, as suggested in different teacher training experiences which incorporate this skill-based dimension of history (Lévesque, 2008; Monte-Sano & Budano, 2013; Pollock, 2014; Lévesque & Zanazanian, 2015) or geography (Cos, de Meer, & Reques, 2011; Delgado & Rodrigo, 2012; Jerez, 2012; De Miguel, 2015; Martínez, 2017; Araya & Cavalcanti, 2018; Jerez, Morales, & Souto, 2019).

The field of studies known as History Education coincides with research focused on the concepts of "historical thinking" and "historical consciousness" two parallel and complementary courses of study (Seixas, 2017). This field of study has been the subject of numerous international academic papers recently, clearly highlighted in the Spanish context (Rodríguez, Gómez & López-Facal, 2020). On this basis, we can consider the existence of a series of skills as methodological content that a teacher must learn and introduce in their classroom practice. We are referring to second-rate strategic content or meta-concepts which indicate how we research and construct the social past and the meaning we attribute to it. These categories are clearly linked to the professional skills of a historian. As such, learning history also involves learning how we build our knowledge of the past, how it is represented, and the significance attributed to it. This entails starting with the construction of historical knowledge and using it to understand the social reality and focus on the present and the future from an understanding of the meaning of the past. As a result, this involves acquiring historical thinking and consciousness, which are neither intuitive nor natural, but rather must be learnt.

When it comes to exploring these historical thinking skills, according to Seixas (2017) we must differentiate between two schools of thought: on the one hand, the Anglo-American angle stemmed from innovative practices and research in the United Kingdom (Lee, 2005) that were further researched in the United States, Canada, Australia and New Zealand (Wineburg, 2001; Lévesque, 2008; VanSledright, 2011; Seixas & Morton, 2013), which employs a more practical and empirical approach to the term "historical thinking"; and on the other hand, the Germanic perspective (in Germany and with influence in Sweden and Brazil), which is more philosophical and pedagogical and prefers the term "historical consciousness" perceived as the ability to use historical information to focus on the present and the future (Rüsen, 2005; Eliasson et al., 2015). This dual tradition means that the typology of said methodological concepts remains open and in discussion. That being said, there is a consensus regarding four basic historical skills: "source and evidence" (the use of sources to obtain evidence), "cause and consequence" (multi-causal explanations), "historical empathy" (understanding the historicity of the past and avoiding presentism through a contextualized historical explanation) and "change and continuity" (historical time). In Canada, Seixas suggested more popular concepts (Seixas & Morton, 2013) where he converges both schools of thought and includes "historical relevance or importance" and "ethical dimension" in addition to the four aforementioned concepts.

As regards geographical education, some authors demonstrate that "in some way, perhaps somewhere deep inside, we are all geographers" in being conscious of the "importance of spatiality in human existence" (Gould, 2000, p. 1). In other words, we can think geographically by revealing aspects related to places, landscapes, and regions (Cavalcanti, 2017), perceived, and thought of as experienced and felt space as a local (Souto, 2018). From there, geographical education becomes a fundamental resource for an overall understanding of problems which affect us on a global scale (Demirci, De Miguel, & Bednarz, 2018) with the aim of understanding the processes of glocalization and the impact of our actions from a socio-environmental perspective. Other authors such as Graves study the way in which the objectives, content, approaches and epistemological questions relevant to this subject can come into play in the comprehensive education of people (Graves, 1985). As such, there is a concern because geographical didactics "allow us to develop reflective opinions about our own teaching practices, in such a way as to be able to make informed decisions" (Souto, 1998).

The teaching and learning of geography has been addressed by the Commission on Geographical Education at the International Geographical Union (IGU) through various international declarations published in relation to: education for development (1992), cultural diversity (2000), sustainable development (2007), research in geographical education (2015) and global understanding (2016). De Miguel, Claudino and Souto (2016) highlight the utopia of these declarations throughout their historical development, as they are politically correct but do not benefit from a large background unless there is a plan of action to further develop it. The 2016 declaration referred to the need for teachers to develop "reflective and critical engagement with educational practices and professional 'habits of mind' that demand improvement in the quality of geographical education" (Comisión de Educación Geográfica de la Unión Geográfica Internacional, 2016) and suggested five points through which to implement the International Plan of Action, calling on education policy makers and geography educators to: enhance the value of geography and its place in the school curriculum; set minimum requirements for geography teachers; promote (inter)national exchanges of meaningful teaching-learning practices; develop a research agenda for geographical education research and its publication in the teaching-learning sector; and create and maintain a professional network.

All of this leads us to reflect on teaching skills related to the geographical education of teachers with the aim of contributing to the development of geographical thinking and consciousness, while appreciating the difference between both concepts as outlined by De Miguel (2015) and the impact on the comprehensive education of pupils in a personal and socially significant way. In this vein, there is concern regarding the geographical education of teachers (Tonda, 2009; Marrón, 2011; De Miguel, De Lázaro, & Marrón, 2013; Moreno & Cely, 2018) and the development of geographical skills among teachers (Cos, de Meer, & Reques, 2011; Jerez, 2012; Martínez, 2017; Araya & Cavalcanti, 2018; Jerez, Morales, & Souto, 2019).

Thus, Cos, de Meer and Reques (2011) make reference to the skills outlined in the Tuning Project and the White Paper on Geography and Territorial Management to explain that university education requires the acquisition of a number of generic skills, to know how to be and how to act, which are classified as personal, instrumental and systemic. Similarly, the development of specific skills, to know and know how to do, are defined as disciplinary, professional, and academic skills. Jerez (2012) outlines that a teacher's geographical skills are conceptual, procedural, and attitudinal in nature. Conversely, Martínez (2017) suggested, using a table to compare with the aforementioned points, the following: knowledge (knowing), skills (knowing how to do), geographical education (knowing how to be) and global skills related in this case to "learning how to teach school geography" (Martínez, 2017, p. 19). Other authors, such as (Araya, 2013, Zwartjes, 2012 and Bednarz et al., 2013 in De Miguel, 2015) cite to concur on the importance of: observing/perceiving, analyzing, interpreting/structuring and acting/applying geographical knowledge to the surroundings; and cite some relevant papers to summarize the aforementioned with: formulating geographical questions, processing information and explaining/communicating geographical facts. Araya and Cavalcanti (2018) presented a case study of geography teachers in training, analyzing the work done by students through teaching processes, in order to construct the concepts suggested by Vygotsky, with the aim of seeing the development of geographical thinking in teachers.

In this regard, research has been done on all elements related to knowing, knowing how to be, knowing how to act and knowing how to think to know how to do (Franquesa, 1996 in Santana, 2019) in connection with teacher training. That's why it's important to consider the contributions of the global and interdisciplinary approach in order to understand the reality from a complex thinking perspective (Zabala, 1999), with the aim of developing, in this case, complex geographical thinking. This question justifies our interest in encouraging a critical reflection of social representations that pupils may have about this skills-based knowledge (Moscovici, 1961; Jodelet, 1986) which is entrenched in their own understanding of the world and the epistemological tradition used to construct it (Souto & García-Monteagudo, 2019). We are concerned that the academic approach to geography perceived hinders the "skills of learning and learning to teach" (Morales & Santana, 2019, p. 100) by not taking into account the emotional link created by the geography we experience through active participation. In this vein, future teachers help to redefine skills and sub-skills through analysis of their own social characterizations. To do this, we have to address their personal school geographies, with the aim of creating frameworks which help to profile teaching skills related to school geography education, as we have demonstrated with our work on cartographic skills (Jerez, Morales, & Souto, 2019).

Tools to coordinate historical and geographical thinking education for teachers in training

The learning and use of historical skills requires the development of linguistic skills insofar as the representation of the past takes a narrative form, creating stories or texts in different formats, in the same way as spatial representation can be found in a pictorial or map format. As a result, the best tool to analyze the extent to which teachers in training learn historical skills is to construct different profiles of historical narratives, especially those presented as a summary of a past phenomenon.

Together with the study of narratives, the analysis of social representations (Moscovici, 1961; Jodelet, 1986) has opened a field of research in which various specialists have worked (Pagés, 1996; Reigota, 2001; Hernández & Pagés, 2016). This theory has allowed us to connect image and social practice (Abric, 2001; Araya; 2002), Kosslyn (1981) analyzed the environment and mental images. In medicine, Hernando (2012) used drawings to study the social representations of nurses. Children's drawings have been studied by many authors (Ricci, 1887; Luquet, 1927; Lowenfeld, 1958; Arnheim, 1979; Goodnow, 1979; Duborgel, 1981; Quiroga, 2007; Picard & Baldy, 2012; Machón, 2016) with the aim of analyzing their meaning. But we want to focus on how to interpret children's personal and social representations (Casas, 1992) and those of the teachers involved in their education, using pictorial representations related to geographical space. Some authors support drawings as a research method because they are useful in showing how pupils perceive or interpret reality, making it a pedagogical resource to develop learning skills by learning from what we depict (Jiménez & Martínez, 2011). In fact, this approach was championed and used as a method to learn and represent everyday surroundings (Barraza, 1999, Comes, 2002; Arto, 2010; Ortiz, Prats, & Baylina, 2011). Seidmann et al. (2014) explored the use of graphic techniques in research about social representations. Piza and Peña (2014) designed an instrument to analyze, through drawing, social representations of the environment held by biology teachers. Cadavid, Vásquez and Botero (2014) also used drawings for their work on social representations of English learning. Gómez and Gavidia (2015) extended it to science education. Fernández (2016) used drawings for perceptions of tourism and Santana (2019) and García-Monteagudo (2019) both interpreted drawings in their respective theses to analyze social representations in school related with natural and rural landscapes.

Empirical analytical studies of historical and geographical thinking

From here on, we will describe important studies carried out under the umbrella of the University of Valencia, which have allowed us to apply the aforementioned theoretical framework, as well as the methodological tools for understanding the skills acquired during teacher training. As regards the research done using narratives among future teachers, they allowed us to identify different levels of historical skills, particularly the use of causality, historical time and historical relevance (Sáiz & López-Facal, 2015; Gómez & Sáiz, 2017; Sáiz, Gómez, & López-Facal, 2018). In this vein, we noticed a clear contrast between the two groups of teachers examined. On the one hand, students in the final year of their Degree in Primary Education, future primary school teachers, revealed very basic, if not non-existent, skill levels. On the other hand, students taking the master's degree in Secondary Education, future secondary school teachers, often had a degree in History thus boasting higher levels of historical skills. To a certain extent, these two pathways reflect novice and expert levels.

Furthermore, we identified a very poor level of understanding and use of historical skills among the future primary school teachers, based on a brief account of the Iberian Peninsula in medieval times (Gómez & Sáiz, 2017; Sáiz, Gómez, & López-Facal, 2018). In the case of research carried out at the University of Valencia, future primary school teachers do not take any specific modules on disciplinary history throughout their university course, but they do study modules on Social Science Didactics. The scant historical education they receive by the end of their compulsory secondary education, at baccalaureate level, is not furthered at any point in their university degree for primary school teachers. The sample studied allowed us to document an overwhelming majority (91 %) of future primary school teachers who did not use methodological historical skills to plan their discourse: this is a consequence of their scarce understanding of substantive historical knowledge and the weak history education they receive on a factual, conceptual and methodological level. In fact, we noticed that the complexity of their accounts and the presence of historical thinking markers grow proportionally to the increase in substantive content itself: those who contribute a larger quantity of historical information do so with greater organizational quality. This scarce historical education shows us that they would hardly be in the right conditions to teach history with the minimum disciplinary and pedagogical levels required, at least according to the accepted standards for research in social science didactics and history education. To this end, the training given to future primary school teachers in history education is clearly deficient and does not improve on the already low levels obtained during the completion of their baccalaureate. It is thus urgent to implement radical change if we want to give future primary school teachers the minimum expertise required to provide history education to primary school pupils.

Conversely, a large majority of the future secondary school history teachers had graduated with a degree in history. We noted progress in the essential learning of knowledge and use of methodological historical skills, which we identified using an account of the history of Spain in the 20th century (Sáiz & López-Facal, 2015). Most of their narratives fell back on, albeit to a low or medium level, the use of methodological historical skills and also added richer substantive content: they didn't limit themselves to the lineal description of political change, but rather showed the interrelationship with other sectors (social, economic) and some also presented problems and interpretations. All in all, we could say that they learned these historical skills by osmosis or by copying the history lectures they attended during their history degree, but not by carrying out pedagogical sequences of historical knowledge building. It also reflects the fact that some history content in higher education is very similar to that at baccalaureate level, just with a change in the intensity and quantity of historical content: they also focused on memorizing, consuming and reproducing academic narratives but not on creating critical and interpretative historical essays or studies, with the use of sources and problem solving as the only possible practices to ensure learning and the broad and sufficient use of historical skills. As such, the history education of history graduates, future secondary school history teachers, also doesn't provide them with critical epistemological interpretation and sufficient skills for the use of historical thinking abilities; that's why they use them in a basic and limited way. They cannot adequately teach topics of which they have basic or non-existent knowledge. This reflects the large majority of future history teachers, who don't go beyond positivist and relativist epistemological views – the hegemonic ones – and don't benefit from a critical perspective, according to the relevant matrix of experts and novices provided by VanSledright (2011, p. 72).

As regards geographical education, in the context of Master's studies in the Faculty of Education and within the lines of research of the SOCIAL(S) group– a research and innovation group in geography and history education – conclusions of their research have allowed us to identify a link between geographical thinking and analyzed social representations.

These pieces of research have made progress in the construction of the skills of learning to learn how to teach from different perspectives. García (2015) addressed the existing correlation between social representations and pedagogical proposals, demonstrating the need for globalized approaches, from the angle of environmental education, for the development of complex geographical thinking. In this vein, Sánchez (2016) revealed the relevance of skills focused on the emotional side by linking environmental education and emotional education. Together with the emotional impact, aspects related to school participation were also deemed relevant (Canet, 2017; Santana, 2019), understood as a necessary component for the development of geographical skills. Parra (2017), on the other hand, underlined the importance of the globalized approach of solidarity and sustainability in teacher training, as related to the goals of the 2030 Agenda. Finally, studies about geographical thinking developed within our research group have demonstrated the convenience of analyzing personal maps and pupils' social representations in order to link the experienced space with their academic knowledge (Jerez, 2018; González, 2019) and school-taught geography (Jerez, 2018; García-Monteagudo, 2019; González, 2019).

In this selection of research, special concern was shown for the social representations of teachers; of pupils (Sánchez, 2016; Canet, 2017; Jerez, 2018;

González, 2019) or of both (García-Monteagudo, 2019; Santana, 2019). The results obtained with students reveal the need to tackle personal geography, in the scope of formal education, due to its educational value in the social construction of experienced spaces and its resulting impact on the development of skills among learners and teachers.

Conclusions and improvement

Drawing from the studies carried out, we noted a significant deficit in teacher training (both for primary and secondary school teachers) due to a lack of basic knowledge for teaching historical and/or geographical skills, given that in their previous training they had no theoretical-practical experience introducing them to learn how to use these skills.

As regards the learning and development of historical thinking, from here on it is necessary to change the role of teachers in training when confronted with historical texts. Rather than encouraging memorization which involves merely consuming and reproducing academic texts which summarize events, we need to foster creative and qualitative practices which encourage the writing of essays or texts which involve tackling historical events in a creative way. In this regard, we suggest three main lines of historical writing practice which should be included in the training of teachers on the Degree in Primary Education or on the postgraduate degree in Secondary Education (Sáiz, 2015; López, Sáiz, & López-Facal, 2018). First, we suggest historical reasoning essays (Chapman, 2011) or explanatory texts based on solving a problem or historical question which involves analyzing sources (Monte-Sano, 2011). Second, we recommend texts summarizing narratives over a period or process in the past (summarize or tell us what you know about. . .) which involve selecting and organizing substantive knowledge possessed by the student, thus pushing them to resort to historical skills such as cognitive abilities to organize their discourse. Finally, our third suggestion is to write historically empathetic texts which provide a writing context in the form of a previous historical event or episode including character profiles they can identify with, accompanied by historical theories and sections which should be written from the character's point of view in the form of open questions. In this way, we can encourage students to write texts in the first person where the student acts as a witness narrator or main character narrator and approximates their understanding through historical empathy of certain periods in the past. Ultimately, the three methods involve creating qualitative historical texts, assuming an active role and undertaking a metacognitive reflection about how you use your own historical knowledge as discourse about the past. Only in this way, if they have experienced and used them, could future primary and secondary school teachers introduce these historical skills into their classroom practices with pupils.

With regard to geographical education, the most emphasized aspect of the studies done is, in our opinion, overcoming the geometric and volumetric spatial limitation of constructivism in order to introduce other variables which identify predisposing factors for certain activities from an emotional perspective. By doing this, we reconsider the cognitive and sensory limitations of the way in which we perceive space to challenge the standards of traditional academics. This question justifies our wish to explore personal and social representations, based on the contributions of geography to perception and behavior, because they link the observation of space with the predisposition to proceed or to act in a certain way. Research related to neuroscience and social representations confirms some of the previous findings of geographical perceptions, especially through the so-called behavioral shift by Horacio Capel in the early seventies.

The results obtained to date corroborate the need to develop complex geographical thinking and the skills and sub-skills needed to learn to learn how to teach. The feelings derived from these personal geographies can contribute to rethinking the skills related to teacher training in geographical education. We are unable to achieve the goals set out in the Paris Agreement and the Sustainable Development Goals of the 2030 Agenda without the participation of citizens who have not developed their predisposition for eco-social transformation. In this vein, it is necessary, but not sufficient, to include sustainability in the curriculum because this emotional link is the foundation of acting and getting involved in the various projects which could contribute to achieving the ecological transition. Without the active participation of individuals, plans such as the SUMP (Sustainable Urban Mobility Plan) and SECAP (Sustainable Energy and Climate Action Plan) lack meaning and could be forgotten like many local 21 Agendas. That is why it is vital for pupils to reflect, based on what they feel, on the geographical space where they want to live, both now and in the future. In this way, teachers could use this as a basis to contribute to the development of skills which make this possible. In this case, geographical education is a fundamental tool for the reasonable construction of these vital spaces, due to its contribution to sustainable interaction and peaceful cohabitation on a local and global scale.

References

Abell, Sandra K./Park, Meredith A./Hanuscin, Deborah L./Lee, Michele H./Gagnon, Mark J.: "Preparing the Next Generation of Science Teacher Educators: A Model for Developing PCK for Teaching Science Teachers". *Journal of Science Teacher Education* 20(1), 2009, pp. 77–93, retrieved 28.10.2020, from DOI: 10.1007/s10972-008-9115-6.

Abric, Jean Claude: *Prácticas sociales y representaciones*. Coyoacán: México. 2001.

Araya, Fabián/Cavalcanti, Lana de Souza: "Desarrollo del pensamiento geográfico: un desafío para la formación docente en Geografía". *Revista de Geografía Norte Grande* 70, 2018, retrieved 28.10.2020, from DOI 10.4067/S0718-34022018000200051.

Araya, Sandra: "Las representaciones sociales: ejes teóricos para su discusión". *Cuaderno de Ciencias Sociales* 127, 2002, retrieved 28.10.2020, from: https://links.uv.es/uES0H0q.

Arnheim, Rudolf: *Arte y percepción visual, psicología del ojo creador*. Alianza Editorial: Madrid. 1979.

Arto, Mónica: "El cambio climático narrado por alumnos de educación primaria y secundaria: propuesta de análisis para dibujos y textos". In: *Investigar para avanzar en la Educación Ambiental. Doctorado Interuniversitario de Educación Ambiental*. Organismo Autónomo Parques Nacionales. Ministerio de Medio Ambiente y Medio Rural y Marino: Madrid. 2010, pp. 11–29.

Barraza, Laura: "Children's Drawing about the Environment". *Environmental Education Research* 5(1), 1999, pp. 49–67.

Cadavid, Isabel, C./Vásquez, Juan C./Botero, Juan D.: "El uso de juegos dramáticos y dibujos para explorar las representaciones sociales de los niños y las niñas acerca del aprendizaje y la enseñanza del inglés". *Íkala, Revista de Lenguaje y Cultura* 19(3), 2014, pp. 287–304, retrieved 28.10.2020, from DOI 10.17533/udea.ikala.v19n3a05.

Canet, Sandra: *Pensar geográficamente en el aula de Educación Infantil: un estudio de casos con alumnado de segundo ciclo* (master's final project). Universitat de València: València. 2017.

Casas, Ferran: "Las representaciones sociales de las necesidades de niños y niñas, y su calidad de vida". *Anuario de Psicología* 53, 1992, pp. 27–45.

Cavalcanti, Lana de Souza: "El análisis de la espacialidad y la comprensión del mundo: llave para la relevancia de la geografía escolar". In: Sebastià, Rafael/Tonda, Emilia (coord.): *Enseñanza y aprendizaje de la geografía para el siglo XXI*. Universidad de Alicante: Alicante. 2017, pp. 97–116.

Chapman, Arthur: "Taking the Perspective of the Other Seriously? Understanding Historical Argument". *Educar em Revista* 42, 2011, pp. 95–106.

Comes, Pilar: La interpretación de las imágenes espaciales. In: *La Geografía y la Historia, elementos del medio*. Ministerio de Educación, Cultura y Deporte: Madrid. 2002, pp. 209–244.

Comisión de Educación Geográfica de la Unión Geográfica Internacional: *Declaración Internacional sobre Educación Geográfica*, 2016, retrieved 28.10.2020, from http://www.igu-cge.org/wp-content/uploads/2018/02/IGU_2016_spanish1.pdf.

Cos, Olga/de Meer, Ángela/Reques, Pedro: "Formación de geógrafos y aprendizaje basado en competencias: balance de un plan de innovación docente". *Biblio3W* XVI(920), 2011, retrieved 28.10.2020, from http://www.ub.edu/geocrit/b3w-920.htm.

Cuesta, Raimundo: *Sociogénesis de una disciplina escolar: la historia*. Pomares Corredor: Barcelona. 1997.

Cuesta, Raimundo: *Clío en las aulas. La enseñanza de la historia en España entre reformas, ilusiones y rutinas*. Akal: Madrid. 1998.

De Miguel, Rafael: "Del pensamiento espacial al conocimiento geográfico a través del aprendizaje activo con tecnologías de la información geográfica". *Giramundo* 2(4), 2015, pp. 7–13.

De Miguel, Rafael/Claudino, Sérgio/Souto, Xosé M.: "La utopía de la educación geográfica en las declaraciones internacionales de la UGI". In: Arroyo, Mercedes/Chaparro, Jeffer (eds.): *XIV Coloquio Internacional de Geocrítica. Las utopías y la construcción de la sociedad del futuro*. Universitat de Barcelona: Barcelona. 2016.

De Miguel, Rafael/De Lázaro, Mª Luisa/Marrón, Mª Jesús (coord.): *Innovación en la enseñanza de la geografía ante los desafíos sociales y territoriales*. Institución Fernando el Católico: Zaragoza. 2013.

Delgado, José Jesús/Rodrigo, Jesús: "El trabajo de campo y las competencias geográficas en el estímulo para el estudio de la geografía: aplicación en un aula de 2º de Bachillerato". *Didáctica Geográfica* 13, 2012, pp. 35–56.

Demirci, Ali/De Miguel, Rafael/Bednarz, Sarah W.: *Geography Education for Global Understanding (International Perspectives on Geographical Education)*. Springer: Cham. 2018.

Duborgel, Bruno: *El dibujo del niño. Estructuras y símbolos*. Paidós: Barcelona. 1981.

Eliasson, Per/Alvén, Fredrik/Yngvéus, Cecilia A./Rosenhud, David: "Historical Consciousness and Historical Thinking Reflected in Large-Scales Assessment in Sweeden". In: Ercikan, Kadriye/Seixas, Peter (eds.): *New Directions in Assessing Historical Thinking*. Routledge: New York. 2015, pp. 171–182.

Fernández, Ana M. "Representaciones del turismo desde los dibujos y explicaciones de las jóvenes generaciones". *Via Tourism Review* 10, 2016, retrieved 28.10.2020, from DOI: 10.4000/viatourism.1431.

García, Dafne: *Las representaciones sociales sobre Educación Ambiental del profesorado de Educación Secundaria: estudio de casos* (master's final project). Universitat de València: València. 2015.

García-Monteagudo, Diego: *La representación social del medio rural: un análisis desde la geografía escolar* (doctoral thesis). Universitat de València: València. 2019.

Gómez, Cosme/Ortuño, Jorge/Miralles, Pedro: *Enseñar ciencias sociales con métodos activos de aprendizaje: reflexiones y propuestas a través de la indagación*. Octaedro: Barcelona. 2018.

Gómez, Cosme/Sáiz, Jorge: "Narrative Inquiry and Historical Skills. A Study in Teacher Training". *Revista Electrónica de Investigación Educativa* 19(4), 2017, pp. 19–32.

Gómez, Víctor/Gavidia, Valentín: "Describir y dibujar ciencias. La importancia del dibujo en las representaciones mentales del alumnado". *Revista Eureka sobre Enseñanza y Divulgación de las Ciencias* 12(3), 2015, pp. 441–455.

González, Alicia. *Percepción crítica y emocional del entorno urbano en la infancia: estudio de campo a través de las salidas de campo y el mapeo digital* (master's final project). Universitat de València: València. 2019.

Goodnow, Jacqueline: *El dibujo infantil*. Ediciones Morata: Madrid. 1979.

Gould, Peter: "Pensar como un geógrafo una exploración a la geografía moderna". *Scripta Nova. Revista Electrónica de Geografía y Ciencias Sociales* 78, 2000.

Graves, Norman J.: *La enseñanza de la geografía*. Visor: Madrid. 1985.

Hashweh, Maher Z.: "Teacher Pedagogical Constructions: A Reconfiguration of Pedagogical Content Knowledge". *Teachers and Teaching: Theory and Practice* 11(3), 2005, pp. 273–292.

Hernández, Lorenia/Pagés, Joan: "¿Cómo enseñar historia y Ciencias Sociales en la educación preescolar?". *Revista Mexicana de Investigación Educativa* 21(68), 2016, pp. 119–140.

Hernando, María: *Cofias y manguitos. Un análisis descriptivo de las representaciones sociales de la enfermera a través de dibujos* (doctoral thesis). Departamento de Psiquiatría. Facultad de Medicina: Madrid. 2012.

Jerez, María: *El aprendizaje de la cartografía personal: estudio de casos en el primer ciclo de Educación Primaria* (master's final project). Universitat de València: València. 2018.

Jerez, María/Morales, Antonio/Souto, Xosé M.: "Las competencias docentes para favorecer el aprendizaje cartográfico en Educación Primaria". In: Puig-Gutiérrez, María/Ferreras-Listán, Mario/Moreno-Fernández, Olga (eds.): *Innovación e investigación en la formación inicial del profesorado de Infantil y Primaria desde las Didácticas de las Ciencias Experimentales y Sociales*. Octaedro: Barcelona. 2019, pp. 31–52.

Jerez, Óscar: "Competencias geográficas del profesorado en Educación Básica". In: Nieto, Emilio/Callejas, Ana I./Jerez, Óscar (coord.): *Las competencias básicas. Competencias profesionales del docente*. Universidad de Castilla-La Mancha: Ciudad Real. 2012, pp. 221–231.

Jiménez, César/Martínez, Yessica: "Visiones y representaciones de estudiantes a través del dibujo". *REMO* VII(21), 2011, pp. 24–31.

Jodelet, Denise: *La representación social: fenómeno, concepto y teoría*. In: Moscovici, Serge (ed.): *Psicología Social. II*. Paidós: Barcelona. 1986, pp. 470–494.

König, Johannes: "First Comes the Theory, Then the Practice? On the Acquisition of General Pedagogical Knowledge during Initial Teacher Education. International". *Journal of Science and Mathematics Education* 11(4), 2013, pp. 999–1028, retrieved 28.10.2020, from DOI: 10.1007/s10763-013-9420-1.

Kosslyn, Stephen M.: "The Medium and the Message in Mental Imagery: A Theory". *Psychological Review* 88(1), 1981, pp. 46–66, retrieved 28.10.2020, from DOI 10.1037/0033-295X.88.1.46.

Lee, Peter: "Putting Principles intro Practice: Understanding History". In: Donovan, M. Suzanne/Bransford, John D. (eds.): *How Students Learn: History in the Classroom*. National Academies Press: Washington. 2005, pp. 31–77.

Lévesque, Stéphane: *Thinking Historically. Educating Students for the Twenty-First Century*. University of Toronto Press: Toronto. 2008.

Lévesque, Stéphane/Zanazanian, Paul: "History Is a Verb: We Learn It Best When We Are Doing It!" *Revista de Estudios Sociales* 52, 2015, pp. 32–51.

López, María/Sáiz, Jorge/López-Facal, Ramón: "Educación histórica y competencia lingüística". In: Miralles, Pedro/Gómez, Cosme (coord.): *La educación histórica ante el reto de las competencias. Métodos, recursos y enfoques de enseñanza*. Octaedro: Barcelona: Octaedro. 2018, pp. 15–26.

Lowenfeld, Viktor: *El niño y su arte*. Kapelusz: Buenos Aires. 1958.

Luquet, Georges H.: *Les dessins d'un enfant*. Librairie Felix Alcan: París. 1927.

Machón, Antonio: *Los dibujos de los niños. Génesis y naturaleza de la representación gráfica. Un estudio evolutivo*. Madrid: Editorial Fíbulas. 2016.

Marrón, Mª Jesús: "Educación geográfica y formación del profesorado. Desafíos y perspectivas en el nuevo Espacio Europeo de Educación Superior (EEES)". *Boletín de la Asociación de Geógrafos Españoles* 57, 2011, pp. 313–341.

Martínez, Luis Carlos: "La geografía en los estudios universitarios de Magisterio: evolución histórica, situación actual y significado docente". *Biblio3W* XXII(1207), 2017, retrieved 28.10.2020, from http://www.ub.edu/geocrit/b3w-1207.pdf.

Monte-Sano, Chauncey: "Learning to Open Up History for Students. Pre-Service Teachers' Emerging Pedagogical Content Knowledge". *Journal of Teacher Education* 62(3), 2011, pp. 260–272.

Monte-Sano, Chauncey/Budano, Christopher: "Developing and Enacting Pedagogical Content Knowledge for Teaching History: An Exploration of Two Novice Teachers' Growth Over Three Years". *The Journal of the Learning Sciences* 22(2), 2013, pp. 171–211.

Morales, Antonio/Santana, Diana: "Las representaciones de la Didáctica de las Ciencias Sociales en los futuros docentes de educación infantil". In: Parra, David/Fuertes, Carlos (coords.): *Reinterpretar la tradición, transformar las prácticas. Ciencias Sociales para una educación crítica*. Tirant Lo Blanch: València. 2019, pp. 99–121.

Moreno, Nubia/Cely, Alexander: *Educación geográfica, formación docente y vida cotidiana*. Universidad Pedagógica Nacional: Bogotá. 2018.

Moscovici, Serge: *La psychanalyse, son image et son public: étude sur la représentation sociale de la psychanalyse*. Presses Universitaires de France: París. 1961.

Ortiz, Anna/Prats, María/Baylina, Mireia, "Métodos visuales y geografías de la infancia: dibujando el entorno cotidiano". *Scripta Nova, Revista Electrónica de Geografía y Ciencias Sociales* XVI(400), 2011.

Pagés, Joan: "Las representaciones de los estudiantes de maestro sobre la enseñanza de las Ciencias Sociales: ¿cuáles son?, ¿cómo aprovecharlas?". *Investigación en la Escuela* 28, 1996, pp. 103–114, retrieved 28.10.2020, from DOI: 10.12795/IE.1996.i28.08.

Parra, Arantxa: *Las representaciones sociales sobre la solidaridad y la sostenibilidad en docentes con formación en mediación y convivencia: análisis de 9 casos* (master's final project). Universitat de València: València. 2017.

Picard, Delphine/Baldy, René: "Le dessin de l'enfant et son usage dans la pratique psychologique". *Développements* 10, 2012, pp. 45–60.

Piza, Karoll J./Peña, Sergio E.: "Diseño de un instrumento para identificar las representaciones sociales sobre el concepto de medio presentes en estudiantes de licenciatura en biología de la Universidad Distrital Francisco José de Caldas". *Bio-grafía Escritos sobre la Biología y su Enseñanza.* Edición Extra. Memorias del VII Encuentro Nacional de Experiencias en la Enseñanza de la Biología y la Educación Ambiental y II Congreso Nacional de Investigación en la Enseñanza de la Biología, 2014, pp. 205–214, retrieved 28.10.2020, from DOI: 10.17227/20271034.vol.0num.0bio-grafia205.214.

Pollock, Scott A. "The Poverty and Possibility of Historical Thinking: An Overview of Recent Research into History Teacher Education". In: Sandwell, Ruth/Heyking, Amy von (eds.): *Becoming a History Teacher.* University of Toronto Press: Toronto, 2014, pp. 60–74.

Quiroga, M.ª Pilar: *Estructura y temática en el dibujo infantil.* Salamanca: Papeles Salmantinos de Educación. 2007.

Reigota, Marcos: *Meio ambiente e representação social.* Cortez Editora: São Paulo. 2001.

Ricci, Corrado: *L'arte dei bambini*, 1887.

Rodríguez, Jairo/Gómez, Cosme/López-Facal, Ramón/Miralles, Pedro: "Tendencias emergentes en la producción académica de educación histórica". *Revista de educación* 389, 2020, pp. 211–242, retrieved 28.10.2020, from DOI 10.4438/1988-592X-RE-2020-389-460.

Rüsen, Jörn: *History: Narration, Interpretation, Orientation.* Berghahn: Nueva York. 2005.

Sáiz, Jorge: *Educación histórica y narrativa nacional.* (doctoral thesis) Universitat de València: València. 2015.

Sáiz, Jorge/Gómez, Cosme/López-Facal, Ramón.: "Historical Thinking, Causal Explanation and Narrative Discourse in Trainee Teachers in Spain". *Historical Encounters: A Journal of Historical Consciousness, Historical Cultures, and History Education* 5(1), 2018, pp. 16–30.

Sáiz, Jorge/López-Facal, Ramón: "Competencias y narrativas históricas. El pensamiento histórico de estudiantes y futuros profesores". *Revista de Estudios Sociales* 52, 2015, pp. 87–101.

Sánchez, Laura: *Las representaciones sociales del concepto de medio ambiente y su incidencia emocional en el tercer ciclo de Educación Primaria. Estudio de casos.* (master's final project) Universitat de València: València. 2016.

Santana, Diana: *Participación escolar y gestión ambiental: una dialéctica educativa.* (doctoral thesis) Universitat de València: València. 2019.

Seidmann, Susana/Di Lorio, Jorgelina/Azzollini, Susana/Riguieral, Gustavo: "El uso de técnicas gráficas en investigaciones sobre representaciones sociales". *Anuario de Investigaciones* XXI, 2014, pp. 177–185.

Seixas, Peter/Morton, Tom: *The Big Six Historical Thinking Concepts.* Nelson: Toronto. 2013.

Seixas, Peter: "Historical Consciousness and Historical Thinking". In: Carretero, Mario/Berger, Stefan/Grever, María (eds.): *Palgrave Handbook of Research in Historical Culture and Education.* Palgrave MacMillan: Londres. 2017, pp. 59–72.

Shulman, Lee S.: "Those Who Understand: Knowledge Growth in Teaching". *Educational Researcher* 15(2), 1986, pp. 4–14.

Souto, Xosé M.: *Didáctica de la Geografía. Problemas sociales y conocimiento del medio.* Ediciones del Serbal: Barcelona. 1998.

Souto, Xosé M.: "La geografía escolar: deseos institucionales y vivencias de aula". *Boletín de la Asociación de Geógrafos Españoles* 79, 2018, retrieved 28.10.2020, from DOI 10.21138/bage.2757.

Souto, Xosé M./García-Monteagudo, Diego: "Conocer las rutinas para innovar en la geografía escolar". *Revista de Geografía Norte Grande* 74, 2019, pp. 207–228, retrieved 28.10.2020, from DOI 10.4067/S0718-34022019000300207.

Tonda, Emilia: "La Geografía en el grado de maestro de Educación Infantil en España". In: *IV Congreso Ibérico de Didáctica da Geografía. A Inteligência Geográfica na Educação do Século XXI.* Associaçao de Professores de Geografia: Lisboa, 2009, pp. 1–23.

VanSledright, Bruce: *The Challenge of Rethinking History Education. On Practice, Theories, and Policy.* Routledge: New York. 2011.

Viñao, Antonio: *Sistemas educativos, culturas escolares y reformas: continuidades y cambios.* Morata: Madrid. 2002.

Wineburg, Sam: *Historical Thinking and Other Unnatural Acts: Charting the Future of Teaching the Past.* Temple University Press: Philadelphia. 2001.

Zabala, Antoni: *Enfoque globalizador y pensamiento complejo: una respuesta para la comprensión e intervención en la realidad.* Graó: Barcelona. 1999.

Cosme J. Gómez Carrasco, Álvaro Chaparro Sáiz, María del Mar Felices de la Fuente and Juan Antonio Inarejos Muñoz

Chapter 8 Perceptions regarding historical competences in trainee geography and history teachers[1]

Abstract: The changes that have taken place in our society require new teaching strategies and new purposes that build citizenship on a broad and multicultural world. The main objective of this research is to analyze the opinions and perceptions of Secondary Education teachers in training on the development of historical competences on History classroom. The research has been carried out in the specialty of Geography and History of the Teacher Training Postgraduate. 646 students from master's degrees in geography and history teacher training from 22 different universities in the 2019/2020 academic year have participated in this research. A quantitative non-experimental design was employed, and data was collected via a questionnaire with a Likert-type scale (1–5). The data revealed by the questionnaire assays the unanimous preference for competences of a procedural and methodological nature and also surmises a clear preference of women for second-order contents.

Keywords: History Education, Historical competences, Questionnaire, Teacher Education, Historical Thinking

Introduction

Many studies have shown the existence of conflicting objectives in the teaching of history in schools today (Barton, 2008; Hammack, 2011; Carretero, Asensio & Rodríguez, 2012), which are resolved by seeking to educate in terms of the identity and nationalism for which it was conceived (Hobsbawm, 1997), thereby legitimising the founding past of the nations (Berger, 2012), and forming

1 This study is the result of the research Project entitled "Methodological concepts and active learning methods for the improvement of teaching competences" (PGC2018-094491-B-C33), funded by the Spanish Ministry of Science, Innovation and Universities, co-funded by the ERDF and the project entitled "Teaching competences and active learning methods. An evaluative research with the teaching staff in social sciences" (20638/JLI/18), funded by the Seneca Foundation to CJGC (Agency of Science and Technology of the Region of Murcia).

historical thinking from a critical perspective which assists in understanding the past and locating our place in the present. From the present-day point of view, it cannot be denied that history teaching in schools must play a fundamental role in forming a critical, participative citizenship, which shows an interest in the world in which it lives (Barton & Levstik, 2004) and, therefore, cannot be based on the romantic and positivist ideals with which it has been conceived as a school subject, but rather it should be characterised by its social function (Rüsen, 2005).

On the basis of this epistemological approach, in England, from the 1960s onwards, different initiatives arose focusing precisely on how to improve the teaching and learning processes of history in order to include the construction of concepts, skills and methodological knowledge and not only memorisation and the reproduction of contents. The greatest proponents of these proposals were projects such as the *History Project 13–16*, which later became known as the *School History Project* (SHP), *Concepts of History and Teaching Approaches, 7–14* (CHATA) and *Constructing History 11–19*, the research and results of which opened up possibilities for new models for the teaching of history (Lee, 1992, 2016; Lee & Ashby, 2000; Cooper & Chapman, 2009; Chapman, 2011; Cooper, 2013; Chapman & Wilschut, 2015) which had a great impact in countries such as the United States (Barton, 1994; Barton & Levstik, 2004; Monte-Sano, 2010; VanSledright, 2014) and Canada (Lévesque, 2008; Éthier, Demers & Lefrançois, 2010; Lévesque & Clark, 2018), where the importance of *The Historical Thinking Project* is particularly worthy of mention. This project is based on six concepts which are essential for the development of historical thinking: historical relevance, the use of primary sources, the analysis of changes and continuities, of causes and consequences, historical perspective, and the ethical and moral dimension of history (Seixas & Morton, 2013).

Since that time, the research lines established by these points of reference have encouraged a great deal of scientific output on an international level (Carretero, Berger & Grever, 2017; Metzger & Harris, 2018), which has led to progress both in the conceptualisation and development of historical thinking skills and concepts (Seixas & Morton, 2013; Van Drie & Van Boxtel, 2018), and in the formation of historical consciousness via the analysis of historical narratives, which is one of the most popular fields of research in countries such as Germany (Rüsen, 2005; Wertsch, 2004; Kölbl & Konrad, 2015; Körber, 2015), Portugal (Barca, 2000; Melo & Lopes, 2004) and Brazil (Schmidt & García, 2005).

Along with these two great fields of research in the framework of historical thinking, there has been an abundance of research on the skills and concepts of historical thinking, such as argumentation and reasoning (Van Drie & Van Boxtel, 2008; Cooper & Chapman, 2009; Monte-Sano, 2010; Wineburg, Martin & Monte-sano, 2013); the use of historical sources (Plá, 2005; VanSledright, 2011; Henríquez & Ruiz, 2014); and historical relevance, which implies the questioning of what and who is deserving of being remembered and studied from the perspective of social and cultural history, the study of mentalities, everyday history and the plurality of histories far removed from ethnocentrism (Lévesque, 2008; Peck, 2009).

Other concepts, such as the analysis of causes and consequences, the understanding of changes and continuities, and historical empathy have been dealt with in detail in the context of the previously-mentioned projects, such as the *School History Project* and CHATA (Lee & Ashby, 2000), in England and in Canada, where authors such as Bryant and Clark (2006) and Lévesque (2008) have worked on them. Furthermore, the uses of history for the formation of identities and of citizenship is another of the fields of study in the framework of the formation of historical thinking which has made it possible to reflect on how history can be used for different purposes and how it can be used as a tool in politics, memory and the construction of national identities (Clark, 1996; Aronsson, 2015; Nordgren, 2016).

Taking into account the results of the research, historical thinking is currently conceived of as a creative process based on the practice of the historian (Levesque, 2011). Thus, the teaching of history cannot be founded on traditional approaches based on the memorisation and reproduction of knowledge. Rather, it must be based on methodologies and strategies which bring together both substantive (or first-order) contents and procedural (or second-order) contents which foster skills such as historical comprehension, argumentation and the interpretation of facts by way of historical sources (VanSledright, 2011), and which focus the learning of history on processes of investigation, research and the critical analysis of social problems (Falaize, 2014).

In the context of Spain, where the present study is set, in spite of the abundance of studies focusing on the formation of historical thinking (Santisteban, 2010; Domínguez, 2015), traditional contents and school routines persist in the classroom (Parra & Fuertes, 2019), focusing on the transmission of mainly conceptual contents, the widespread use of textbooks (Bel, 2017; Valls, 2019) and a low level of cognitive development and historical competences in comparison with other countries such as France and England (Gómez & Chapman, 2017; Martínez & Gómez, 2018).

Along the lines of other recent research, we are totally in agreement with the fact that the transformation of this type of school practices must necessarily pass through, among other requirements, adequate initial teacher training which helps teachers to build their identities as education professionals. In this regard, knowing the prior conceptions of future teachers regarding the teaching of history is essential if we wish to deconstruct and transform their pre-established ideas and models with the aim of improving the teaching and learning processes of this subject (Virta, 2002; Darling-Hammond & Bransford, 2005; Fives & Buehl, 2012). Thus, beginning with the premise that carrying out a diagnosis of the perceptions and conceptions of students teachers in their initial training will make it possible to lay the foundations for improving their competences for the teaching of history, with their initial training being a key aspect in this context (González & Skultely, 2018). Of course, as pointed out by Gómez-Carrasco, López-Facal and Castro-Fernández (2020), there is an abundance of bibliography regarding the perceptions and beliefs of trainee teachers, above all, from the perspective of the components of teachers' knowledge proposed by Shulman (1989). However, there is a paucity of research examining the pedagogical content knowledge (PCK) of trainee history teachers. As some authors have stated, there is no solid empirical support which makes it possible to establish comparisons between different spatial contexts. In this regard, the studies by Westhoff and Polman (2008), Korthagen (2010), Pollock, (2014) and Lévesque (2014), among others, stand out for examining the analysis of the relationships between epistemological conceptions and historical thinking concepts, and the aims of teaching history, although they do not offer holistic perspectives which make it possible to generalise.

From the context of Spain, some studies have been oriented towards an in-depth analysis of this object of research, demonstrating that trainee teachers, on the whole, do not have a critical epistemological interpretation of history, that they are not aware of historical thinking skills and do not possess the necessary metacognition regarding what historical knowledge is and what purpose it should serve (Gómez, Rodríguez & Mirete, 2018; Sáiz, Gómez & López-Facal, 2018). It is also noted that they maintain traditional representations of history and that their conceptions are contradictory as, in spite of being in favour of innovation and methodological renovation, on the whole, they lack the skills which enable a critical interpretation of the curriculum and continue to consider the subject of history as a set of data which provides a certain degree of general culture.

Consequently, this research examines in detail the pedagogical/teaching content knowledge (PCK) possessed by trainee secondary teachers in Spanish universities, specifically in their representations of history and the objectives they confer upon its teaching. Within the framework of these objectives, we focus on analysing which competences they believe their future pupils should acquire in history classes. In our opinion, approaching how future teachers understand history is fundamental, due to the fact that this influences the construction of their teaching identity and the teaching and learning processes which they will establish in their classes. The results of this research will allow us to approach initial teacher training from a perspective which stresses the transformation of the trainees' epistemological conceptions regarding what it means to teach and learn history, in such a way as to change traditional models of history education.

Method

Participants

646 students from master's degrees in geography and history teacher training from 22 different universities have participated in this research: Murcia, Alicante, Valencia, Jaume I, Barcelona, La Rioja, Zaragoza, Oviedo, Basque Country, Santiago de Compostela, Complutense University of Madrid, Autonomous University of Madrid, Valladolid, Extremadura, Castilla-La Mancha, Huelva, Seville, Córdoba, Málaga, Almería, Jaén and Granada. Participation in this research has involved 70 % of the universities which offer master's degrees in geography and history teacher training in Spain.

Design of the research

A quantitative non-experimental design was employed, and data was collected via a questionnaire with a Likert-type scale (1–5). This type of design was chosen due to the fact that it is capable of responding to problems both in descriptive terms and in relation to variables when data is collected systematically, thereby guaranteeing the rigour of the data obtained (Hernández & Maquilón, 2010). Designs employing surveys are extremely common in the field of education as they can be applied to multiple problems and make it possible to gather information on a large number of variables.

Data collection tool

The data were collected from the questionnaire entitled "Teaching approaches in history education" which consists of two parts. The first part is the previously validated ATI questionnaire in its version with 22 items (Trigwell, Prosser & Ginns, 2005) which refer to teaching approaches. The second part of

Tab. 1: Items from Block II-2 of the questionnaire: What competences do you believe should be acquired in history classes?

Item	Name
40	Chronology and dating skills of historical facts.
41	Describing and listing the main historical facts relating to the origin and consolidation of the nation.
42	Arguing and discussing the historical relevance of a topic.
43	Learning to analyse historical sources.
44	Explaining the causes and consequences of historical facts and processes.
45	Explaining the changes and continuities of historical processes.
46	Acquiring historical perspective in order to give contextualised explanations.
47	Arguing the ethical dimension of history which provides pupils with an orientation for life.
48	Problematising our identities, values and collective consensuses.
49	Educating in social values and for a critical citizenship.

the questionnaire is concerned with opinions relating to the most appropriate topics, resources, competences and methods for teaching history. This questionnaire has five blocks and is designed with a Likert-type evaluation scale (1–5). In this research, block two was used (Tab. 1), in which participants were required to answer the question "What competences do you believe should be acquired in history classes?," and in which there are 10 competences which the participants were required to evaluate from 1 (extremely irrelevant) to 5 (extremely relevant). These items correspond to competences related with chronological and first-order knowledge (items 1 and 2), the competences of historical thinking proposed by Seixas and Morton (2013) (from item 3 to item 8), and two competences related with the themes of identity and civic values (items 9 and 10).

The content was validated by seven experts: four from the area of the teaching of the social sciences, one an expert in the field of teaching and school organisation, another from the field of research methodology and diagnostics in education, and one from the field of educational psychology. This validation was carried out by way of a questionnaire in which the experts were required to evaluate the sufficiency, clarity and relevance of the items with a rating scale of 1–4. In the case of the items from this block, the results were satisfactory (Tab. 2) in terms of sufficiency (mean 3.57), clarity (mean 3.85) and relevance (mean 4).

Tab. 2: Evaluation of the items from the block according to the experts

Statistics	**Sufficiency**	**Clarity**	**Relevance**
Number of observations	7	7	7
Minimum	3.000	3.000	4.000
Maximum	4.000	4.000	4.000
Frequency of the minimum	3	1	7
Frequency of the maximum	4	6	7
Range	1.000	1.000	0.000
1st quartile	3.000	4.000	4.000
Median	4.000	4.000	4.000
3rd quartile	4.000	4.000	4.000
Mean	3.571	3.857	4.000
Variance (n)	0.245	0.122	0.000
Typical deviation (n)	0.495	0.350	0.000

The reliability of the block chosen for this research was analysed. Reliability can be defined as the degree of accuracy offered by a measure. In order to be considered reliable, a scale must have the capacity to exhibit consistent results in successive measurements of the same phenomenon. The objective of this is to probabilistically determine the degree of variation attributable to random or causal errors which are not linked to the construction of the tool. It guarantees the consistency expressed in the determination of the degree of error contained in the application of a scale and, therefore, in the measurement of the phenomenon. The reliability of the block has been studied making use of two indices: Cronbach's alpha and the Guttman scale. In the case of Cronbach's alpha, an index of .70 was obtained, which can be considered to be acceptable. As far as the Guttman scale is concerned, indices oscillating between .63 and .82 (Tab. 3) were obtained. If the data of Cronbach's alpha and the Guttman scale are observed, the values improve with the elimination of item 2 ("Describing and listing the main historical facts relating to the origin and consolidation of the nation") (Tab. 4).

a. Procedure and Data Analysis

The research was approved by the ethics committee of the University of Murcia. In order to gather data, contact was made with the coordinators of the geography and history speciality of master's degrees in teacher training and with teachers of this speciality. A letter was sent explaining the objectives of the

Tab. 3: Guttman scale

Guttman L1	Guttman L2	Guttman L3	Guttman L4	Guttman L5	Guttman L6
0.631	0.723	0.701	0.821	0.695	0.743

Tab. 4: Cronbach's alpha and Guttman scale values with the elimination of each of the items

Variable	Cronbach's alpha <scale/ eliminated item >	Guttman L6 <scale/ eliminated item >
Chronology and dating skills of historical facts (creating and interpreting timelines)	0.688	0.731
Describing and listing the main historical facts relating to the origin and consolidation of the nation	0.733	0.759
Arguing and discussing the historical relevance of a topic	0.697	0.744
Learning to analyse historical sources	0.664	0.716
Explaining the causes and consequences of historical facts and processes	0.659	0.707
Explaining the changes and continuities of historical processes	0.653	0.701
Acquiring historical perspective in order to give contextualised explanations (empathy or historical understanding of the context)	0.656	0.707
Arguing the ethical dimension of history which provides pupils with an orientation for life	0.667	0.714
Problematising our identities, values and collective consensuses	0.689	0.731
Educating in social values and for a critical citizenship	0.674	0.720

research project and providing a link to the webpage via which the students could participate. In the first page of this link there was an informed consent document for the participants to sign before beginning the survey. Twenty-two Spanish universities replied and provided trainee teachers with the link from November 2019 to February 2020.

For the analysis of the data, descriptive analyses, contingency tables and inferential analysis (Student's *t* and one-way ANOVA) were used. All of these analyses were carried out employing the *R* lavaan library (Rosseel, 2012) and the XLSTAT program in its 2020.3.1 version.

Results

Descriptive statistics

Tab. 5 shows the descriptive statistics of Block II-2 of the questionnaire, in which the trainee teachers evaluated the different competences which, in their opinion, should be acquired in history classes. Thus, if we take as a guideline the mean evaluation attributed by the participants in the survey to each one of the items, the competences which are better considered and which, therefore, should be developed by pupils during a history class are: *Acquiring historical perspective in order to give contextualised explanations* (item 46) and *Educating in social values and for a critical citizenship* (item 49), given that both of these items received a mean evaluation of 4.61. The outcome of this index is determined by the better consideration of women for these items, with a mean value of 4.72, compared to the opinion of men, whose scoring established mean values of 4.54 for item 46 and 4.52 for item 49.

Following the same pattern, as can be observed in Tab. 5, on the contrary, the competences which trainee teachers consider less important and, therefore, should not be given priority in history classes are: *Chronology and dating skills*

Tab. 5: Descriptive statistics of the items from Block II-2 of the questionnaire

	Mean	%1	%2	%3	%4	%5	Deviation
Item 40	3.75	1.56	7.00	28.60	40.10	22.70	0.93
Item 41	3.16	8.71	17.40	34.50	27.80	11.50	1.11
Item 42	4.32	0.78	1.40	11.00	38.80	48.00	0.78
Item 43	4.42	0.47	1.09	9.63	33.50	55.30	0.75
Item 44	4.54	0.00	0.94	5.15	33.40	60.50	0.64
Item 45	4.51	0.16	0.31	6.36	34.60	58.60	0.64
Item 46	4.61	0.31	0.31	5.77	24.80	68.80	0.64
Item 47	4.37	0.93	0.93	11.60	32.90	53.60	0.80
Item 48	4.17	1.87	1.56	17.60	35.90	43.10	0.90
Item 49	4.61	0.93	1.24	5.12	21.70	71.00	0.73
Total	4.25	1.57	3.22	13.53	32.35	49.31	0.79

of historical facts (item 40), which received a mean score of 3.75 and *Describing and listing the main historical facts relating to the origin and consolidation of the nation* (item 41), which received a mean score of 3.16. It must be pointed out that both of these competences are linked with traditional learning processes based on memorisation, in which the participation of the pupil in his/her own learning is secondary to that of the teacher. In both cases, these items refer to first-order contents which develop, in a minor way, the competences of historical thinking (VanSledright, 2014). This would explain why they are considered less relevant than the other items, which are related with second-order contents of a procedural and methodological nature.

Fig. 1 shows the complete scale of evaluation of the items to be developed in the learning processes of historical contents. It shows how the participants, along with items 46 and 49 mentioned above, also positively evaluated item 44 (*Explaining the causes and consequences of historical facts*) and item 45 (*Explaining the changes and continuities of historical processes*), to which they attributed values of more than 90 %, with the majority selecting options 4 and 5 of the closed Likert-type evaluation scale, thereby demonstrating a certain degree of unanimity in the perceptions of those surveyed.

As mentioned above, item 40, relating to skills of chronology and the dating of historical facts, is one of the worst-considered items by the participants. However, in spite of its low score, it is true that the students attributed it with certain importance as a necessary competence in history classes, as it obtained 8.56 % of negative opinions (values 1 and 2), 28.6 % of intermediate values (value 3) and 62.84 % of values of 4 and 5, with the most represented score being 4 ("somewhat relevant"). This result shows how, despite it being one of the least valued items, the trainee teachers consider it important for pupils to know how to locate the historical facts taught in chronologies and timelines. Along the same lines, item 48, referring to the competence of *Problematising own collective identities, values and consensuses*, was viewed positively, obtaining 79 %, with only 17.6 % of those surveyed considering it neither relevant nor irrelevant (response value 3). Indeed, a similar consideration can be found for items 42 (*Arguing and discussing the historical relevance of a topic*) and 47 (*Arguing the ethical dimension of history which provides pupils with an orientation for life*), with many positive evaluations and intermediate values of approximately 11 %. However, taking into account all of the items, it must be noted that both 42 and 47, along with item 43 (*Learning to analyse historical sources*), have intermediate percentages of evaluation. They are above the average, albeit with scores somewhat lower than other items relating to the explanation of causes and consequences, changes and continuities, the acquisition of historical

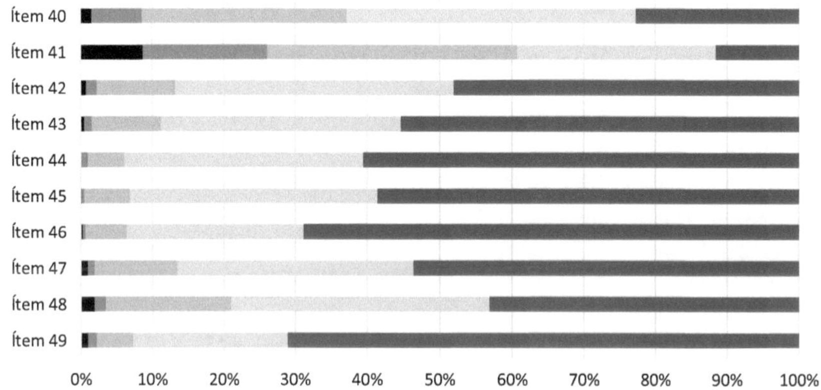

Fig. 1: Evaluation of the items regarding the competences to be developed in history classes. Red: Extremely irrelevant; orange: Irrelevant; grey: Neither relevant nor irrelevant; light green: Somewhat relevant; dark green: Extremely relevant

perspective for giving contextualised explanations, and the aim of educating in social values and the construction of a critical citizenship. In these results, there is a lesser (albeit slight) consideration of history as a subject which should favour the acquisition of skills regarding argumentation, the meaning of historical relevance, the recognition of the ethical dimension of history and the analysis of historical sources for their interpretation. This fact can perhaps be explained by the lack of knowledge or confusion of those surveyed regarding what is implied by these competences or how they can be developed on a methodological or procedural level

Item 41, relating to the competence of *Describing and listing the main historical facts related with the origin and consolidation of the nation*, is that which offers the greatest concentration of negative (26.1 %) (extremely irrelevant or irrelevant) and intermediate responses (34.50 %), and which, therefore, generates the worst opinion among the participants, who only awarded it 39.40 % of positive values (between 4 and 5).

Differences according to university of origin

A one-way ANOVA test was carried out in order to identify statistically significant differences according to the university in which the participants were carrying out their master's degree studies. There are significant differences for items 1 and 2 (first-order contents), and items 8, 9 and 10, referring to the ethical

Tab. 6: Results of the one-way ANOVA test according to university

	Item 1	Item 2	Item 3	Item 4	Item 5	Item 6	Item 7	Item 8	Item 9	Item 10
University of the Basque Country	3.714	3.381	4.429	4.571	4.619	4.714	4.762	4.524	4.524	4.857
University of Barcelona	4.111	3.500	4.333	4.667	4,.778	4.722	4.611	4.611	4.278	4.444
University of Málaga	4.081	3.649	4.135	4.649	4.703	4.486	4.784	4.595	4.437	4.622
University of Oviedo	3.778	2.667	4.556	4.556	4.667	4.667	4.624	4.667	4.222	4.667
University of Jaén	3.778	3.037	4.222	4.778	4.741	4.704	4.667	4.630	4.222	4.593
University of La Rioja	3.900	2.850	4.300	4.500	4.550	4.550	4.650	4.750	4.150	4.900
University of Granada	4.059	3.314	4.451	4.490	4.627	4.510	4.627	4.373	4.196	4.588
Autonomous University of Madrid	3.580	2.773	4.227	4.455	4.591	4.591	4.864	4.273	4.500	4.773
University of Castilla-La Mancha	3.778	3.833	4.556	4.444	4.556	4.556	4.556	4.333	3.833	4.611
University of Extremadura	3.970	3.788	4.455	4.303	4.576	4.485	4.564	4.273	3.909	4.576
Jaume I University	3.889	3.889	4.667	4.444	4.444	4.667	4.556	4.222	3.556	4.111
University of Murcia	3.675	3.229	4.291	4.300	4.577	4.600	4.700	4.475	4.075	4.540
University of Huelva	3.467	3.267	4.467	4.333	4.467	4.467	4.467	4.400	4.333	4.800
University of Córdoba	4.143	3.571	4.321	4.393	4.643	4.429	4.500	4.179	4.107	4.500
University of Santiago de Compostela	3.375	2.583	4.417	4.292	4.542	4.708	4.875	4.292	4.375	4.500
University of Zaragoza	3.517	2.690	4.172	4.552	4.517	4.759	4.676	4.345	4.172	4.207
University of Seville	3.652	3.130	4.435	4.217	4.435	4.304	4.522	4.435	4.435	4.739
University of Almería	3.514	3.114	4.257	4.457	4.429	4.457	4.675	4.514	3.943	4.714
University of Valencia	3.182	2.359	4.214	4.212	4.365	4.450	4.709	4.605	4.431	4.852
University of Valladolid	4.226	3.190	4.048	4.429	4.524	4.524	4.429	4.000	4.000	4.571
Complutense University of Madrid	3.957	3.239	4.283	4.391	4.533	4.522	4.391	4.043	4.134	4.522
University of Barcelona	4.200	3.400	4.000	4.400	4.200	4.200	4.400	4.200	4.200	4.600
Rovira i Virgili University	3.857	3.436	4.524	4.381	4.311	4.143	4.333	3.762	3.286	4.429
University of Alicante	3.667	3.375	4.333	4.292	4.333	4.083	4.375	4.083	3.958	4.375

dimension, the problematisation of identities and educating in social and civic values. Tab. 6 shows the differences between universities for each item.

Item 41 is that which received the highest score in value 2 with 17.40 %, receiving in some universities a mean score never higher than 3, as is the case of the Autonomous University of Madrid (2.77), the University of La Rioja (2.85), the University of Oviedo (2.67), the University of Santiago de Compostela (2.58), the University of Valencia (2.35) and the University of Zaragoza (2.69). It can also be observed that it is the item with the greatest typical deviation in the whole of the responses. The difference in relation to the typical deviation of the responses in item 41 can be observed in Fig. 2, which shows the different alternatives selected by the participants in the questionnaire according to the university of origin. Thus, in the University of Oviedo, 55.6 % of the participants consider the development of this competence to be irrelevant or extremely irrelevant, compared to 53.6 % of those from the University of Valencia and 40.9 % of participants from the Autonomous University of Madrid. On the other hand, at other universities, this item received a higher evaluation. For example, 64.3 % of participants from the University of Córdoba believed this competence to be relevant or extremely relevant, as did 64.9 % of those from the University of Malaga, 66.7 % from Jaume I University and 72.2 % from the University of Castilla-La Mancha.

At the opposite end of the analysis, item 45 (*Explaining the changes and continuities of historical processes*) stands out due to the fact that it is the statement which received the least number of negative responses (0.47 %) and has, as mentioned above, the best evaluation (93.2 %). This item, along with items 44 and 46, shows an extremely low typical deviation in its responses, with only 0.64, which demonstrates an extremely homogeneous opinion of the students surveyed with hardly any disparities between universities.

Differences according to gender

The Student's *t* test was performed, and contingency tables were created in order to analyse differences according to sex. Fig. 3 shows that, in three of the ten items, there are statistically significant differences according to the Student's t test. In these three items, female participants scored significantly higher than male participants: acquiring historical perspective (4.7 compared to 4.5); arguing the ethical dimension (4.4 compared to 4.2); and educating in social and civic values (4.7 compared to 4.5).

It is possible to examine these differences in more depth with the data collected in Tab. 7. It is notable how item 49 (*Educating in social values and for a critical citizenship*), one of the best valued items, is that which presents the greatest difference in responses between male and female participants. In this case, there is a divergence of 0.20 between the mean score attributed by men (4.52) and that given by women (4.72); a difference which can also be observed in the standard deviation, given that men present a deviation of 0.79 compared

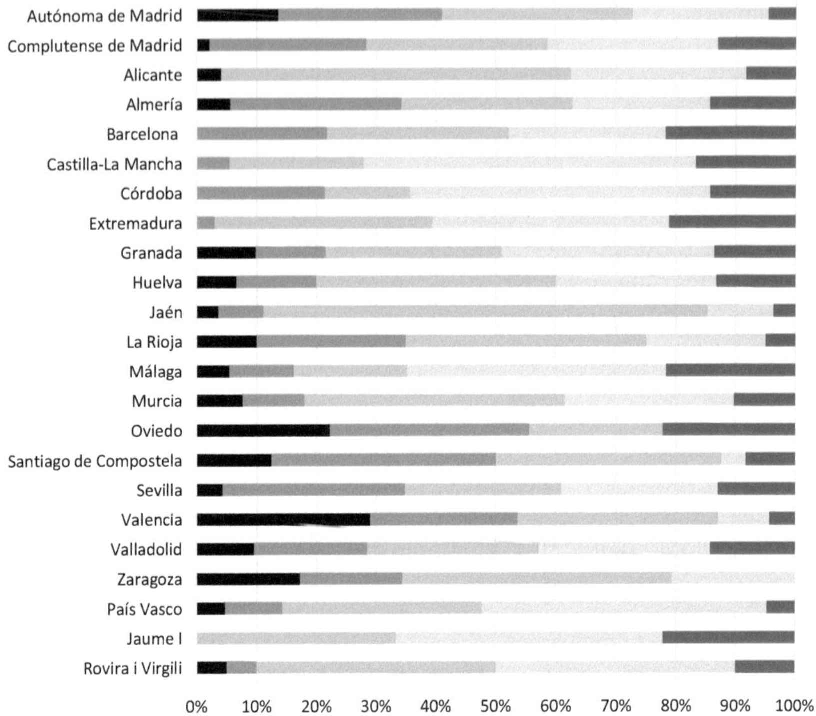

Fig. 2: Results of item 41 for each university. Red: Extremely irrelevant; orange: Irrelevant; grey: Neither relevant nor irrelevant; light green: Somewhat relevant; dark green: Extremely relevant

to 0.63 for women. Therefore, it can be stated that women valued it more highly than men and that they present a greater unanimity when judging this item.

It can be appreciated how women gave worse evaluations to items such as 40 and 41, relating to the development of first-order contents while, at the same time, they generally offer higher values to items relating to the development of second-order concepts. In this regard, items 46, 47 and 49 particularly stand out as the evaluation given by women compared to that of men shows a considerable divergence due to the fact that there is a numerical difference of several points between the valuations, and not of decimal points as occurs in the cases in which men score more highly than women (Tab. 7).

Taking into account another variable, in this case academic training, an interesting trend can be discovered, particularly for item 41, for which 50 % of

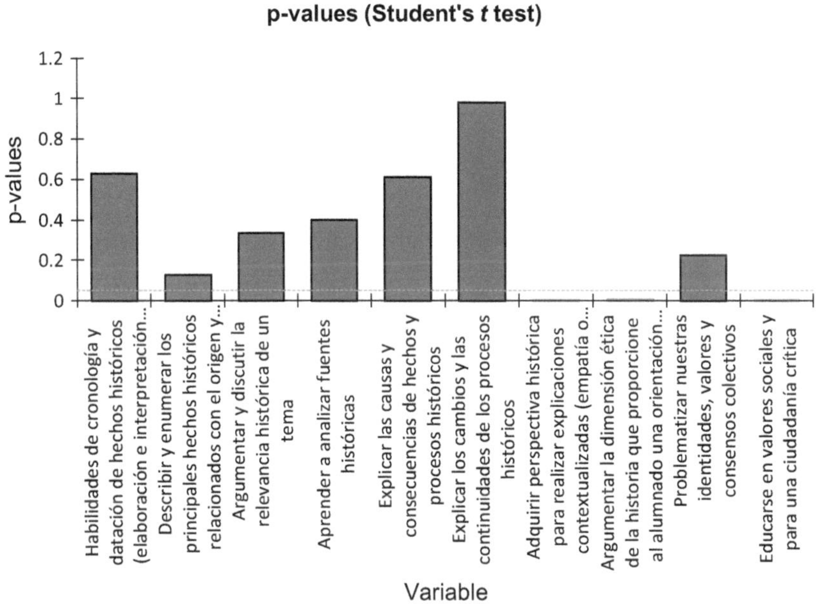

Fig. 3: Results of the Student's *t* test *for each item according to sex*

those participants with postgraduate studies considered the development of the competence *Describing and listing the main historical facts relating to the origin and consolidation of the nation* to be "extremely irrelevant" or "irrelevant." This percentage is considerably higher than that offered by those with only graduate studies, of whom 25.91 % considered it to be "irrelevant" or "extremely irrelevant."

Another equally interesting trend is that observed in relation to the variable linked to teaching innovation training. In this regard, for item 48, referring to the competence of *Problematising our collective identities, values and consensuses*, it can be observed that those participants who had done a teaching innovation course valued this item worse than those who had not. Thus, 10.9 % of those who had done such training awarded values of between 1 and 2 in their responses to item 48, compared to 2.6 % of those who had not done such courses, who considered this item to be irrelevant or extremely irrelevant, thereby establishing a difference of more than eight points in the responses. In fact, 18.8 % of those who had not done such training concentrated their responses in the option "Neither relevant nor irrelevant," compared with 6.25 % of those

Tab. 7: Responses to the items of the questionnaire according to the variable of sex

Item	Responses	Women	Men	Item	Responses	Women	Men
40	1–2	**8.6**	8.12	**45**	1–2	0.714	0.279
	3	29.7	27.7		3	6.79	6.15
	4–5	61.9	**64.1**		4–5	92.5	**93.6**
41	1–2	**29.3**	23.9	**46**	1–2	0.358	0.845
	3	33.9	34.6		3	4.3	7.04
	4–5	36.8	**41.6**		4–5	**95.3**	92.1
42	1–2	2.5	1.97	**47**	1–2	1.79	1.96
	3	11.1	11		3	10	13.1
	4–5	86.4	**87.1**		4–5	**88.2**	84.9
43	1–2	1.43	1.4	**48**	1–2	3.96	3.07
	3	8.57	10.4		3	17.6	17.3
	4–5	**89.6**	88.2		4–5	78.4	**79.6**
44	1–2	1.43	0.563	**49**	1–2	1.79	2.52
	3	5.38	5.07		3	3.21	6.72
	4–5	93.2	**94.4**		4–5	**95**	90.8

surveyed who had completed a teaching innovation course. Again, this divergence in the response clearly indicates how the influence of initial or ongoing training can have a great impact on the conceptions of future teachers and on their epistemological and methodological representations regarding teaching and learning processes.

Discussion and conclusions

The results yielded from this research are unequivocal. In order to achieve the objective of fostering historical thinking from a critical perspective, which helps to understand the past and situate oneself in the present, it is necessary to implement a series of changes in the learning of history. The teaching of this subject based on the conceptualisation of the skills and concepts of historical thinking makes it possible to deploy its social function and contributes towards forming critical citizens and relegating essentialist and nationalistic perspectives based on wide-ranging memorisation. In order to achieve this aim it is relevant to carry out transformations on different levels. One of the most important changes concerns, without doubt, the knowledge of the prior conceptions of future teachers regarding the teaching of history. By revealing the pre-established ideas and models of the future teachers, it is possible to detect traditional methodologies, routines and contents in order to transform ossified processes of teaching and learning in this subject. In summary, it is essential to scrutinise the initial training of teachers and link it to their perceptions and to the pedagogical and didactic knowledge of the content which they possess.

In this regard, this research has been focused on unravelling which competences trainee teachers believe should be acquired by their future pupils in history classes. This analysis is believed to be essential in calibrating the processes of teaching and learning which they will later implement in the classroom. The methodology employed has made it possible to achieve the objectives proposed at the outset, based on a quantitative non-experimental design which has rigorously collected the data provided by 646 students via a questionnaire with a Likert type (1–5) scale regarding the competences which should be acquired in history classes. The evaluation made of the 10 selected competences (two related with chronological and first-order knowledge, the six historical thinking skills proposed by Seixas & Morton and two competences related with issues of identity and civic values) has made it possible to draw robust conclusions.

The two first competences were those which received the worst evaluations from the future teachers. They are irremediably associated to a traditional method of teaching history based on the memorisation of dates and national milestones, in which the participation of the pupil in his/her own learning is eclipsed by that of the teacher. At the opposite extreme, the data revealed by the questionnaire assays the unanimous preference for competences of a procedural and methodological nature and also surmises a clear preference of women for second-order contents (particularly in the item *Educating in social values and for a critical citizenship*). At the same time, within the methodological competences, a preference could be observed for their explanation of causes and consequences, the observation of changes and continuities, the acquisition of historical perspective in order to give contextualised explanations and the aim oriented towards education in social values and the construction of a critical citizenship. Of secondary importance, perhaps due to the greater degree of complexity in their development, were the consideration of history as a subject which should favour the acquisition of skills for argumentation, the meaning of historical relevance, the recognition of the ethical dimension of history and the analysis of historical sources for their interpretation.

Neither was there room for doubt regarding the impact of academic training and teaching innovation courses. The participants surveyed who had completed postgraduate studies were assured in attributing less value to the competence of *Describing and listing the main historical facts related with the origin and consolidation of the nation* than those who only possessed graduate studies. On the other hand, the participants surveyed who had carried out a teaching innovation course awarded a lower evaluation to the item *Problematising our collective identities, values and consensuses* than those who had not received such training.

To conclude, the results obtained in this research are encouraging due to the preference of those surveyed towards a method of teaching of history based on methodological competences which casts aside the mere memorisation of contents linked to first-order concepts. Future research should corroborate that these conceptions of future teachers do not remain a mere declaration of intentions and that an innovative form of teaching history which is adapted to the needs of the 21st century is implemented.

References

Aronsson, Peter: "Shaping Lives: Negotiating and Narrating Memories". *Etnográfica*, 19(3), 2015, pp. 577–591, retrieved on 7.21.2020, from DOI https://doi.org/10.4000/etnografica.4125.

Barca, Isabel: *O pensamento histórico dos jovens: idéias dos adolescentes acerca da provisoriedade da explicação histórica*. Universidade do Minho, Instituto de Educação e Psicologia, Centro de Estudos em Educação e Psicologia: Braga 2000.

Barton, Keith C.: *Historical Understanding among Elementary Children*. Unpublished doctoral thesis. University of Kentucky, Lexington 1994.

Barton, Keith C: "Research on Students' Ideas about History". In: Levstik, Linda S./Thyson, Cynthia A. (eds.): *Handbook of Research on Social Studies Education*. Taylor & Francis: New York 2008, pp. 239–258.

Barton, Keith C./Levstik, Linda S.: *Teaching History for the Common Good*. Lawrence Erlbau: New Jersey 2004.

Bel Martínez, Juan C.: "Imagen y libros de texto de Historia en Educación Primaria: estudio comparativo a partir de un análisis cualitativo". *Revista de Educación*, 377, 2017, pp. 82–112, retrieved on 7.21.2020, from DOI: 10.4438/1988-592X-RE-2017-377-354.

Berger, Stefan: "De-Nationalizing History Teaching and Nationalizing it Differently! Some Reflections on How to Defuse the Negative Potential of National(ist) History Teaching". In: Carretero, Mario/Asensio, Mikel/Rodríguez-Moneo, María (eds.): *History Education and the Construction of National Identities*. Information Age Publishing: Charlotte, NC 2012, pp. 33–47.

Bryant, Darren A./Clark, Penney: "Historical Empathy and Canada: A People's History". *Canadian Journal of Education* 29(4), 2006, pp. 1039–1064.

Carretero, Mario/Asensio, Mikel/Rodríguez-Moneo, María (eds.): *History Education and the Construction of National Identities.* Information Age Publishing: Charlotte, NC 2012.

Carretero, Mario/Berger, Stefan/Grever, Maria (eds.): *Palgrave Handbook of Research in Historical Culture and Education.* Palgrave McMillan: London 2017.

Chapman, Arthur: "Historical Interpretations". In: Davies, Ian (ed.): *Debates in History Teaching.* Routledge: London, England 2011, pp. 96–108.

Chapman, Arthur/Wilschut, Arie (eds.): *Joined-up History: New Directions in History Education Research.* Information Age Publishing, Inc: Charlotte, NC 2015.

Clark, Penney: *Take It Away, Youth: Visions of Canadian Identity in British Columbia Social Studies Textbooks, 1925-1989.* Unpublished doctoral thesis. University of British Columbia: Vancouver 1996.

Cooper, Hilary (ed.): *Teaching History Creatively.* Routledge: London 2013.

Darling-Hammond, Linda/Bransford, John (eds.): *Preparing Teachers for a Changing World: What Teachers Should Learn and Be Able to Do.* Jossey-Bass: San Francisco, CA 2005.

Domínguez, Jesús: *Pensamiento histórico y evaluación de competencias.* Graó: Barcelona 2015.

Éthier, Marc A./Demers, Stéphanie/Lefrançois, David: "Las investigaciones en Didáctica sobre el desarrollo del pensamiento histórico en la enseñanza primaria. Una panorámica de la literatura publicada en francés e inglés desde el año 1990". *Enseñanza de las Ciencias Sociales,* 9, 2010, pp. 61–74.

Falaize, Benoit: "O ensino de temas controversos na escola francesa: os novos fundamentos da história escolar na França?". *Revista Tempo e Argumento, Florianópolis,* 6(11), 2014, pp. 224–253, retrieved on 7.21.2020, from DOI: 10.5965/2175180306112014224.

Fives, Helenrose/Buehl, Michelle: "Spring Cleaning for the 'Messy' Construct of Teachers' Beliefs: What Are They? Which Have Been Examined? What Can They Tell Us?". In: Harris, Karen R. et al. (eds.): *APA Handbooks in Psychology. APA Educational Psychology Handbook, Vol. 2. Individual Differences and Cultural and Contextual Factors.* American Psychological Association: Washington, DC 2012, pp. 471–499, retrieved on 21.7.2020, from DOI: https://doi.org/10.1037/13274-019.

Gómez Carrasco, Cosme J./Chapman, Arthur: "Enfoques historiográficos y representaciones sociales en los libros de texto: un estudio comparativo,

España-Francia-Inglaterra". *Historia y Memoria de la Educación*, 6, 2017, pp. 319–361, retrieved on 7.21.2020, from DOI: 10.5944/hme.6.2017.17132.

Gómez-Carrasco, Cosme J./López-Facal, Ramón/Castro-Fernández, Belén M.: "Trainee Teachers' Perceptions of History Teaching and the Critical Education of Citizenship". In: Delgado Algarra Emilio J./Cuenca López, José M. (eds.): *Handbook of Research on Citizenship and Heritage Education*. Information Science Reference, IGI Global: Hershey, PA 2020, pp. 239–263, retrieved on 7.21.2020, from DOI: 10.4018/978-1-7998-1978-3.ch012.

Gómez Carrasco, Cosme J./Rodríguez Pérez, Raimundo A./Mirete Ruiz, Ana B.: "Percepción de la enseñanza de la historia y concepciones epistemológicas. Una investigación con futuros maestros". *Revista Complutense de Educación*, 29(1), 2018, pp. 237–250, retrieved on 7.21.2020, from DOI: https://doi.org/10.5209/RCED.52233.

González, Gloriana/Skultety, Lisa: "Teacher Learning in a Combined Professional Development Intervention". *Teaching and Teacher Education*, 71, 2018, pp. 341–354, retrieved on 7.21.2020, from DOI: 10.1016/j.tate.2018.02.003.

Hammack, Phillip L.: *Narrative and the Politics of Identity: The Cultural Psychology of Israeli and Palestinian Youth*. Oxford University Press: New York 2011.

Henríquez, Rodrigo/Ruiz Zúñiga, Marcela: "Chilean Students Learn to Think Historically: Construction of Historical Causation through the Use of Evidence in Writing". *Linguistics and Education*, 25(1), 2014, pp. 145–157, retrieved on 7.21.2020, from DOI: 10.1016/j.linged.2013.10.003.

Hernández, Fuensanta/Maquilón, Javier J.: "Introducción a los diseños de investigación educativa". In: Nieto Santiago (ed.): *Principios, métodos y técnicas esenciales para la investigación educativa*. Dykinson: Madrid 2010, pp. 109–126.

Hobsbawm, Eric J.: *Nations and Nationalism since 1780: Programme, Myth, Reality*. Cambridge University Press: Cambridge 1997.

Kölbl, Carlos/Konrad, Lisa: "Historical Consciousness in Germany: Concept, Implementation, Assessment". In: Ercikan, Kadriye/Seixas, Peter (eds.): *New Directions in Assessing Historical Thinking*. Routledge: New York 2015, pp. 17–28.

Körber, Andreas: *Historical Consciousness, Historical Competencies – And Beyond? Some Conceptual Development within German History Didactics*. Deutsches Institut für Internationale Pädagogische Forschung (DIPF): Open Access, 2015, pp. 1–56, retrieved on 7.21.2020, from DOI: 10.13140/RG.2.1.1524.9529.

Korthagen, Fred A.J.: "Situated Learning Theory and the Pedagogy of Teacher Education: Towards an Integrative View of Teacher Behavior and Teaching Learning". *Teaching and Teacher Education*, 26(1), 2010, pp. 98–106, retrieved on 7.21.2020, from DOI: 10.1016/j.tate.2009.05.001.

Lee, Peter: "History in School: Aims, Purposes and Approaches. A Reply to John White". In Lee Peter/Slater John/Walsh Paddy/White John P./Shemilt Denis (eds.): *The Aims of School History: The National Curriculum and Beyond*. Tufnell Press: London, England 1992, pp. 20–34.

Lee, Peter: "Historical Literacy and Transformative History". *Educar em Revista*, 60, 2016, pp. 107–146, retrieved on 7.21.2020, from DOI: 10.1590/0104-4060.45979.

Lee, Peter/Ashby, Rosalyn: "Progression in Historical Understanding among Students Ages 7–14". In: Stearns, Peter N./Seixas, Peter/Wineburg, Sam (eds.): *Knowing, Teaching and Learning History: National and International Perspectives*. New York University Press: New York 2000, pp. 199–222.

Lévesque, Stéphane: *Thinking Historically: Educating Students for the 21st Century*. University of Toronto Press: Toronto 2008.

Lévesque, Stéphane: What Is the Use of the Past for Future Teachers? A Snapshot of Francophone Student Teachers in Ontario and Québec Universities. In: Sandwell Ruth/Von Heyking Amy (eds.): *Becoming a History Teacher: Sustaining Practices in Historical Thinking and Knowing*. University of Toronto Press: Toronto 2014, pp. 115–138. DOI: 10.3138/9781442619241-006.

Lévesque, Stéphane/Clark, Penney. "Historical Thinking: Definitions and Educational Applications". In: Metzger Scott Alan/Harris Lauren McArthur (eds.): *The Wiley International Handbook of History Teaching and Learning*. Wiley-Blackwell: New York 2018, pp. 119–148.

Martínez Hita, María/Gómez Carrasco, Cosme J.: "Nivel cognitivo y competencias de pensamiento histórico en los libros de textos de Historia de España e Inglaterra. Un estudio comparativo". *Revista de educación* 379, 2018, pp. 145–169. DOI: 10.4438/1988-592X-RE-2017-379-364.

Melo, Maria do C./Lopes, José M.: *Narrativas Históricas e Ficcionais. Recepção e Produção para Professores e Alunos. Actas do 1º Encontro sobre narrativas históricas e ficcionais. Actas do 1º Encontro sobre Narrativas e Ficcionais.* Centro de Investigação em Educação, Universidade do Minho: Braga 2004, pp. 29–39.

Metzger, Scott Alan y Harris, Lauren McArthur (eds.): *The Wiley International Handbook of History Teaching and Learning*. Wiley-Blackwell: New York 2018.

Monte-Sano, Chauncey: "Disciplinary Literacy in History: An Exploration of the Historical Nature of Adolescents' Writing". *The Journal of the Learning Sciences* 19(4), 2010, pp. 539–568.

Nordgren, Kenneth: "How to Do Things with History: Use of History as a Link between Historical Consciousness and Historical Culture". *Theory & Research in Social Education* 44(4), 2016, pp. 479–504. DOI: 10.1080/00933104.2016.1211046.

Parra, David/Fuertes, Carlos: *Reinterpretar la tradición, transformar las prácticas. Ciencias Sociales para una educación crítica.* Editorial Tirant Lo Blanch: Valencia 2019.

Peck, Carla: *Multi-Ethnic High School Students' Conceptions of Historical Significance: Implications for Canadian History Education.* Unpublished doctoral thesis. University of British Columbia: Vancouver, BC 2009.

Plá, Sebastián: *Aprender a pensar históricamente. La escritura de la historia en bachillerato.* Plaza y Valdés: México 2005.

Pollock, Scott A.: "The Poverty and Possibility of Historical Thinking: An Overview of Recent Research into History Teacher Education". In: Sandwell Ruth/Von Heyking Amy (eds.): *Becoming a History Teacher.* University of Toronto Press: Toronto 2014, pp. 60–74. DOI: 10.3138/9781442619241-006.

Rosseel, Yves: "Lavaan: An R Package for Structural Equation Modelling". *Journal of Statistical Software* 48(2), 2012, pp. 1–36. Retrieved from http://www.jstatsoft.org/v48/i02/.

Rüsen, Jörn: *History: Narration, Interpretation, Orientation.* Berghahn Books: New York, NY 2005.

Sáiz, Jorge/Gómez, Cosme J./López-Facal, Ramón: "Historical Thinking, Causal Explanation and Narrative Discourse in Trainee Teachers in Spain". *Historical Encounters: A Journal of Historical Consciousness, Historical Cultures, and History Education* 5(1), 2018, pp. 16–30.

Santisteban, Antoni: "La formación de competencias de pensamiento histórico". *Clío & Asociados* 14, 2010, pp. 34–56.

Schmidt, Maria A./García, Tania M.: "A formaçao da consciência histórica de alunos e professores e o cotidiano em aulas de História". *Cad. Cedes, Campinas* 25(67), 2005, pp. 297–308.

Seixas, Peter/Morton, Tom: *The Big Six Historical Thinking Concepts.* Nelson: Toronto 2013.

Shulman, Lee S.: "Paradigmas y programas de investigación en el estudio de la enseñanza: una perspectiva contemporánea". In: Wittrock Merlin C. (ed.): *La investigación de la enseñanza. Vol. 1. Enfoques, teorías y métodos.* Paidós/MEC: Barcelona 1989, pp. 9–91.

Trigwell, Keith/Prosser, Michael/Ginns, Paul: "Phenomenographic Pedagogy and a Revised Approaches to Teaching Inventory". *Higher Education Research & Development*, 24(4), 2005, pp. 349–360.

Valls, Rafael: "Enseñanza escolar de la historia y uso de los manuales en las aulas". In: Parra, David/Fuertes, Carlos (eds.): *Reinterpretar la tradición, transformar las prácticas. Ciencias Sociales para una educación crítica.* Tirant Humanidades: Valencia 2019, pp. 125–148.

Van Boxtel, Carla/Van Drie, Jannet: "Historical Reasoning: Conceptualizations and Educational Applications". In: Metzger, Scott A./Harris, Lauren M. (eds.): *The Wiley International Handbook of History Teaching and Learning.* Wiley & Blackwell: New York 2018, pp. 149–176, retrieved on 7.21.2020, from DOI: https://doi.org/10.1002/9781119100812.ch6.

Van Drie, Jannet/Van Boxtel, Carla: "Historical Reasoning: Towards a Framework for Analyzing Students' Reasoning about the Past". *Educational Psychology Review*, 20(2), 2008, pp. 87–110, retrieved on 7.21.2020, from DOI:https://doi.org/10.1007/s10648-007-9056-1.

VanSledright, Bruce A.: *The Challenge of Rethinking History Education: On Practice, Theories, and Policy.* Routledge: New York 2011.

VanSledright, Bruce A: *Assessing Historical Thinking and Understanding: Innovate Designs for New Standards.* Routledge: New York 2014.

Virta, Arja: "Becoming a History Teacher: Observations on the Beliefs and Growth of Student Teachers". *Teaching and Teacher Education*, 18(6), 2002, pp. 687–698, retrieved on 7.21.2020, from DOI: 10.1016/S0742-051X(02)00028-8.

Wertsch, James V.: "Specific Narratives and Schematic Narrative Templates". In: Seixas, Peter (ed.), *Theorizing Historical Consciousness.* University of Toronto Press: Toronto 2004, pp. 49–62.

Westhoff, Laura M./Polman, Joseph L.: "Developing Preservice Teachers' Pedagogical Content Knowledge about Historical Thinking". *International Journal of Social Education*, 22(2), 2008, pp. 1–28, retrieved on 7.21.2020, from https://files.eric.ed.gov/fulltext/EJ818470.pdf.

Wineburg, Sam/Martin, Daisy/Monte-Sano, Chauncey (eds.): *Reading Like a Historian: Teaching Literacy in Middle and High School History Classrooms.* Teachers College Press: New York 2013.

Francisco Javier Trigueros Cano, José María Campillo Ferrer, Pedro Miralles-Martínez, and Jesús Molina Saorín

Chapter 9 The acquisition of historical competences in Spain: Perceptions of primary and secondary education teachers[1]

Abstract: This study aims to examine the perceptions of primary and secondary education teachers on the acquisition of skills through the subject of History. The research data of this study have been collected through a questionnaire that has been designed ad hoc. The sample finally selected is made up of 332 teachers, most of whom are on a permanent or contract basis (66.8 %). All the participants are Spanish and work in various parts of the country. The data collected through the questionnaires were analysed with the Statistical Package for the Social Sciences (SPSS v26.0) and descriptive analyses and statistical tests were carried out. The results indicate that teachers are concerned with addressing and promoting key competences in this area, particularly in understanding the causes and consequences of historical processes and in improving civic education as a basis for promoting critical thinking and meaningful learning among students.

Keywords: History, Competences, Primary education, Secondary education, Perception, Teachers

Introduction

This chapter offers an in-depth examination of educators' perceptions of the acquisition of historical competencies in primary and secondary education. Firstly, it highlights the history and development of key competences taking into account some of the different theoretical and epistemological positions adopted when addressing these core elements in education. Next, and with the objective of clarifying the initial conceptual frameworks, the definition of historical competence is presented, taking note of the – relatively recent – paradigm shift on which

1 This research was funded by the "Spanish Ministry for Science, Innovation and Universities. Secretary of State for Universities, Research, Development and Innovation," grant number PGC2018-094491-B-C33, and "Seneca Foundation. Regional Agency for Science and Technology," grant number 20874/PI/18.

the teaching of history is being based. In this sense, the objectives of this study are explicitly defined, which allows a clearer idea of the focus of this research. Thirdly, the design and procedure of this study are described in detail, as well as the instruments, sampling and data collection and analysis. Subsequently, the main conclusions that help to achieve the objectives set are presented, and all this is done in an orderly sequence that makes use of tables and illustrative figures as well as text. In the final part of this work, the main conclusions of the study are examined, interpreting them in the light of what has been identified in the specialised literature on competence-based teaching approaches. Finally, the limitations of this study are presented along with some basic guidelines, so that they can serve as real catalysts for further progress in this field of knowledge.

Brief description of the key competences

In recent years, the notion of competence has become widespread in education in a polysemic way; today, this term refers to both an educational priority and a central element in relation to the quality of reference texts (such as education laws), as well as other normative texts, educational regulations, curricula or programmes. These texts serve as a basis for all teaching and learning procedures in their search for consistency with existing educational legislation. In this sense, in the framework of the project Definition and Selection of Competences: Theoretical and Conceptual Foundations (DeSeCo), competence is defined as the ability to respond successfully to complex demands by activating mental prerequisites (Weinert, 2001).On this basis, each competence is structured around a demand and involves a combination of interconnected skills, knowledge and cognitive attitudes that together can be organised to achieve progress in a contextual and appropriate manner. To better understand them, three key factors are identified: their contribution to positive lifelong outcomes, their successful application in different contexts, and their effectiveness in meeting complex demands (Rychen & Salganik, 2003). This broad conceptual basis covers the development of students' problem-solving capacity, their motivation, as well as their social skills and willingness to learn in different contexts. If we move away from the perspective of the school system, the acquisition of competences is also applicable (and commonly used) for adequate personal performance in life and in work, cultural and social contexts.

To describe the characteristics of these key competences, we will use the descriptors proposed by Mateo (2010):

- They promote the development of skills rather than the acquisition of content (although this is essential in the learning process).

- They have the characteristic of being multifunctional and flexible enough to respond to dynamic and diverse circumstances.
- They include knowledge, skills and attitudes that can be applied in response to human needs; they can be seen as applications of that knowledge, involving a critical understanding of how those needs arise.
- They are dynamic, enabling individuals to adapt to the needs of a highly changing society; they can be acquired formally, non-formally and informally (in different contexts).
- They are interdisciplinary in nature, which reinforces the convergence between subjects in the curriculum, as well as the connectivity of what is learned both in academia and externally (from business, sports or work).
- They facilitate high quality education by providing opportunities for individuals to actively participate in society; they encourage effective transition into the labour market by promoting pathways to enhance students' skills. They also promote equity by providing a common framework for interaction, learning and practice.
- They form the basis for lifelong learning by promoting both validation of prior learning and life experiences or integration of skills (academic, social and vocational).

Formal, non-formal and informal contexts

Competences can be acquired through various educational pathways and in various contexts, whether formal, non-formal or informal (Council of Europe, 2018). In formal contexts, they can be developed through different curriculum areas, which contribute to their acquisition in a multifaceted way. In such school contexts, educators should adopt a broad approach aimed at promoting the acquisition of basic knowledge in existing curricular areas. This (competency-based) subject teaching is critical for both educators and learners, as it determines the learning opportunities for the latter in designing what they learn and how they learn it. In this sense, the appropriate selection and distribution of tasks in academic contexts creates favourable conditions for students to participate in meaningful learning experiences, so developing their thinking and reasoning abilities. Non-formal learning is also an integral part of competence development as it encourages the acquisition of the maximum number of interpersonal, communication and cognitive skills in non-academic (traditional) contexts. Its effectiveness is developed through experiences gained through voluntary work, as well as from other youth organisations. A further advantage of non-formal education is that it caters for people of all ages (from children to adults) through short courses, seminars or workshops run by employers, social partners or guidance professionals, which undoubtedly tends to reduce existing educational disparities by supporting the literacy of adults and young people. In addition, the contribution of informal contexts to the development of these complex skills also needs to be taken into account; emphasis is placed on information exchange and training, as it helps people acquire skills for life and work outside the formal school system. While informal education is often seen as a less attractive option than formal education, it should be noted that it can promote high quality education as much as that provided in formal contexts. Informal education can be even more interactive and attractive than formal education and represents an opportunity to push the boundaries of knowledge and experience. Given the cost-effectiveness of these practices, there is a need to further explore new forms of teaching and learning in order to establish better cooperation between social and academic stakeholders (Bingham & Conner, 2010). For this reason, appropriate responses to new adaptive learning networks have to be developed based on the adoption of a variety of teaching approaches in a wide range of contexts (Davidson-Hunt, 2006).

Competency-oriented approaches in different contexts

Although there are various learning contexts, most competency-based education models occur within the formal school system, promoting the creation of new teaching approaches that require the participation of students in classroom work (Éthier, Demers, & Lefrançois, 2010). It should be noted that these acts of teaching are no longer based on memorization or repetition, as developed by traditional methods, but rather on interactive practice based on the learner, focusing – primarily – on the learner's needs, interests, background and prior knowledge as a starting point (Friesenhahn et al., 2013). The role of the teacher is to guide students and monitor their progress as they complete planned tasks, and where they apply previously acquired knowledge and skills to practical situations in order to progressively acquire greater competence and capacity (Bauer-Marschallinger, 2016). The application of this knowledge still needs to be monitored by educators, as students are on their way to working more autonomously without professional support. Students learn in a social framework in which interaction and participation are core factors that positively influence the effectiveness of their actions. This collaborative learning atmosphere is crucial for presenting and defending ideas, for talking with colleagues, for sharing experiences and for working cooperatively. In this way, active learning occurs naturally among classmates and on a continuous basis, which facilitates the acquisition of knowledge and skills (Nordentoft & Wistoft, 2012). In this respect, the definition of key competences includes integral elements of quality education other than those related to the academic field, such as social inclusion, personal fulfilment, health or employability. Therefore, teachers should promote the exchange of personal, social and learning experiences through more emotionally oriented tasks related to the arts, sports or leisure activities, as they can enhance the development of multifaceted skills by reinforcing positive behaviour and attitudes towards learning. In addressing these issues from a lifelong learning perspective, social and economic criteria are considered in an attempt to support individuals in developing their competences throughout their lives. Therefore, both the promotion of academic learning processes as well as social and emotional learning practices encourage the emergence of different learning scenarios in which individuals can improve their skills and assess their abilities. As a clear example, in work environments, people can assess performance while employed, or examine careers by exploring different occupations, which can facilitate the transition from work to education (or from education to work) throughout life. In this regard, it should be mentioned that the incorporation of competency-based approaches (providing students with the tools to analyse the environment), opens the door to the reconceptualization of roles and responsibilities in the teaching and learning process. In formal educational settings, these new frameworks help to create new profiles of learners who discuss content, analyse ideas and transform them into values for themselves and others. It is therefore essential that learners focus on their autonomy in order to move beyond the learning provided by the exclusive use of memory as the main applied strategy, so that they can move away from the passive roles that – traditionally and during the last decades – had been assigned to them. To eliminate such role assimilation, teachers must encourage students to create knowledge rather than simply accept it without trying to go beyond it. This objective is paramount in competency-based teaching, which is clearly defined by its holistic vision and by virtue of which students are placed at the centre of the learning process. In this respect, competences are well suited to articulating more active teaching approaches in formal education with a view to strengthening connectivity between curriculum subjects.

The teaching of history and the definition of historical competences

In particular, history is a clear example of the paradigm shift in teaching and learning. It has been considered – mainly – as a subject in which the teaching of content was based on the repetition of facts, dates and events in the context of primary or secondary education systems or in any other educational context beyond compulsory schooling. However, the teaching of history is much more than just rote learning; and this change is becoming more evident in the formal school system. A more thorough assessment of the nature of history itself is now encouraged, which favours the participation of students in conceptual reasoning based on their own discipline (Gómez & Miralles, 2016). This reasoning facilitates historical thinking, allowing students to think as historians do, and consequently, to deepen their understanding of historical events and situations (Seixas & Morton, 2013). As historical thinking is based on the creative process of interpreting past events and producing stories about history, its development is essential to encourage students to construct a meaningful account of what happened in the past and its significance over time (with its impact on the present) (Lévesque, 2008). This historical construction is based on the use of two different types of content: substantive (or first order) content, which refers to both historical concepts and specific events, dates and characters. And, on the other hand, a second-order content that includes those strategic contents and corresponds to methodological concepts (such as strategies or skills) that help individuals to analyse, synthesise and evaluate information, thus facilitating the conceptualisation of history as a construction (VanSledright, 2010). When analysing the first-order content, deeper questions arise about the procedures, agents of change or reasons that led them to make crucial decisions in the past. Among the competences in this area is the ability to ask such questions (Ercikan & Seixas, 2015). Thus, these historical competences are applications of this methodology that aim to answer these questions by activating complex cognitive abilities from a multivariate approach (Yeager, 2001). These skills with which historians conduct their research must be also developed by history students because it is crucial – for them – to be able to discover the disciplinary and procedural processes that define historical thought, so that they can better understand the changes caused by human activity over time (Seixas & Morton, 2013). To this end, the authors present a list of six key concepts that can be used to achieve significant progress in historical thinking linked to the educational field: historical relevance, historical evidence, change and continuity, cause and consequence, historical perspective and, finally, the ethical dimension. Below, to facilitate their understanding, we offer a brief summary of each of these dimensions.

When analysing historical relevance, students should determine which historical events are worth remembering and explain the reasons for such decisions; a key question is whether history fits into current historical narratives and, if not, what legitimate constructions around these significant events should be produced. In addressing historical evidence, primary sources are considered to be first-hand evidence that was created in the same period of time that historians are studying. These traces of the past can provide students with meaningful information, provided that they are established in their historical contexts (Levstik & Barton, 2011); in other words, history students should not simply accept the source as it is, but should learn to read it more critically and to contextualise it in a more meaningful way. With regard to the dimension of change and continuity, students must be trained to describe substantial changes in people's lives and explain why they occurred; these explanations can be made by comparing two points in the past or past and present situations, which facilitates a complex analysis of both diachronic and synchronic breaks and continuities. The work around the cause and consequence dimension involves the analysis of historical events taking into account that there are important contextual conditions that must be addressed in order to have a clearer idea of the reasons for and effects of the relevant content under study; for example, the combination of political, economic, religious, social and cultural factors may vary from country to country, from region to region, from person to person and – indeed – from period to period. Similarly (within this dimension) the authors also identify the specific approach adopted by the historian as another variable that can affect the type of historical thinking developed. With regard to the category of historical perspective, in order to avoid the application of current ideas and norms when interpreting the past (what is sometimes called "presentism"), students need to handle relevant contextual information that will help them to better understand the historical actors, actions, achievements, failures and plans that took place, so facilitating the creation of a critical historical perspective. Finally, the ethical dimension is based on the development of questions of an ethical nature: how can students judge and evaluate past actions, how can they address historical injustices (whose social damage remains evident), and how can they commemorate those who participated in or were affected by wars or military operations, and whose sacrifices have meant we live in more just and democratic societies? Taken together, the development of these six dimensions aims to develop historical thinking that will enable students to develop their historical skills, to analyse historical sources more critically, to master historical discourse and, consequently, to participate more actively in the construction and reconstruction of history by enjoying their comprehensive and detailed learning.

As we know, recent years have seen a proliferation of studies aimed at understanding the links between the teaching of history and the acquisition of historical skills (Dallinger et al., 2016). Within this scope, some studies aimed to explore the effectiveness of strategies and resources that foster historical thinking and reasoning (Gestsdóttir, van Boxtel & van Drie, 2018). Other studies focused on the development of competences in the field of history education under particular learning scenarios, such as the Content and Language Integrated Learning approach (Dalton-Puffer & Bauer-Marschallinger, 2019). In Spain there is also a thriving academic literature on this subject, where many research groups (from all over the country) have analysed the change of paradigm from traditional teaching methods to approaches more oriented towards the acquisition of historical skills (Gómez, Vivas & Miralles, 2019). These studies give us a clear idea of the essential changes that have been promoted in education and that have favoured the replacement of the traditional teaching method with more effective instructional practices (Santisteban, 2010). In particular, some research on the development of historical competences has focused on the application and evaluation of projects and programmes within this innovative framework (Molina & Egea, 2018). Notwithstanding, some studies have recently analysed the implications of historical thought in the teaching of history, and conclude that there are still relevant educational challenges that continue to be an important issue in the teaching of history (Fordham, 2017). In other words, further research is required to investigate the effectiveness of this new educational paradigm (Thorp & Persson, 2020). For example, there are still few studies that analyse this discipline-based conceptual reasoning from different perspectives (such as those presented by teachers). Only a few studies have been carried out on this subject in Spain (Sáiz & López, 2015); in particular, the models of national narrative assumed by history teachers in training have been examined, and it has been concluded that there is a need to continue training teachers in narrative skills due to the prevalence of simple representations of national identity. Another study analysed the perceptions of secondary school teachers in training on the teaching of cultural heritage and the acquisition of historical knowledge and found a close link between the positive appreciation of heritage and the development of historical knowledge (Miralles, Gómez & Rodríguez, 2017). More recently, the historical thinking skills of primary school teachers in training were examined in this country, in terms of their professional skills in organizing historical knowledge and integrating second-order concepts; the results of the study confirmed the limited skills of teachers in training in applying historical thinking processes, suggesting that more training is needed to consolidate specific skills in the teaching of history (Sáiz, Gómez & López, 2018). In short, these studies agree on the need for more specific training in competency-based skills that can promote more creative and dynamic teaching experiences, which is one of the greatest demands of teachers (Monteagudo-Fernández et al., 2020).

Method

Objectives

This work aims to carry out a comparative study of the perceptions of primary and secondary school teachers regarding the teaching of historical skills in the subject of History, and more specifically analyse which skills should be acquired when teaching History. In this sense, a series of more specific objectives has been formulated for the development of this main objective:

- To analyse the different perceptions of teachers regarding the acquisition of historical competences according to the Spanish region in which they work.
- To analyse the different perceptions of the teaching staff regarding the acquisition of historical skills according to the gender of the participants.
- To analyse the different perceptions of the teaching staff regarding the acquisition of historical competences according to their academic profile.
- To examine the different perceptions of the teaching staff regarding the acquisition of historical competences according to the type of school where they work.
- To examine the different perceptions of the teaching staff regarding the acquisition of historical competences according to the educational level (primary or secondary).
- To analyse the different perceptions of teachers regarding the acquisition of historical competences according to the educational innovation projects in which they have participated.

Sample

A quantitative methodology has been used, suitable for a non-experimental study of a transversal nature in which descriptive and relational studies are combined, which has allowed us to obtain the information we wanted from the subjects in the sample. In order to achieve the proposed objectives, a questionnaire was drawn up and validated by experts in the field of social sciences, as well as by teachers of primary and secondary education. Teachers were asked to respond to nine questions related to first and second order historical concepts (Gómez &Miralles, 2019). Tab. 1 shows the proposed items and their correspondence to these contents.

Tab. 1: Relationship between competences and historical concepts

ITEMS RELATED TO THE FIRST ORDER CONCEPTS	
Items	**Concepts**
1. Skills in chronology and dating of historical events (elaboration and interpretation of time lines)	Chronology
2. Describe and list the main historical facts related to the origin and consolidation of the nation	Knowledge of a concept or fact
ITEMS RELATED TO THE SECOND ORDER CONCEPTS	
Items	**Concepts**
3. Argue and discuss the historical relevance of an issue	Historical significance
4. Learning to analyse historical sources	Primary source evidence
5. Explain the causes and consequences of historical events and processes	Cause and consequence
6. Explain the changes and continuities of historical processes	Continuity and change
7. Acquire historical perspective to make contextualized explanations (empathy or historical understanding of the context).	Historical perspective
8. Argue an ethical dimension to history that provides students with a direction for life	Ethical dimension
9. Problematise collective identities, values and consensus	
10. Educate in social values and for a critical citizenship	

The sample is made up of participants from a wide geographical area of Spain, as well as teachers and professors with different profiles according to the administrative situation, academic training, educational level, gender or type of educational centre (state or private). Sample selection procedures were carried out by means of a non-probabilistic sampling process of an accidental or casual type (Thomas & Nelson, 2007). 332 teachers from all over Spain participated, distributed in various autonomous communities (such as Andalusia, Asturias, the Canary Islands, Castile and Leon, Extremadura, Galicia, Madrid, Basque Country, Region of Murcia and Community of Valencia). The completion of the questionnaires was in two ways: over the Internet and in paper format (at the choice of the participants); the printed option was the most used by the participants. Over 67 % of the participants stated that they had participated (throughout their initial or ongoing training period) in various educational innovation projects, which could indicate that they are concerned about

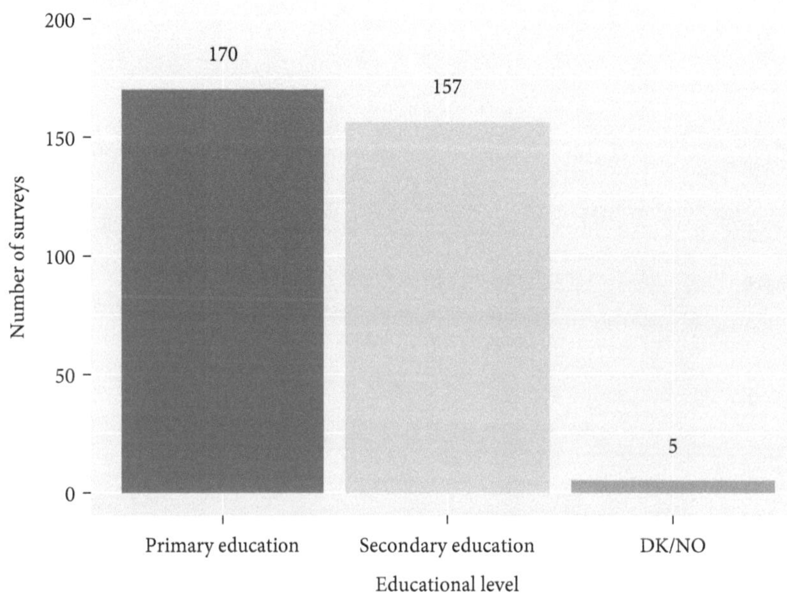

Fig. 1: Distribution of participants according to educational level

improving the teaching of history. As for the educational level at which they work, slightly more than 51 % of the respondents worked in primary education, while 47.2 % perform their teaching duties at the secondary level (the rest did not specify this information -cf. Fig. 1-). In relation to the sex of the participants, the sample can be considered balanced since 52 % were female and 48 % male (Fig. 2).

In terms of the administrative situation, permanent teachers in state schools predominate, representing 67 % of the respondents, followed by the group of teachers in private schools with a participation of 17 %; and the group formed by temporary or trainee teachers, 16 % of the sample. In terms of the academic profile of the participants (Fig. 2), 58 % are graduates, 34 % have only studied for a diploma, just over 5 % have completed a master's degree and almost 3 % have a doctorate. Taking into account the work centre of the participants (state or private), it should be noted that the number of teachers working in a state centre is almost 82 %, and that 17 % are teachers from private centres (only 1 % do not identify the type of centre in which they work).

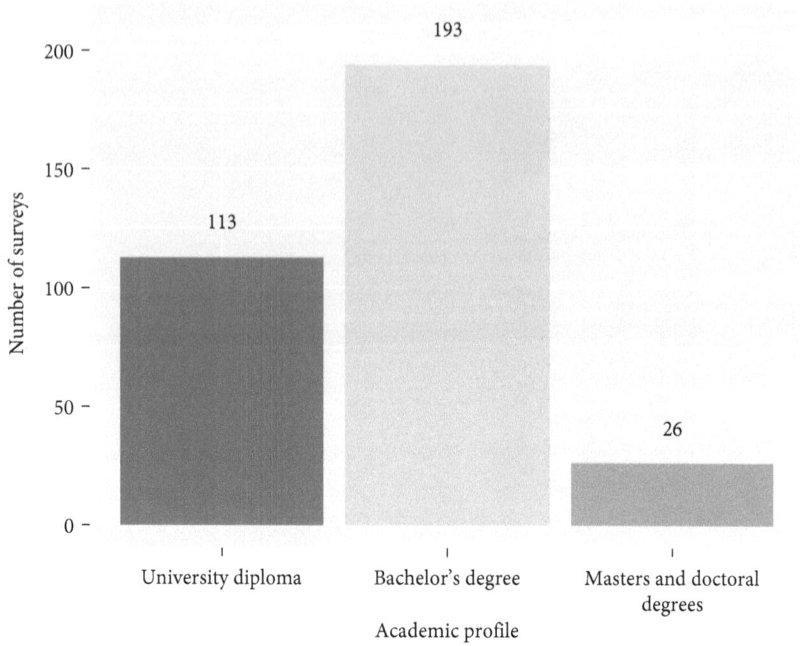

Fig. 2: Distribution of participants according to academic profile

Results

After analysing the perception shown by the teaching staff participating in this study with regard to which historical competences should be acquired, a clear and very high trend is observed in the answers obtained, given that more than 90 % consider that the acquisition of competences (in terms of historical relevance, historical sources or evidence – historical evidence –, change and continuity, causes and consequences, historical perspective and ethical dimension), are very important, although the item referring to problematising our collective identities, values and consensus only offers a coincidence of 75 %. There is also broad agreement (76 %) on the importance of acquiring skills in chronology and dating of historical events (elaboration and interpretation of time lines), as well as in describing and listing the main historical events related to the origin and consolidation of the nation (63 %). Discriminating by the Spanish region in which they work, the results show that there is full agreement on the causes and consequences of historical events and processes, as well as on the changes and continuities of historical processes, as they all coincide. It should be stressed that these are aspects of considerable relevance, and there is not a single questionnaire whose response is little or not very relevant. Only in each of the Autonomous Communities of Valencia, the Basque Country, Madrid and the Region of Murcia is there a group of teachers (around 10 %) who give little importance to the work of the historical perspective in order to carry out contextualised explanations (empathy or historical understanding of the context), as well as to the work of the ethical dimension of history in order to provide students with an orientation for life, while for the rest (both in these regions and in the others) they consider it very important. In terms of problematising our collective identities, values and consensus, except in the Basque Country and the Region of Murcia (where – again – 10 % of teachers do not consider it a relevant aspect), the rest do consider it as a transcendental issue, although it is worth noting that in Asturias 75 % of those surveyed maintain a neutral position (considering it neither a relevant nor a less relevant aspect). However, when talking about educating in social values and for a critical citizenship more than 90 % – in all regions – perceive that this is a relevant issue.

As regards learning to analyse historical sources, it once again has a high positive percentage in the responses in favour of its relevance, and it can be seen that the load of these reveals that for the participants it is an important or very important subject in equal proportions, with both questions adding up to a total of responses of over 75 % (highlighting – in this item – that only Extremadura and Castilla y León show indifference in this respect). With regard to the need to argue and discuss the historical relevance of a topic, the importance of this issue is once again generally predominant. And with regard to the competences related to the first-order concepts, the responses of those who consider them to be of little relevance or who show indifference – in general – increase. Specifically, when referring to describing and enumerating the main historical facts related to the origin and consolidation of the nation, the responses that give little relevance to this subject range from 30 % to 50 %. In this case, only Extremadura, the Canary Islands and Asturias stand out, where for 75 % of the participants this is an important issue to be dealt with (it should be noted that in Galicia the responses are divided almost equally between irrelevant, indifferent and relevant). Finally, when talking about the treatment of skills in chronology and dating of historical events (elaboration and interpretation of time lines), only the Canary Islands, Castile and Leon, and Extremadura consider it a very important aspect in its entirety, while the rest return percentages close to 30 % in which they give little importance to this competence.

It has been found that the gender variable has little influence on the trend of responses, since a similarity can be seen in all of them. We can only highlight differences of around 10–12 % in the answers on the competences related to the first-order concepts, with the teachers giving more importance to these questions.

As regards the influence of the academic profile of the interviewees, these were grouped into three categories: graduates, undergraduates (or graduates) and teachers with postgraduate qualifications (with a master's degree, advanced studies diploma or doctorate). In this sense, there are again significant differences in the competences of the concepts of chronology or knowledge of a historical fact or concept; that is, with the competences linked to the development of skills in chronology and dating of historical facts (elaboration and interpretation of time lines) as well as in the description and enumeration of the main historical facts related to the origin and consolidation of the nation. In this sense, it can be observed that the higher the level of education studies (or the higher the degree), the less importance is given to these competences, which shows that training beyond the minimum required for teaching gives greater importance to the treatment of competences related to second-order concepts: precisely those that allow the development of a more appropriate, more critical and more reflective historical thinking among students. However, it should be noted that in relation to the ethical dimension, in the answers about the importance of arguing for the ethical dimension of history in order to provide students with an orientation for life, differences are also observed according to the training received, with more value being placed on this aspect by the members of the group of participants with a postgraduate degree – followed by the group of graduates – and less importance being placed on it among those with a diploma or degree.

As regards the ownership of the centre of work (state or private), there are hardly any differences in the trend of the answers, except for those related to the ethical dimension where they are more significant; specifically, when answering the question about problematising our collective identities, values and consensus, the teaching staff coming from private centres consider it relevant in 56 %, while those working in state centres raise the percentage to 80 % among those who consider it quite necessary to work in that sense.

Performance at the primary or secondary stage only offers notable differences in the answers to questions 2 and 9. In this sense, although in the answers to the question related to the first-order concept of knowledge of a concept or a fact, teachers who do not reveal whether it is relevant for them – or not – range from 25 % (in primary) to 30 % (in secondary), the same does not happen among those who give little relevance to this type of question (half in primary with respect to secondary – 8 % versus 15 %, respectively). Furthermore, the graph again highlights the answers given in relation to the question about problematising our collective identities, values and consensus, considering it less important at the primary stage, with just over 32 % among those who consider it to be an irrelevant or unimportant aspect compared to the answers given by secondary school teachers which barely reach 15 % (with it being an important aspect for 85 %).

Finally, it can be seen that participation (or not) in educational innovation projects influences some of the responses, especially those concerning first-order concepts. In this sense, those who have participated in this type of training attach less importance to the first two questions of the questionnaire, with a difference of 20 percentage points in the question on describing and listing the main historical facts related to the origin and consolidation of the nation. Once again there is a striking difference in the answers to question 9, which deals with problematising our collective identities, values and consensus, in which just over 30 % of the answers from teachers who have not taken part in innovation projects give little relevance to or show indifference to this question, while just 20 % of the group that has participated in these training projects give little importance or show their indifference.

Conclusions

This study examines educators' perceptions of history teaching with the aim of developing meaningful practices and promoting competency-based learning, as well as fostering mutual awareness and cooperation. To this end, respondents were asked whether their teaching practices focused on the demonstration – by students – of problem-solving skills and critical knowledge. It follows that educators agree on the progress that still needs to be made and the fundamental changes in direction that still need to be made in this discipline (Sáiz & Gómez, 2016). In this sense, the general objective of teaching history should be to design, apply and guarantee the best and most relevant learning experience possible for the student. However, curricular frameworks that propose the teaching of content based on competences (as suggested by the LOMCE in the case of Spain) tend to provide vague guidelines for educators, which can lead to confusion when these guidelines are followed in real practice (Gómez, Miralles & Molina, 2015). Therefore, their clarification should be considered a basic prerequisite to delimit and classify these pedagogical elements. Being more explicit, in curricular terms, about the acquisition of skills and competences, the development of independent thinking and critical analysis could progressively improve the teaching of this subject. It is therefore desirable that – in addition to these educational initiatives – research efforts be continued with a view to identifying the basic needs of students and teachers, whose input should be necessary for the implementation of the new regulations. As can be seen from this work, it is essential that teachers, as well as policy makers (and society at large), gather meaningful and comparable information that will allow a complete and comprehensive picture to be drawn of students' learning outcomes during the stages of compulsory education. This consultative way of developing a common curriculum rationale (focused on practice) would enrich the new competence-based teaching approaches with a clear focus on skills development and quality of work. This would be highly relevant both for building teachers' capacity and skills, and for including the teaching of historical thinking in their future teaching units. In short, both teachers and students must move towards a vision of history teaching that puts historical skills to the fore, thus ensuring that the implementation of the central concepts of historical thought in the classroom must be at the heart of history teaching (Páez, Bobowik & Liu, 2017). As can be seen, the participating teachers give special value to the fact that the students acquire the skills related to historical relevance, and also to change and continuity, with a high consensus identical to that existing in this respect regarding the high importance for students of being educated in social and critical values towards the construction of a committed and active citizenship (in both cases with a 90 % consensus). As presented in the previous section (dedicated to the results), it would be interesting to continue with other studies that allow the scope of the findings to be widened with respect to the influence that postgraduate training may have (among teaching staff) on the treatment of competences related to second-order concepts, insofar as the data indicate that such concepts allow for the development of more critical and reflective historical thinking among students, and that for this to happen, postgraduate training of teaching staff is an essential variable (Sáiz & López, 2016). And this same position would be extended to address the ethical dimension of history, not only because its application offers students a direction for life but also because the importance given to this dimension is much higher among those teachers who have postgraduate studies (hence our interest in suggesting continuity to this line of research). Similarly, it would also be useful to continue investigating the perceptions that have been raised in terms of collective identities, values and consensus, to the extent that teaching participation in innovation projects becomes the touchstone of a balance that is sometimes somewhat volatile. Without a doubt, and in an educational scenario dotted with so many events that structure the life of the citizen (of an economic, cultural, political, legal nature...), it would seem sensible to give the teaching of historical competences the importance they should never have lost, insofar as their good exercise, design, development and execution will end up affecting – by capillarity – the development of both the character and personality of students and – in the future – their exercise (more or less active, more or less critical) in a society that is absolutely sifted by the economic, political and legal. And probably – and by virtue of all this – the educational ecosystem has much to say in this sense, and good proof of this would be the relevance that, for this purpose, both teacher training and their perception of these and other variables involved – directly – in this scenario can attain. Hence, we encourage both the continuity of studies of this nature and the monitoring of the findings (and humble productions) of this research group.

References

Bauer-Marschallinger, Silvia: *Acquisition of historical competences in the CLIL history classroom.* (doctoral thesis). University of Vienna: Vienna, 2016, retrieved 20.4.2020, from DOI 10.25365/thesis.42043.

Bingham, Tony/Conner, Marcia: *The new social learning: A guide to transforming organizations through social media.* American Society for Training & Development: San Francisco, 2010.

Council of Europe, *Council Recommendation of 22 May 2018 on key competences for lifelong learning Text with EEA relevance.* Luxembourg: *Official Journal of the European Union, 2018. OJ* C 189 4.6.2018. Retrieved 20.6.2020, from https://eur-lex.europa.eu/legal-content/EN/TXT/PDF/?uri=CELEX:32018H0604(01)&from=LT.

Dallinger, Sara/Jonkmann, Kathrin/Hollm, Jan/Fiege, Christiane: "The effect of content and language integrated learning on students' English and history competences–Killing two birds with one stone?". *Learning and Instruction* 41, 2016, pp. 23–31.

Dalton-Puffer, Christiane/Bauer-Marschallinger, Silvia: "Cognitive discourse functions meet historical competences: Towards an integrated pedagogy in CLIL history education". *Journal of Immersion and Content-Based Language Education* 7(1), 2019, pp. 30–60. Retrieved 14.5.2020, from DOI https://doi.org/10.1075/jicb.17017.dal.

Davidson-Hunt, Iain J.: "Adaptive learning networks: Developing resource management knowledge through social learning forums". *Human Ecology* 34(4), 2006, pp. 593–614.

Ercikan, Kadriye/Seixas, Peter (eds.): *New directions in assessing historical thinking.* Routledge: New York, 2015.

Éthier, Marc-André/Demers, Stéphanie/Lefrançois, David: "Las investigaciones en didáctica sobre el desarrollo del pensamiento histórico en la enseñanza primaria. Una panorámica de la literatura publicada en francés e inglés desde el año 1990". *Enseñanza de las Ciencias Sociales. Revista de Investigación* (9), 2010, pp. 61–73, retrieved 28.06.2020, from https://www.raco.cat/index.php/EnsenanzaCS/article/view/191359.

Fordham, Michael: "Tradition, authority and disciplinary practice in history education". *Educational Philosophy and Theory* 49(6), 2017, pp. 631–642, retrieved 28.06.2020, from DOI 10.1080/00131857.2015.1135777.

Friesenhahn, Günter J./Schild, Hanjo/Wicke, Hans-Georg/Balogh, Judit/Plantz, Christina: *Learning mobility and non-formal learning in European contexts. Policies, approaches and examples.* Council of Europe Publishing: Strasbourg, 2013.

Gestsdóttir, Súsanna/van Boxtel, Carla/van Drie, Jannet: "Teaching historical thinking and reasoning: Construction of an observation instrument". *British Educational Research Journal* 44(6), 2018, pp. 960–981, retrieved 22.03.2020, from DOI https://doi.org/10.1002/berj.3471.

Gómez, Cosme J./Miralles, Pedro: "Historical skills in compulsory education: Assessment, inquiry based strategies and students' argumentation". *Journal of New Approaches in Educational Research (NAER Journal)* 5(2), 2016, pp. 130–136.

Gómez, Cosme J./Miralles, Pedro/Molina, Sebastián: "Presentación Evaluación, competencias históricas y educación ciudadana". *Revista de Estudios Sociales* (52), 2015, pp. 9–14, retrieved 14.06.2020, from DOI http://dx.doi.org/10.7440/res52.2015.01.

Gómez, Cosme J./Vivas, Verónica/Miralles, Pedro: "Competencias históricas y narrativas europeas/nacionales en los libros de texto". *Cadernos de Pesquisa, São Paulo* 49(172), 2019, pp. 210–234, retrieved 15.04.2020, from DOI https://doi.org/10.1590/198053145406.

Lévesque, Stephane: *Thinking historically: Educating students for the twenty-first century.* University of Toronto Press: Toronto, 2008.

Levstik, Linda. S./Barton, Keith. C.: *Doing history: Investigating with children in elementary and middle schools.* Routledge: New York, 2011.

Mateo, Lourdes: "Origen y desarrollo de las Competencias Básicas en Educación Primaria". *Temas para la educación* 7(1), 2010, pp. 1–17.

Miralles, Pedro/Gómez, Cosme J./Rodríguez, Raimundo A.: "Patrimonio, competencias históricas y metodologías activas de aprendizaje: Un análisis de las opiniones de los docentes en formación en España e Inglaterra". *Estudios pedagógicos (Valdivia)*, 43(4), 2017, pp. 161–184, retrieved 12.05.2020, from DOI http://dx.doi.org/10.4067/S0718-07052017000400009

Molina, Sebastián/Egea, Julián: "To assess the acquisition of historical skills in secondary education: A case study focused on historical empathy". Ensayos: *Revista de la Facultad de Educación de Albacete* 33(1), 2018, pp. 1–22.

Monteagudo-Fernández, José/Rodríguez, Raimundo/Escribano-Miralles, Ainoa/ Rodríguez-García, Antonio: "Percepciones de los estudiantes de Educación Secundaria sobre la enseñanza de la historia, a través del uso de las TIC y recursos digitales". *Revista Electrónica Interuniversitaria de Formación del Profesorado* 23(2), 2020, pp. 67–79, retrieved 23.05.2020, from DOI https://doi.org/10.6018/reifop.417611

Nordentoft, Helle M./Wistoft, Karen: "Collaborative learning and competence development in school health nursing". *Health Education* 112(5), 2012, pp. 448–464, retrieved 28.07.2020, from DOI http://dx.doi.org/10.1108/09654281211253452

Páez, Darío/Bobowik, Magdalena/Liu, James: "Social representations of the past and competences in history education". In: Carretero, Mario/Berger, Stefan/Grever, María (eds.): *Palgrave handbook of research in historical culture and education*. Palgrave Macmillan: London, 2017, pp. 491–510, retrieved 28.05.2020, from DOI https://doi.org/10.1057/978-1-137-52908-4_26.

Rychen, Dominique S./Salganik, Laura H.: *Highlights from the OECD project definition and selection competencies: Theoretical and conceptual foundations (DeSeCo)*. Hogrefe & Huber Publishers: Cambridge MA, 2003.

Sáiz, Jorge/Gómez, Cosme J.: "Investigar pensamiento histórico y narrativo en la formación del profesorado: fundamentos teóricos y metodológicos". *Revista Electrónica Interuniversitaria de Formación del Profesorado* 19(1), 2016, pp. 175–190, retrieved 15.05.2020, fromhttps://revistas.um.es/reifop/article/view/206701.

Sáiz, Jorge/Gómez, Cosme J./López, Ramón: "Historical thinking, causal explanation and narrative discourse in trainee teachers in Spain". *Historical Encounters* 5(1), 2018, pp. 16–30.

Sáiz, Jorge/López, Ramón: "Competencias y narrativas históricas: el pensamiento histórico de estudiantes y futuros profesores españoles de educación secundaria". *Revista de Estudios Sociales* (52), 2015, pp. 87–101.

Sáiz, Jorge/López, Ramón: "Narrativas nacionales históricas de estudiantes y profesorado en formación". *Revista de Educación* 374, 2016, pp. 118–141.

Santisteban, Antoni: "La formación de competencias de pensamiento histórico". *Clio & asociados* (14), 2010, pp. 34–56, retrieved 22.06.2020, from https://www.clio.fahce.unlp.edu.ar/article/view/clion14a03.

Seixas, Peter/Morton, Tom: *The big six historical thinking concepts*. Nelson Education: Toronto, 2013.

Thomas, Jerry/Nelson, Jack: *Métodos de investigación en actividad física*. Paidotribo: Barcelona, 2007.

Thorp, Robert/Persson, Anders: "On historical thinking and the history educational challenge". *Educational Philosophy and Theory* 52(8), 2020, pp. 891–901, retrieved 23.06.2020, from DOI https://doi.org/10.1080/00131857.2020.1712550.

VanSledright, Bruce: *The challenge of rethinking history education: On practices, theories, and policy*. Routledge: New York, 2010.

Weinert, Franz E.: "Concept of competence: a conceptual clarification". In: Rychen, Dominique S./Salganik, Laura H. (eds.): *Defining and selecting key competencies*. Hogrefe: Göttingen, Germany, 2001.

Yeager, Elizabeth Anne: "Historical thinking and other unnatural acts: Charting the future of teaching the past". *Social Education* 65(5), 2001, pp. 307–307.

Helena Rausell, Álvaro-Francisco Morote, Rafael Valls, and Moisés Domingos

Chapter 10 Social representations and teaching of cross-cutting topics for history and geography teachers in training: Gender and climate change[1]

Abstract: This research focuses on the importance of initial training when addressing the representations that future teachers have about socially relevant topics in the scope of geographical and historical education. These include gender and socio-environmental problems, which are the focus of this piece of work. To this end, this study will try to identify the degree to which teachers in training understand the concept of gender, in order to determine to what extent initial gender-related education has an influence on them. As regards climate change, we will explore participants' perceptions of the education they received during their schooling and analyze the didactic proposals they would present to their pupils. The sample includes a total of 124 students from the Degree in Primary Education and 69 students taking the Master's Degree in Secondary Education (specialized in geography and history; University of Valencia). The results indicate that the previous education received by students results in significant differences when it comes to defining and, consequently, understanding the concept of gender. However, when it comes to climate change, although participants stated that they had received specific training on the subject, the activities they proposed were not sufficiently problematized or innovative to be introduced into the classroom with any possibility of success.

Key words: Teaching training, Gender, Climate change, History, Geography

Introduction

The theory of social representations (Moscovici, 1978) appeared as a paradigm which could explain the workings of academic standards in social science education. Attempts to improve history, geography and art history education in schools are affected by the spontaneous perceptions that pupils, teachers,

1 This research is part of the "Social representations of school content in the development of teaching skills" project run by the Ministry of Science, Innovation and Universities (PGC2018-094491-B-C32) and co-financied with EU FEDER funds.

textbook editors and legislators hold. These perceptions have been developed in a historical context dominated by a hegemonic culture which influences all actors in the school setting (Jodelet, 1994; Audigier, 2002; Borg, 2003). It is also important to underline that social representations allow us to understand the distinctive features and effects of placing individuals in specific cultural contexts and the process of integration and re-signification of the hegemonic ideas of the social environment they are part of. In this way, we can emphasize their function as a symbolic obstacle to integrating and constructing more modern attitudes which incorporate new knowledge (Domingos & Lira, 2019). In terms of gender, for example, it may be one of the factors that explains why the progress and contributions made in the last few decades by female academics is not reflected among teachers nor in the specific field of social science didactics research (Donoso & Velázquez, 2013). This is despite the fact that introducing this topic into research broadens and modifies scientific knowledge.

When the European Union implemented the Horizon 2020 program at the end of 2013, they identified 137 science and technology sectors in which gender analysis could be beneficial for rcsearch. These fields included nanotechnology, geosciences, oceanography, organic chemistry, aeronautics, ecology and biophysics, but they also mentioned social sciences, arts and humanities and all disciplines which are focused on curriculum development, educational policy and educational research. As highlighted in the program itself, gender equality concerns the entirety of Horizon 2020: "Integrating the gender dimension in research and innovation is an added value in terms of excellence, creativity, and business opportunities."[2]

As regards teaching of the weather and climate, this topic has become one of the branches of geography that receives the most academic attention and social interest due to its importance in the current context of climate change (Martínez & Olcina, 2019). Knowledge and teaching of its content is a complex task (Olcina, 2017). Even now we observe how controversial it is when the scientific community presents a simple explanation for the evolution of climatic conditions on a global scale and its current status, due to the amalgamation of

2 European Commission: *Funding and tender opportunities*; https://ec.europa.eu/research/participants/docs/h2020-funding-guide/cross-cutting-issues/gender_en.htm.European Commission, Stanford University, National Science Foundation: *Gendered Innovation*. https://genderedinnovations.stanford.edu/.

factors involved, the social representations generated and the stereotypes and fake news created by the media (Olcina, 2009; Morote, 2019).

In Spain, explanations for climate change carry more weight in the educational setting due to the requirement to address these topics in classrooms, as established both for Primary Education (Royal Decree 126/2014, 28th February) and Secondary and Baccalaureate Education (Royal Decree 1105/2014, 26th December). Equally, it is a pedagogical challenge for teachers who bear the responsibility of educating the youngest members of society on understanding and potentially adapting to this phenomenon (Eklund, 2018). In this vein, we should highlight that education is one of the non-structural measures which allows us to tackle this phenomenon, but is traditionally an ignored factor (Morote, 2019).

Although gender is starting to be included in educational research, studies focused specifically on gender and teacher training remain few and far between (Hahn, 1996; Crocco, 2006; Vavrus, 2009). Furthermore, the absence of gender in official curricula in Primary and Secondary Education is of great concern. These issues (gender and climate change) are barely mentioned, if at all, in the curricula of Teacher Training Colleges and Postgraduate qualifications in Secondary School Education (Rebollar, 2013; Ibeas & Garrido, 2018).

Despite the high proportion of female teachers involved in teaching kindergarten and primary and the growing number in secondary and university education, teachers receive very little or no training on gender, which practically condemns them to unconsciously reproducing stereotypes, inequalities and gender roles in their future practice and discourse (Jaramillo, 1999; Vavrus, 2009; Anguita, 2011). As regards climate change, in the last decade several international studies have demonstrated the importance of dedicating time to studying this topic in the school environment (Ahmad & Numan, 2015; McWhirter & Shealy, 2018). In Spain, it must be noted that there is no consolidated field of research about the teaching of this topic in geography and/or social science didactics (Morote, 2019). However, there is a long-standing record of scientific research about how to teach climatology (Martínez & López, 2016; Martínez & Olcina, 2019).

Although progress toward gender equality in the last few decades is undeniable, inequalities between the genders are still considerable. As is well known, the role of the family and education in the transferal of stereotypes and gender roles is decisive (Rausell & Talavera, 2017). Meanwhile, this research on climate change stems from interest in several factors: (1) the current status of the problem and the existing and future effects on society. This aspect is further accentuated in the Mediterranean region ("at-risk region") (Calvo, 2001) due to the increase, both in frequency and intensity, of extreme weather events (IPCC, 2018). (2) The importance of studying climate change in primary, secondary and baccalaureate education, as per regulation; (3) despite the longstanding record of scientific research about climatology didactics, both in this study's area (the region of Valencia) and at a national level, relatively little is known about teacher training on climate change; and (4) the importance of raising awareness among and teaching the youngest members of society about the causes and effects of this phenomenon given that they will teach and raise awareness among future generations.

For all of these reasons, the objectives of this research are, on the one hand, understanding to what extent teachers in training recognize the concept of gender to try to determine how much of an impact initial training has on them and, on the other hand, exploring participants' perception of the education related to climate change that they received during their school years and analyzing the didactic proposals they would present to their pupils. Ultimately, we will explain the similarities observed between the responses given by students taking the Degree in Primary Education and those on the Master's Degree in Secondary Education as regards every day and socially relevant phenomena (in this case gender and environmental problems), where an "ad hoc" explanation derived from practicality tends to be predominant instead of the rationality of academic theories.

With regard to the hypotheses of this study, on the topic of gender we expect to see a difference between the responses given by students taking the Master's Degree in Secondary Education and those on the Degree in Primary Education, as the former have received some initial training (specific or crosscutting) on gender. Concerning climate change, the teachers in training will not have received adequate training during their previous school education. As for their didactic proposals, we believe that they do not have sufficient didactic complexity to be able to present or create pedagogical materials, which could contribute to future teachers being unable to address this topic in the classroom.

Methods

In line with the aforementioned objectives, we carried out a number of surveys about gender and climate change in the format of a case study of the future primary, secondary and baccalaureate teachers at the Faculty of Teaching (University of Valencia, Spain).

Research design

This research is of a mixed nature (quantitative and qualitative). It is based on a mixed descriptive and exploratory study (non-experimental). We decided to use this type of methodology because it allows us to carry out a contextualized description of the object of study, its non-hypothetical approach and the gives us the possibility of reflecting on its singularity using a dense and accurate description of the researched phenomenon (Álvarez & San Fabián-Maroto, 2012). Moreover, it adopts a cross-cutting approach given that the information analyzed was gathered during a specific period of time (2019–2020 academic year) in the format of a case study: students taking the Degree in Primary Education and the Master's Degree in Secondary Education (hereafter MAES) at the University of Valencia. The latter had previously studied a Bachelor's degree in History or Art History or, in a smaller proportion, Geography or the Environment.

Context and participants

With regard to the context and the participants, we used a non-probabilistic sample (available or convenient sample) for the selection procedure. In the case of students taking the Degree in Primary Education, groups were chosen from the 3rd and 4th years of the course, taking modules in social science didactics. For 3rd year students, this was Basic Aspects and Social Science Didactics, whereas 4th year students studied the Applied Aspects module (2019–2020 academic year). Concerning the representativeness of the sample, we took into account the number of registered students (a total of 155 students). On the basis of a reliability level of 95 % and a margin of error of 5 %, the representative sample had to include 111 students. In the end, the total number of students who participated increased to 124 (53 students in their 3rd year – gender study – and 71 students in their 4th year – climate change study).

As for students taking the Master's Degree in Secondary Education, we selected the group taking the Learning and Teaching of Geography and History course and the Supplementary Disciplinary Training module. In terms of the representativeness of the sample, we took the number of registered students into account (a total of 41 students). With a reliability level of 95 % and a margin of error of 10 %, the representative sample had to include 29 participants. In the end, a total of 35 students participated in the gender study and 34 participated in the climate change study.

With reference to the demographic characteristics of the participants, a very high proportion of students taking the Degree in Primary Education were female

(70.1 %). This is representative of the normal range of the type of students at the Teaching Faculty of the University of Valencia (70.2 % of students registered in the 2019–2020 academic year were women). As regards age, participants ranged from 19 to 23 years old with an average of 21.9. For the postgraduate degree, although there was greater parity, women still outnumbered men (45.7 % men, 54.2 % women), with a higher age range between 20 and 26 years old and an average of 25.5 years old.

Analysis tool

In order to gather information, we used the survey method (Alaminos & Castejón, 2006; Ruíz, 2009), which involved designing and carrying out a series of questionnaires given to the participants at different points in the academic year. For the gender study, we used two different questionnaires for undergraduate and postgraduate students. The first questionnaire contained five questions (four open questions and a multiple-choice question) which was filled out by students at the beginning of the year for an initial evaluation. The second included five open questions. It was also a pre-test questionnaire which was used prior to sessions dedicated to gender. In both cases, we asked participants to define the concept of gender.

As regards the tool used to gather information about climate change, we created a questionnaire separated into 5 sections with a total of 19 questions. Taking into account the proposed objectives of this research, we analyzed the data obtained from section 1 (demographic characteristics; items 1 and 2), section 3 (education about climate change during their schooling; items 3 and 4) and section 5 (teacher training about climate change; items 15, 16, 17 and 19) (Tab. 1)

These questionnaires were validated by three teachers in the departments of Experimental and Social Science Didactics (University of Valencia) and Regional Geographical Analysis and Physical Geography (University of Alicante). As for the procedure, the undergraduate questionnaire was completed on two occasions, initial (gender) and intermediate (climate change), during the third week of September and the last week of November (first semester). The postgraduate questionnaire was completed during the last week of March (second semester of the 2019–2020 academic year). Both questionnaires took around 20–30 minutes to finish. With regard to the response rate, it should be noted that all the participants answered. Finally, we must underline that the whole procedure was carried out anonymously, through the creation of lists using student numbers and providing written confirmation as to the confidentiality of the data collected.

Tab. 1: Analyzed items and sections from the questionnaire on climate change

Section 1. Demographic characteristics
-Item 1. Gender
-Item 2. Age
Section 3. Climate change education during schooling
-Item 3. During your school education, did you receive any information about climate change? Can you tell us what information you were given? Answer: Yes/No/Don't remember
-Item 4. If the answer to the previous question was 'yes', during which educational stage did you receive this information? Answer: Primary school/Secondary school/ Baccalaureate
Section 5. Teacher training on climate change
-Item 15. As a future teacher, do you feel you are qualified to be able to teach this content? Select answer: Yes/No/Do not know, no answer
-Item 16. During your university education, have you received any training on this content? Answer: Yes/No/Do not remember
-Item 17. Is studying this content at school worth the effort? Answer: Yes/No/Do not know, no answer
-Item 19. As a future teacher, what type of activities, information and other materials would you provide to your pupils in order to further their knowledge about climate change?

Source: Prepared by the authors.

Results

What do we mean by gender?

In the questionnaire on gender, we asked participants in both groups to define the concept of gender. It's worth recalling that the sample for these two questionnaires in particular included 53 students from the third year of the Degree in Primary Education and 35 students from MAES. Undergraduate students had had no contact with gender during their initial training, while 57.1 % of postgraduate students confirmed having received some specific or cross-cutting education on gender.

In order to analyze the responses, we established a series of codes which function by representing a unit of meaning (Simons, 2009). We documented a total of 29 codes, with a total recurrence of 211 and an average of 2.4 codes per response. These codes are displayed in Tab. 2 along with the frequency they were used.

Tab. 2: Codes identified in relation with the definition of gender

Code	Nº	Code	Nº	Code	Nº
Biology	2	Male-female binary	33	Non-binary	1
Characteristics	19	Classification	16	Socio-cultural construction	15
Culture	6	To be defined	1	Defining	1
Distinction	13	Education	1	Fluidity	1
Genitalia	5	Historicity	7	Identity	12
Sexual identity	3	Identification	9	Innate	3
Innate (non)	1	Masculinity	1	Perspective	1
Power	3	Roles	12	Emotion	7
Being (way of)	1	Sex	21	Social	12

Source: Survey results. Prepared by the authors.

We can observe that the most frequent codes are, in this order, male-female binary (15.6 %), sex (9.9 %), characteristics (9.0 %), classification (7.6 %), social and cultural construction (7.1 %), differentiation or distinction (6.2 %), identities (5.7 %), social (5.7 %), roles (5.7 %) and identification (4.3 %). Conversely, almost a third of codes are documented on only one or two occasions. We also established a series of categories in which we grouped the codes (Simons, 2009; Álvarez & San Fabián-Maroto, 2012). These categories are: (1) classification, (2) sex and biological difference, (3) relational approach, (4) feminism and post-structural feminism.

The first of these categories indicates a very basic, almost intuitive, approximation of gender. It includes codes which present gender as a way to "classify" or "identify" people, and yet it represents a fairly significant proportion (11.8 %). The second category covers allusions to sex and biological differences, underlining the innate nature and its link with certain attributes or a particular sexual identity, which gender theories specifically fight against (15.6 % of responses). The third category was identified as relational approach and includes references to differences between men and women; characteristics, qualities or differences attributed to men and women; gender roles and stereotypes. This section carried the most proportional weight of the total (44.1 %), followed by the last category which corresponds to feminism and post-structural feminism (28.0 %). This final category includes the definition of gender as a social or cultural construction, reference to identities, rejecting the biological or male-female binary argument and the (scarce) allusions to gender fluidity or masculinity.

Furthermore, the contrast between the responses given by the groups surveyed – undergraduate and postgraduate (MAES) students – allowed us to identify the following differences:

1) The first is based on the number of codes used. While undergraduate students mentioned a total of 16 codes, MAES students mentioned 24 different codes, which represents an increase of 50 %.
2) Postgraduate students also provided more elaborate responses. For undergraduates, the average number of codes used per response was 1.89, while postgraduate students used an average of 3.17 codes per response, more than one point higher than the undergraduate average.
3) Some codes are used only by undergraduate students while others are mentioned only by postgraduate students. The former were the only ones to consider gender as a defining element, explaining it using differences and identification or considering it to be innate or related to our way of being. On the other hand, MAES students introduced twelve new codes, specifically the rejection of male-female binary, the non-innate nature, the idea that it is a concept under construction or a social and cultural construction, the link to culture, education, roles or stereotypes, its fluid nature, historicity or contingency, masculinities and the shaping of perspective or approach.
4) The two groups also used the shared codes differently. Some codes represent a fairly similar proportion within each subgroup, as is the case for male-female binary (relatively similar percentage around 15.0 % in both cases) and gender as a way of being or feeling (3.5–3.0 %, respectively). However, the number of times other codes were mentioned varies greatly, as is shown through the higher percentages of codes such as characteristics (13.0 % versus 5.2 %), classification (11.0 % versus 4.4 %), sex (11.0 % versus 8.8 %) or social (10.0 % versus 1.7 %) with a higher prevalence among those we could consider non-experts (the undergraduate group).
5) Finally, the categories used vary as shown in Tab. 3. Undergraduate students favor relational approach and understand gender as a classification method linked to biological gender, while postgraduate students resort in almost one out of every three answers to codes related with feminism and post-structural feminism (Tab. 3).

Tab. 3: Absolute and relative proportions of categories associated with the definition of gender for undergraduate and postgraduate students

Category	Group	Nº codes	Percentage
Classification	Undergraduate	21	21.0 %
	Postgraduate	5	4.5 %
Relational approach	Undergraduate	51	51.0 %
	Postgraduate	42	37.8 %
Sex and biological argument	Undergraduate	20	20.0 %
	Postgraduate	13	11.7 %
Feminism and post-structuralism	Undergraduate	8	8.0 %
	Postgraduate	51	28.0 %

Source: Survey results. Prepared by the authors.

Teacher training on climate change

Climate change education during schooling

With regard to school education about climate change, for item 3 we asked if the students had received lessons about this topic during their schooling. The results show that most of them had received training on the subject. Among future primary school teachers, 83.1 % (n=59) confirmed that they had lessons on the topic in comparison with 67.6 % (n=23) of future secondary school teachers (Tab. 4). These lessons were mainly given, in the case of the undergraduate students, in secondary school (47.4 %). The same can be noted for MAES students, of whom 50 % reported having received lessons on the topic

Tab. 4: Lessons on climate change during schooling

	Degree in Primary Education	Master's degree in Secondary Education (secondary + baccalaureate)
Item 3	Yes: (n= 59; 83.1 %) No: (n= 12; 16.9 %) Do not remember: 0	Yes: (n= 23; 67.6 %) No: (n= 7; 20.6 %) Do not remember: (n=4; 11.8 %)
Item 4	Primary Education: (n= 38; 33.3 %) Secondary Education: (n=54; 47.4 %) Baccalaureate: (n= 22; 19.3 %)	Primary Education: (n= 7; 18.4 %) Secondary Education: (n=19; 5 %) Baccalaureate: (n=12; 31.6 %)

Source: Survey results. Prepared by the authors.

during secondary school (Tab. 4). However, the difference with undergraduate students is that the training received at baccalaureate is greater due to the fact that MAES students go through this stage, while some undergraduate students get in through modules.

Teacher training and didactic proposals on climate change

In this part (section 5), dedicated to climate change education, the first question is related to how students perceive their ability to explain this topic in the classroom (item 15). The results obtained, both for undergraduate and MAES students, reveals an amalgamation of very diverse opinions (see Tab. 5). For example, 36.6 % (n= 26) of future primary school teachers feel they do not have the necessary skills, 32.4 % (n=23) believe that they do, and the remaining 31.0 % (n=22) don't know. As for future secondary school teachers, 44.1 % (n= 15) feel that they are qualified to teach the topic, 32.3 % (n=11) do not, and the remaining 23.5 % (n=8) do not know. We can draw various conclusions from these results: although the number of participants who feel unqualified is practically the same for undergraduate and postgraduate students (36.6 % and 32.3 % respectively), the data shows that there is a significant difference in terms of believing they are qualified to explain this topic (32.4 % versus 44.1 %). Although the sample is small and doesn't allow us to generalize, we can observe how the training received can have a certain impact, as MAES (History, Art History and Geography) students feel more capable of providing these explanations. The high percentage of undergraduate students who do not know or did not answer also stands out, representing almost a third of the sample.

Another question relates to whether they have received training on climate change during their university education or not (specifically during the undergraduate degree) (item 16). There was a range of answers among the undergraduate students: 49.3 % (n=35) confirmed that they had received lessons on the topic while the rest (50.6 %; n= 36) said they had not. As regards MAES students, the majority (almost 60 %) confirm that they were taught about the topic. Here, once again, their education has an impact on their answers, especially for history students whose science is highly interrelated with geographic content. As for item 17 – "is it worth the effort studying this content during schooling?" – the data indicates that 100 % of students (both undergraduate and postgraduate) agreed.

Finally, the aim of the last question was to identify which educational activities and proposals the future teachers would suggest to address the topic of

Tab. 5: Teacher training on climate change

	Degree in Primary Education	Master's Degree in Secondary Education (secondary + baccalaureate)
Item 15	Yes: (n=23; 32.4 %) No: (n= 26; 36.6 %) Do not know, no answer: (n=22; 31.0 %)	Yes: (n=15; 44.1 %) No: (n=11; 32.3 %) Do not know, no answer: (n= 8; 23.5 %)
Item 16	Yes: (n= 35; 49.3 %) No: (n= 36; 50.7 %) Do not remember: 0	Yes: (n= 20; 58.8 %) No: (n= 13; 38.2 %) Do not remember: (n= 1; 2.9 %)
Item 17	Yes: (n= 71; 100 %) No: 0	Yes: (n= 34; 100 %) No: 0

Source: Survey results. Prepared by the authors.

climate change in the classroom (item 19). As regards future primary school teachers (a total of 116 answers), they mainly highlighted activities related to field trips ("excursions") (n=30; 25.9 %) as well as research and looking for information (n= 27; 23.3 %) (Tab. 6). In third place, students proposed activities explaining the phenomenon of climate change in relation with the local surroundings. Although the number of responses to this effect is low (n= 13; 11.2 %), these proposals are interesting in order to provide meaningful learning, relating academic study to real-life situations so that pupils can see what is explained in the classroom. Traditional responses follow, including lectures, videos, documentaries, analyzing the news and so on. As for MAES students (a total of 41 responses), the main data showed proposals related to research and looking for information on climate change (n=18; 43.9 %). Here, once again, the factor of "education" plays a relevant role, as we must not forget that most of these students have a background in history where this sort of didactic proposal is more frequent. Field trips, for example, are only mentioned by 9.8 % of MAES students, perhaps due to the smaller proportion of students with a background in geography and due to the possibility of more excursions during primary school. Finally, it must be noted that 2 students (4.9 %) revealed that they did not feel qualified to explain the topic. In reality, people who understand the need and are aware of their limitations are in the right mindset to learn. We can positively evaluate these answers in the sense that these students are at the right starting point for learning, which also allows us to reflect on the educational model used to train future teachers.

Tab. 6: Didactic proposals to teach climate change (item 19)

Degree in Primary Education			Master's Degree in Secondary Education (secondary + baccalaureate)		
Proposals	**Nº**	**%**	**Proposals**	**Nº**	**%**
Field trips	30	25.9	Research/looking for information	18	43.9
Research/looking for information	27	23.3	News	6	14.6
Real situations	13	11.2	Real situations	5	12.2
Videos	12	10.3	Field trips	4	9.8
Lectures	10	8.6	Other	4	9.8
Workshops	7	6.0	Videos	2	4.9
Debates	5	4.3	Don't feel qualified	2	4.9
News	5	4.3			
Other	7	6.0			

Source: Survey results. Prepared by the authors.

Discussions and conclusions

Through the exploratory analysis of this research regarding gender, we can confirm the validity of the hypothesis which identified gender education as a fundamental variable. The comparison of the responses given by the two groups of participants – students taking the Degree in Primary Education and postgraduate students with a degree in History or Art History – revealed a number of considerable differences as a result of this initial training.

The observations made by the undergraduate students, who suffer from a lack of any gender education, reveal that they do not master specific terminology, they do not mention gender roles or stereotypes and they are not aware that gender is contingent and related to culture. They are also unable to define gender, beyond understanding it as a way of classifying, identifying and differentiating between men and women according to a number of characteristics related to sexual identity. They do this in a very generic way, in line with the acceptance of social norms using a classification system that they adapt, as explicitly mentioned in some cases, from literature, music or art. Conversely, the responses provided by the postgraduate students introduce new elements. Several participants deny the biological argument, instead contending the fluidity of gender and underlining its contingency. They explain gender in terms of power relations, linking it to culture and defining it as a social, historical and/or cultural construction.

Studies such as those by García-Pérez et al. (2011) reveal similar results to those obtained in this piece of research. They identify a lack of basic gender knowledge among university students, linking this deficiency with a set of educational necessities which, in their opinion, should try to prevent the ignorance of professionals whose future work will involve educating new generations. Jaramillo (1999), Anguita (2011), Rebollar (2013) and Ibeas and Garrido (2018), among others, denounce the lack of gender education among future teachers, a direct consequence of the lack of specification in recent educational laws in Spain and the so-called "trap of gender mainstreaming" (Anguita & Torrego, 2009; Donoso & Velázquez, 2013; Rebollar, 2013).

As a result, we can conclude that undergraduate students see gender from the perspective of hegemonic social representations which naturalize power relations between men and women and conceal the social construction of these relationships. On the contrary, postgraduate students present an understanding of gender which is not based on hegemonic social representations, but rather on the education they have received at university. With regard to climate change, the first hypothesis was not substantiated as most participants (both undergraduate and postgraduate) attested that they had received lessons on the topic during their schooling. However, the second hypothesis (proposals) was corroborated, given that the activities that students proposed for study in the classroom were not very innovative nor did they present any questioning of the issue. In this vein, teaching the causes and consequences of climate change using the perspective of the pupils' immediate surroundings would support pupil engagement to better interpret and understand this phenomenon which is overshadowed by the media. This better understanding would help to raise awareness and educate citizens who would be less vulnerable to the phenomenon.

One method to ensure that pupils have a more real understanding of climate change issues would involve explaining the topic using information shared in the academic sphere and contrasting it, for example, with the information provided by the media. This approach is being studied by the *Geographical Association* in primary schools in the United Kingdom, where for several decades they have been analyzing case studies which use research about natural disasters and climate change for classroom material (Greenwood, 2018).This is a way of surpassing ordinary knowledge and the resulting social representations.

Vulnerability is a variable of considerable importance in order to tackle the current and future realities of climate change. It tends to be a forgotten factor, as explained by Morote and Pérez (2019), and, in most cases, is the main variable in asserting natural disasters, which is why school education on the topic is so important. In comparison with other studies, González & Maldonado's study (2014) on social representations and climate change among university students stands out in the American sphere (Mexico). The authors explain that scientific information shared by various sources, both scientific and popular, does not have a sufficient impact on young people to lead to changes in behavior aimed at putting the brakes on this phenomenon. Furthermore, they contend that new communication and environmental education strategies aimed at pupils are needed to promote environmentally friendly behavior.

Meanwhile, López and Oller (2019) explain that there is fairly ambiguous conceptual knowledge, influenced by the media, in relation to environmental problems, as well as a limited capacity to propose didactic activities which encourage necessary social change and holistic and comprehensive environmental education. Morote, Campo and Colomer (2019) noted that the main sources (about climate change) that students on the initial teacher training course for primary education use are the media. This is thus dangerous given the impact that the media has on our current society with so-called "fake news" (Brisman, 2018), as well as the limited capacity for conceptual interpretation shown in the analysis of information received.

With regard to the activities suggested by teachers in training, Souto, Morote and García (2019) mainly highlight activities involving how to tackle flooding (emergency protocols), analyzing past events and going on field trips. However, a model to explain hydrographic systems, which could help pupils understand the phenomenon of flooding, was not observed, thus making understanding of these events more difficult. Furthermore, the students' proposals put too much emphasis on tackling these natural disasters from the perspective of the consequences (once the disaster has happened) rather than the causes (climatic or worsened by human activity).

Starting with education, pupils should be taught to develop a critical perspective about the interpretation of information, thus giving them a well-founded understanding of gender and climate change issues. Excluding gender from the field of social studies implies not only deficient education of citizens, but also a lack of understanding of the concealed power systems which shape our world (Crocco, 2006). If we accept that the term gender is key to contemporary feminists' approach to assert a specific defining territory (Aguilar, 2009), the difficulties revealed by many teachers in training to define gender appear to be particularly worrying.

Meanwhile, in the field of climate change, teacher training is faced with the considerable challenge of cultivating pupils' critical perspectives, so that they: (1) are able to interpret the territory; (2) possess the tools needed to explain this problem in relation to vulnerability and sustainability, as well as the involvement of citizen participation and (3) prefer consulting academic texts, casting aside (or contrasting with), insofar as possible, certain information provided by the media. All of this is of greater interest if we take into account the fact that climate change is included in the curriculum for primary and secondary education, as well as the responsibility to educate the youngest members of society and the challenge that climate change represents as one of the greatest socio-environmental problems of the 21st century.

References

Ahmad, Saima/Numan, Sharker: "Potentiality of disaster management education through open and distance learning system in Bangladesh open university". *Turkish Online Journal of Distance Education* 16 (1), 2015, pp. 249–260.

Aguilar, Consol: "¿Por qué es importante el género en la pedagogía crítica?" *Revista interuniversitaria de formación del profesorado* 64, 2009, pp. 121–138.

Alaminos, Antonio/Castejón, Juan Luis: *Elaboración, análisis e interpretación de encuestas, cuestionarios y escalas de opinión*. Universidad de Alicante: Alicante 2006, 119 pages.

Anguita, Rocío: "El reto de la formación del profesorado para la igualdad". *Revista Electrónica Interuniversitaria de Formación del Profesorado* 14 (1), 2011, pp. 43–51, accessed 11.6.2020, DOI 10.14198/fem.2013.21.07.

Anguita, Rocío/Torrego, Luís: "Género, educación y formación del profesorado. Retos y soluciones". *Revista interuniversitaria de formación del profesorado* 64, 2009, pp. 17–26.

Álvarez, Carmen/San Fabián-Maroto, José Luis: "La elección del estudio de caso en la investigación cualitativa". *Gazeta de antropología* 28 (1), 2012, pp. 1–12.

Borg, Simon: "Teacher cognition in language teaching. A review of research on what teachers think, know, believe, and do". *Language Teaching* 36, 2003, pp. 81–109, accessed 1.6.2020, DOI 10.1017/S0261444803001903.

Brisman, Avi: "Representing the 'invisible crime' of climate change in an age of post-truth". *Theoretical Criminology* 22(3), 2018, pp. 468–491.

Calvo, Francisco: *Sociedades y territorios en riesgo*. Ediciones del Serbal: Barcelona, Spain 2001.

Crocco, Margareth Smith: "Gender and social education: What's the problem?". In: Ross, E. Wayne (ed.): *The social studies curriculum. Purposes, problems, and possibilities*. State University of New York Press: Albany 2006, pp. 171–193.

Domingos, Moisés/Lira, André Augusto Diniz: "Representações Sociais, Sens Pratique, Poder Simbólico e o Processo de Construção do Ser Docente". *Arxius de Ciències Socials*, 41 (1), 2019, pp. 43 - 64.

Donoso, Trinidad/Velázquez, Anna: "¿Por qué una propuesta de formación en perspectiva de género en el ámbito universitario?" *Profesorado. Revista de currículum y formación del profesorado* 17 (1), 2013, pp. 71–88.

Eklund, Sheri: Climate change education with a bright horizon? Pedagogical reflections on teacher training for climate education that aims to empower students (Stockholm University, Sweden) (doctoral thesis). 2018.

García-Pérez, Rafael et al.: "El patriarcado no es transparente". *Competencias del profesorado para reconocer desigualdad. Cultura y educación* 23 (3), 2011, pp. 385–397, accessed 7.5.2020, DOI: 10.1174/113564011797330298.

González, Edgar/Maldonado, Ana Lucía: "¿Qué piensan, dicen y hacen los jóvenes universitarios sobre el cambio climático?: Un estudio de representaciones sociales". *Educar em revista* 3, 2014, pp. 35–55, accessed 9.6.2020, DOI 10.1590/0104-4060.38106.

Greenwood, Henry: "A climate change assembly". *Primary Geography* 96, 2018, pp. 22–23.

Hahn, Carole L.: "Gender and political learning". *Theory and Research in Social Education* 24 (1), 1996, pp. 8–36.

Ibeas, Nieves/Garrido, Begoña: *La perspectiva de género en la educación aragonesa*. Universidad de Zaragoza: Zaragoza 2018, 61 pages.

Jaramillo, Concepción: *Formación del profesorado: igualdad de oportunidades entre chicas y chicos*. Ministerio de Trabajo y Asuntos Sociales – Instituto de la mujer: Madrid 1999, 61 pages.

Jodelet, Denise (ed.) : *Les représentations sociales*. Presses Universitaires de France: Paris 1989, 454 pages.

López, José Antonio/Oller, Montserrat: "Los problemas medioambientales en la formación del profesorado de educación primaria". REIDICS: Revista de Investigación en Didáctica de las Ciencias Sociales 4, 2019, pp. 93–109, accessed 19.5.2020, DOI https://doi.org/10.17398/2531-0968.04.93.

Martínez, Luis Carlos/Olcina, Jorge: "La enseñanza escolar del tiempo atmosférico y del clima en España: currículo educativo y propuestas didácticas". *Anales de Geografía de la Universidad Complutense* 39 (1), 2019, pp. 125.148, accessed 15.6.2020, DOI 10.5209/AGUC.64680.

Martínez, Ramón/López, José Antonio: "La enseñanza de la climatología en los manuales escolares de Ciencias Sociales en Educación Primaria". In Sebastiá, Rafael/Tonda, Emilia (ed.): *La investigación e innovación en la enseñanza de la Geografía*. Universidad de Alicante: San Vicente del Raspeig, Spain 2016, pp. 245–258.

McWhirter, Nathan/Shealy, Tripp: "Case-based flipped classroom approach to teach sustainable infrastructure and decision-making". International Journal of Construction Education and Research 16 (1), 2018, pp. 1–21, accessed 15.6.2020, DOI 10.1080/15578771.2018.1487892.

Morote, Álvaro-Francisco: "La enseñanza del cambio climático en la Educación Primaria. Exploración a partir de las representaciones sociales del futuro profesorado y los manuales escolares de Ciencias Sociales". *ENSAYOS, Revista de la Facultad de Educación de Albacete* 34 (2), 2019, pp. 213–228.

Morote, Álvaro-Francisco/Campo, Benito/Colomer, Juan Carlos: "La percepción del cambio climático en los futuros docentes de Educación Primaria. Una experiencia de conocimientos previos a partir de la enseñanza de las Ciencias Sociales". In: AGE and Universidad de Valencia. *Crisis y espacios de oportunidad. Retos para la Geografía*. Universidad deValencia: Valencia, Spain 2019, pp. 106–120.

Morote, Álvaro-Francisco/Pérez, Alfredo: "La comprensión del riesgo de inundación a través del trabajo de campo: Una experiencia didáctica en San Vicente del Raspeig (Alicante)". *Vegueta. Anuario de la Facultad de Geografía e Historia* 19, 2019, pp. 609–631.

Moscovici, Serge: *La psychanalyse, son image, son public*. Presses Universitaires de France: Paris 1978, 512 pages.

Olcina, Jorge: "Cambio climático y riesgos climáticos en España". *Investigaciones Geográficas*, nº 49, 2009, pp. 197–220.

Olcina, Jorge: "La enseñanza del tiempo atmosférico y del clima en los niveles educativos no universitarios. Propuestas didácticas". In Sebastiá, Rafael/Tonda, Emilia (eds.): *Enseñanza y aprendizaje de la Geografía para el siglo XXI*. Universidad de Alicante: Alicante, Spain 2019, pp. 119–148.

Rausell, Helena/Talavera, Marta: "Dificultades de la coeducación en la formación del profesorado". *Feminismos* 29, 2017 pp. 329–345, accessed 1.5.2020, DOI 10.14198/fem.2017.29.13.

Rebollar Sánchez, Edith Mariana: *El género en los planes de estudio de los grados de educación de las universidades públicas españolas* (doctoral thesis). Universidad Autónoma de Barcelona: Barcelona 2013, 187 pages.

Ruiz, Antoni: "Método de encuesta: construcción de cuestionarios, pautas y sugerencias". *Reire. Revista d'Innovació i Recerca en Educació* [online] 2009, pp. 96–110.

Simons, Helen: *Case study research in practice*. SAGE: London 2009, 189 pages.

Souto, Xosé Manuel, Morote, Álvaro-Francisco/García, Diego: "Crisis y riesgos naturales en la educación social. El caso del riesgo de inundación en Educación Primaria". En AGE y Universidad de Valencia. *Crisis y espacios de oportunidad. Retos para la Geografía*. Universidad de Valencia: Valencia, Spain. 2019, pp. 171–185.

Universidad de Valencia: *Anuario estadístico*, accessed 05.06.2020 at https://webges.uv.es/MS10/servlet/mstrWeb?evt=3010&Server=MSTRATEGY&Project=RECULL&. 2020.

Vavrus, Michael: "Sexuality, schooling, and teacher identity formation. A critical pedagogy for teacher education". *Teaching and Teacher Education* 25, 2009, pp. 383–390.

Carlos Fuertes, Elvira Asensi, Carlos Fuster, and Sérgio Claudino[1]

Chapter 11 Socially acute questions and critical citizenship in trainee geography and history teachers: From theory to classroom observation

Abstract: This chapter aims to delve into the relationship between theoretical knowledge about "socially relevant" topics or issues provided by university, and the eventual addressing of these topics in practice during the teacher's first experiences in the geography and history classroom. In this sense, the authors expects to identify the complex relationship that will cause changes and continuities in teachers-in-training's perceptions and praxis; and to pinpoint its difficulties and possibilities, that will allow to promote a critical citizen education and to improve teacher training itself. The sample of this research is made up by 28 Master's degree in Secondary Education Teaching students (major in geography and history) and their internship tutors, who are working geography and history teachers. The sample will be completed in the future with multiple Portuguese geography teachers. The information gathered by means of different tools (observation guide, internship report, classroom diaries, quizzes, didactic proposals designed and implemented by the teachers-in-training, teachers-tutor interviews) will be organized in different conceptual categories, namely citizen participation or addressing of socially relevant topics, according to previously designed rubrics.

Keywords: History education, Critical citizenship, Geography education, Teacher education, Classroom observation

Teacher training, curricular models and critical citizenship

Geography and History (GH) teacher training is a common topic of discussion in social sciences teacher education and is understood in many different ways. At least, in as many ways as there are ideal models of curriculum, GH teacher and citizen. Bearing in mind relevant aspects as the selection and justification of content, the role of teachers or the desired model of citizenship and purposes, it has been proposed a classification between three models of

1 The authors participate in the project "Las representaciones sociales de los contenidos escolares en el desarrollo de las competencias docentes" (Ref: PGC2018-094491-B-C32), of the Ministerio de Ciencia, Innovación y Universidades (Government of Spain), co-funded by the ERDF (EU).

curriculum: technical, practical and critical (Rozada 1997). These three ideal models often appear combined in teaching practice, but their different conceptualization is useful both in the initial stage of GH teacher training and in the framework of this research.

The *technical model* is related to traditional GH teaching methods. Focusing on the effectiveness of the teaching-learning process, it is based on the transmission of a wide corpus of conceptual knowledge – often factual or descriptive knowledge – through one-way lectures, the uncritical use of the textbook and passive learning activities. It lies on a supposed objectivity of a school knowledge that is not challenged and generally implies the dissemination of classic narratives, hardly problematized, linked to scholarly, nationalistic purposes and the reproduction of inequalities and social issues. In this way, an "enlightened" and "patriotic" citizen is pursued or promoted (Tadeu da Silva 1999), in line with modern nation-states and a restricted conception of citizenship, in favor of a "personally responsible citizen" (Boyle-Baise 2003), more concerned about uncritically fulfilling duties than exercising or claiming rights.

The *practical model* emerged as a result of the development of psychopedagogy according to Piaget's theory and has become increasingly important since the last third of the twentieth century. Based on a discovery-learning model, the focus shifts from teacher to student, prioritizing their acquisition of skills and competencies, as well as their cognitive development. According to the importance of students in this competence approach, the practical model would promote a "participatory citizen" model (Boyle-Baise 2003). However, the little attention to the selection of conceptual content and, with it, to the analysis and deconstruction of social issues, identities and inequalities, fosters a "competitive person," in line with a capitalism demanding useful skills for the labor market and not for the exercise of a critical citizenship (Tadeu da Silva 1999; Fernández Liria/García Fernández/Galindo Ferrández 2017: 114).

Finally, the *critical model* – the model to follow here – draws on critical pedagogy and pursues the main goal of creating *critical citizens* who are able to understand, challenge and transform the social world (Apple 1979); of making the student a person "with distrustful and critical judgement on the current social dynamics" (Tadeu da Silva 1999); that is, a "justice-oriented citizen" (Boyle-Baise 2003), someone who participates in the public sphere, turning criticism into action in order to solve social and environmental issues and injustices of our time. Thus, a GH teaching promoting critical thinking through an explanatory, analytical, multifactorial, problematizing and open approach is advocated (Jenkins 2009; Seixas/Morton 2013). Although it shares the prominence given to students and competence development with the practical model, it focuses on the conception of the teacher as engaged in contextualization and as a transformative intellectual (Giroux 2015), capable of critically reading the curriculum and selecting and problematizing content. Certainly, from a critical perspective, the central question is to know what knowledge should be taught and, above all, the purpose of such selection.

Epistemological approaches linked to the "new history" are put at the centre – history from below, everyday history, cultural history, gender history, postcolonial studies, and so on – and the "new geographies" or "other geographies" – radical geography, perception geography, feminist geography, urban studies, and so on – (Iggers 1997; Harvey 2001). Perspectives that focus – in a non-stereotyped or altered way – on the difficulties, perceptions and agency of subaltern groups – working classes, women, peasants, dominated cultures, and so on –. Perspectives that prioritize the denaturalization of certain deeply rooted narratives in the educational and social sphere despite having been questioned by historical and geographical research. Hence, they focus on the deconstruction and analysis of conflicts, inequalities and exclusive and essencialized identities, often establishing explicit connections between the topics studied and the socio-environmental issues of the present. In short, approaches to GH teaching that prioritize attention to "open-ended questions" or "socially acute questions" (SAQs) (Legardez/Simonneaux 2006), which have greater reflective and transformative potential, in coherence with the objective of making critical citizens.

Placing ourselves in a critical perspective and facing the technical or practical conceptions of *teacher training*, we understand that the initial training of prospective GH teachers must go beyond the apparently neutral objective of training in techniques and resources to be more "effective" in the transmission or acquisition of a content that is not subject to discussion. Of course, the importance of a disciplinary, methodological and psychopedagogical training is recognized. However, it is considered that a solid epistemological, ethical and sociological training is absolutely essential, which places reflection on content selection and GH teaching purposes at the center. Only in this way will prospective teachers understand the problems of teaching in the broadest sociopolitical and cultural contexts and will be able to promote equality, critical thinking, a democratic public sphere and the making of true critical citizens.

From a critical perspective, it is also understood that in the initial training of teachers it is essential to analyse and deconstruct their representations about GH teaching, since these will really condition their own teaching practice (Adler 2008; Parra 2019). And precisely that is the objective of this article, focused on assessing the representations of a group of GH teacher students after a theoretical training carried out by one of the authors of the article through the subject

"Learning and Teaching Geography and History," which tackles the analysis of the curricular models presented and inspired by a critical perspective. Thus, we will study whether prospective teachers internalize this training and whether they are capable of implementing it to the classroom observation and reflection on the teaching practice of their tutors during their first internship in a secondary education center (Stenhouse 1975). More specifically, the focus will be on the ability to implement the training received on content selection and approach, attending to SAQs and promoting the making of critical citizens.

Method

Research approach

This research is part of a qualitative research model through an ethnographic approach to education (Rockwell 2009; Bagley/Beach/Silva 2018) that tries to tackle the representations of trainee teachers about the daily practices of employed GH teachers. Starting from the reflections and documentation gathered from the classroom observation carried out by GH teacher students, we will deepen the relationship between theoretical knowledge and practical training, assessing the way in which previous representations and notions of prospective GH teachers are undermined or reinforced within the framework of this experience (Pinheiro dos Santos 2019).

Participants and context

A non-probability sampling (opportunity sampling or convenience sampling) was used for research, since prospective GH teachers in Secondary Education were deliberately selected. Specifically, the research was conducted with students of the Master's Degree in Secondary Education Teacher (MAES), specialty Geography and History, of the University of Valencia (Spain), during the 2019/20 academic year. Being in the last stage of the initial training and during and after the internship period were criteria for sampling. The participants of the study were twenty-eight, and the selection tried to guarantee a plural and representative profile out of the total of seventy-nine enrolled students in the Master's. Thus, the sample presents a gender balance (53 % female and 47 % male), averaging twenty-six years of age. Regarding previous training, 54 % were graduates in History (n=15), 35 % were graduates in Art History (n=10) and the remaining 11 % were graduates in Geography and other related degrees (n=3). Likewise, 39 % of the participants had another specialization Master's degree in those disciplines (n=11).

Internships were available in twenty-two secondary and high schools: 77 % were state centres, while the remaining 33 % were private or semi-private centers (religious and seculars). Regarding geographical location, 40 % were located in the city of Valencia, 36 % in the metropolitan area of the city, and 20 % in other municipalities of the province. Regarding tutors – teachers who accompanied the students during the practices in schools, whose teaching practice was subjected to observation –, they were twenty-three. They were also balanced by gender (56 % female and 44 % male), averaging fifty-one years of age and twenty-two years of teaching experience. In terms of academic education, 72 % had a bachelor's degree in History, 8 % in Art History and 20 % in Geography or other disciplines. Most of them were teaching both in compulsory secondary education (12–16 years) and in post-compulsory high school (16–18 years). Regarding pedagogical training, all of the tutors had a post-graduate certificate in education (CAP), an intensive short three month course that contrasts with a demanding official master's degree of 60 ECTS credits currently took by devoted students throughout a full academic year.

Instrumentation and data collection

Data collection has been carried out using several useful tools to analyse students' ability to reflect on tutors' teaching practice based on the recording of daily observation in classrooms and the collection and production of different documents (Moon 2004). First, the *field notes or diaries*, documents in which the participants committed their daily thoughts on what had happened in the classroom to paper, have allowed us to know what the students focused their attention on during the observation period. The diaries analysed, for which no guide was provided, presented a format in which detailed descriptions of simple classroom situations predominated, along with specific reflections (Zabalza 2004).

A second tool used was the *internship reports*. These documents, produced after the end of the internship period, acquire a much more analytical and reflective format on the part of the students and, therefore, become a fundamental source in this research. An orientation guide was provided to elaborate the documents: twenty-page limit and different topics to necessarily reflect on. The analysis of tutors' teaching practice was here the main point, while an attentive analysis of both the planning and the implementation of the classes was requested.

A third and final tool used were *tutor-created materials*, attached to the reports, providing rich information to better contextualize and analyse the students' reflection (González 2018). Among the resources, there were exams, worksheets, teaching units, multimedia presentations or assessment rubrics for the tutors to quantitatively and qualitatively assess the student work during the internship.

For the analysis of the different tools for collecting data, both manifest and latent contents and silences were considered, favoring the analysis of reflections focused on the detection of critical practice by tutors. The information was treated considering three categories. Thus, we processed and classified the information according to the conceptualization and assessment of the tutor; to the relative importance students gave to methodology and content in their analysis; and, finally and more extensively, to the specific reflections they made on content selection, connection with SAQs and contribution to the making of a critical citizenship. In the framework of the analysis, we also assessed the incidence of several differential variables, particularly the background knowledge of the students.

Results and discussion

Methodological reductionism

Analysing the understanding and ability to apply theoretical knowledge to the observation and examination of teaching practice, a first element to assess was the conceptualization of the practice tutors and the relevance given, beyond methodology, to content. Specifically, we are interested in knowing how important the extent of the connection between content selection and approach to SAQs is for the students when a tutor follows a particular curricular model, and if it allows the making of critical citizens. As shown in Tab. 1, in this regard, a balance was found between those students who essentially relied on methodological aspects to conceptualize their tutors (14) and those who also focused on content selection and approach (14).

A first conclusion is that the initial methodological reductionism showed by all the MAES students persist, although attenuated, after the internship among half of the sample – specially among graduates in Art History, Geography and others. That is to say an almost exclusive concern for methodology and the understanding of "innovation" or good teaching practice as something essentially related to methodology and resources. This reductionism was verified in the first theoretical sessions through several activities that showed results

Tab. 1: Degree of the students, dimensions considered and conceptualization of tutors

degree taken by students		dimensions considered in the conceptualization of tutors		conceptualization of tutors by students	
History	5	Methodology	14	Technical or traditional	10
Art History	6	*("Methodologist" Students)*		Practical or technical-practical	4
Geography and other	3			Critical or technical-critical	-
History	10	Methodology and content	14	Technical o traditional	3
Art History	4	*("Critical" students)*		Practical o Technical - practical	2
Geography and other	-			Critical o Technical-critical	9

coinciding with other investigations on the representations of teachers in initial training, closely related to their school memories and their expectations about the master's degree and the teaching profession (Parra 2019: 79–86). After the internship period, their persistence is clearly traceable in diaries and reports, both in the long analysis of the methodology and in the predominance in the main arguments and emphasis used for the conceptualization of tutors and the curricular model they follow.

First, some examples allow us to deduce that, in the case of the students called "methodologists" for research purposes, even the classroom observation was focused on these aspects, as the following fragment of a diary shows: "Mainly I pay a lot of attention to resources used for the classroom dynamics [...] how the teacher managed the class [...] what attention-getting strategies the teacher used" (S17). Second, even if specific examples of attention to content are provided, these students often showed a poor understanding of the critical perspective regarding the treatment of SAQs. This is the case of a student who, focused on methodology, presented in the art history class a mere anecdotal and uncritical reference to a current news story on the restoration of the "Ghent Altarpiece" as an example of the "usual" treatment by her tutor of "what we know from SAQs" (S19).

Third, a relevant issue is that students focusing exclusively on methodological aspects did not detect features of the critical model in the observed teachers, with most tending to classify their tutors as "technical" or "traditional" without any clarification (ten out of fourteen). To this end, some elements – note: not specific to GH – stood out, such as the use of one-sided lectures, the centrality of textbooks, low cognitive activities or the little attention paid to students' interests. In relation to this, a negative vision of tutors' teaching practice was appreciated – that has been previously pointed out for the prospective history teachers in Argentina –, regretting a simplistic and little nuanced practice, focused on the most obvious aspects or on issues related to school inertia: "again and again they refer to teachers' practices by evaluating (negatively, most of the time) and not by interpreting; by inspecting what is repeated, and not what subtly changes or transforms, by judging what is 'routine' and 'tradition'" (González 2018: 26).

Along these lines, some indications in the analysed reports suggested that the conceptualization as "technical" or "traditional" without further clarification could be, in some cases, a consequence of a somewhat rigid and stereotyped insight among these students, perhaps due to prejudices related to age – fifty-one years on average – or to less pedagogical training received by tutors.[2] Of course, it is common to find partial and undeveloped arguments and, particularly, with little attention to content selection and approach.

A student, who qualified her tutor as "technical" based on methodological aspects, reinforced her argument by attaching an exam and questioning the methodological approach – "it prioritizes memorization." However, in a clear example of how indicative silences in reports are, she obviated some content that, together with typical elements of historicism, had a greater critical and reflective potential, such as cause and effect activities or focused on gender inequalities in Ancient Greece (S2). Sometimes, although there were teachers – apparently – methodologically close to a technical model using content with critical potential, this happened very occasionally and was not considered a relevant element to refine the conceptualization. This is the case of a student whose only allusion to content – "focused on making students reflect on historical events and the current impact of these events" – is mentioned without evaluation and ignored when she undeniably places her tutor into a "traditional" model (S17).

Focusing on socially acute questions and identifying critical practice

Beyond this first group, it should be emphasized that the other half of the students observed – who could be considered as "critical" due to their coherence

2 In any case, it is important to emphasize that there is no correlation between age or pedagogical training of tutors and their consideration under one model or another by students.

with the theoretical courses they had taken – did pay balanced attention to the different aspects involved in GH teaching-learning process. These students, among which there were many graduates in History, without ignoring methodological issues, emphasized in their analysis the way in which their tutors selected content and approach in a more or less coherent way in relation to SAQs and the making of critical citizens. It was also revealing that their classification showed a clear contrast with that carried out by the methodologists, as observed in Tab. 1.

Thus, a reversal of trends was detected, with a less tutors considered without clarification under a "technical" or "traditional" model (3) and more typical features of the "practical" model (2) and, especially, the "critical" model (9), in any case predominating the mixed figures, halfway between the technical model and the practical or the critical model. This was a more nuanced and plural representation, fitting with research on current teaching practice (González 2018). It would seem, in this sense, that the students who had best internalized the theoretical knowledge and, therefore, were further away from a methodological insight, were those more able to read between the lines during classroom observation and to reflect later on, getting over rigid, stereotyped and negative opinions about employed teachers. Their reflections, more complex and with more solid and explanatory arguments, are key to identify both teachers' good practice and limits and difficulties in the content selection and approach from a critical perspective, focused on SAQs.

First, when these "critical" students considered their tutors under a technical model, a practical model or an intermediate path between the two, they focused not only on methodological issues but also on the absence of a coherent critical perspective in terms of content selection and approach. In this sense, their conceptualization was more solid and convincing than that carried out by the methodologists. A student who highlighted that "lecture explanations took up most of the session time" offers a good example of the usual class dynamics. Those explanations were based exclusively on a textbook with activities "that involve the simple repetition of knowledge." But, at the same time, she considered her tutor under a technical model by taking into account the predominance of vague description, incomplete problematization and lack of attention to SAQs when choosing content and approach. In an example of the way in which greater curricular knowledge – in this case, in art history – encourages more incisive analyses on the treatment of content, she regretted the predominance of formal and stylistic analysis in art, without a historical contextualization that would provide reflection on society: "for example, it was said that the Gothic altarpieces fulfilled a decorative function and they were hanging mainly on palace walls. The importance of this art piece in relation to a belief in purgatory and the destiny of souls after death in general was not considered" (S28).

Second, students with a more global perspective were the ones who, to a greater extent, identified the use of critical practice in relation to the problematization of content and attention to SAQs when choosing content and approach. Thus, although her tutor was "quite classical, as she followed the textbook," the student added later that "we could consider her critical to a certain extent since she tried to relate the subject to current issues and tried to go gradually creating a critical consciousness in students when it comes to perceiving the world they live in." Among the numerous examples she pointed out to convey her view, she highlighted that, when dealing with Ancient History, "slavery in Rome was related to modern slavery and she asked the students if they thought the situation was fair or unfair," comparing the existing "socio-economic and gender differences between Ancient Rome and modern society" (S24). This and other cases of clarification derived from an in-depth analysis of both the methodology and the content, suggest, as indicated above, that a part of the tutors considered exhaustively and negatively as "technical" by the methodological students are probably teachers more multifaceted that assume critical elements in their practice, to a greater or lesser extent.

Third, the positive interpretation that critical students assumed about those tutors who approached the content coherently connecting with SAQs, did not lead them, however, to sweetened or uncritical views regarding their teaching practice. They frequently showed the predominance of combined practices, questioning the way in which the teaching methodology and particularly the assessment criteria, closer to a technical model, limit or hinder the impact of a critical approach to content. Thus, a student underlined as the first argument to consider her tutor under a critical model "the constant problematization of the issues studied, such as the ethics of colonization." These were addressed "using reflective questions, debates and trying to constantly interact with the students" in such a way that "he tried to make them think for themselves." However, relying on activities and tests made by her tutor and attached to her report showing the predominance of short, rote answers, she pointed out that "it would be positive if the degree of critical reflection created in the classroom environment during the lessons moved on to activities and tests" (S1).

Very similar observations are made by other students who rightly pointed out the importance of the exam as a rote learning memory test, exclusively based on the textbook, and a barrier to critical education (Merchán 2001): "the teacher intended to present a more critical history than the textbook," but "the students only thought about grades and they cared very little about what he said if it was not going to be on the test" (S27).

Geography and citizenship education

The students giving attention to "critical" practice focused on several specific aspects related to content selection and approach, connecting with SAQs and promoting the making of critical citizens. In fact, many students highlighted the way in which the observed teachers exploited, to a greater or lesser extent, the potential of geography content "to educate for citizenship" (Souto 2010).

Often surprised by this, graduates in history and art history tend to consider that "the subject itself" enabled these connections "when dealing with topics such as economics, demography or climatology." These subjects have a critical approach that "may seem more obvious," because the main content of the geography curriculum is largely understood as SAQs. However, both the problematization of these topics that could be developed from neutral approach, and the link of other "less obvious" content to SAQs were underlined. It was the case of a tutor who, in the "soil" unit, highlighted "the relationship between siliceous soil and the problems in coal mines of Asturias and the conflicts that have occurred since then" (S24).

In the same way, they questioned the insufficient use of SAQs with a critical potential in the geography curriculum. Thus, for example, a student appreciated that her tutor had chosen, to deal with the subject "energy sources," an "extremely interesting" documentary on the so-called planned obsolescence. On the contrary, she regretted that the reflective potential of the documentary was hampered by the teacher's inaction and the uncritical use of the source. By merely "playing the whole documentary," without pauses, questions, contextualization or further analysis activities, the students showed "lack of interest" and only a few "paid attention to the video." A confirmed pattern of inhibition and substitution of the teacher for audio-visual aids that is especially questionable in SAQs (González 2014: 165), since they certainly limit the contribution of such content – and sources – to critical thinking (Burke 2001).

Expanding subjects (and their limits) in history education

On the other hand, all students focused on the content taught paid attention to subject expansion. This is particularly noticeable in history topics, in which critical teachers would be promoting a more inclusive history education than the traditional mainstream one. In the first place, women emerge strongly, appreciating both the trend towards their incorporation into the teaching discourse and the shortcomings (Díez/Fernández 2019). A student with a very positive image of her tutor, whose characteristics "would point to the critical model," valued her tutor's particular attention to "the situation of women" in different history and geography topics. However, in her classroom diary, she questioned his uses of language, reproducing typical topics of an androcentric history regarding prehistory: "today I did not like that when he was talking about human evolution and mentioned 'the first man' he said: 'well, we say man, so we do not always have to refer to man and woman'. He did not problematize the issue and hinted that it is normal to do so" (S1).

Two other students, who shared a tutor, positively assessed his attention to "current social issues," explaining that he "did not avoid conflictive topics" and had not used a textbook for years, since making his own class notes. At the same time, they highlighted that a work group made up by the tutor and other teachers reviewed the materials in regular weekly meetings, drawing the conclusion that it was necessary for the notes to "focus more on the role that women have played in history" and in "the historical construction and operation of power relations," including gender (S20, S21).

Second, a particular focus on the problems of the working classes and the most excluded or invisible social sectors was also observed in several cases as a feature of teachers considered critical. Thus, a student who stated that his tutor "always told me that the educational purpose was to think critically about present-day issues and the society we live in," illustrated his tutor's teaching practice with a detailed description of a class on European feudal society for 13–14 year old students:

> There was a pyramid scheme in the textbook, but he made his own [on the blackboard] by adding a sector of population that did not appear in the book, made up of outcasts, criminals, orphans, widows, bandits, vagrants, and so on. In other words, he integrated those minorities excluded from feudal society into the explanation, minorities who were on the fringes of the law and, in short, ignored in the textbook. Thus, he reversed the explanation and placed greater emphasis on these most disadvantaged groups. Consequently, the response of the students was great, since they began to ask many questions about this social group, as why they were excluded or why they were not considered important enough to be in the textbook. In addition, the teacher, solving doubts, put it in relation to the exclusion of many minorities nowadays and the problem it raises. Lastly, my tutor left the classroom saying, "it is worth being a teacher for days like this" (S8).

This also shows – as underlined in other reports – how some "critical" teachers use textbooks as a source that can be problematized, questioning the narratives, connecting past and present and thus promoting critical thinking in students.

Third, other students underlined the greater visibility of ethnic, religious or cultural "others." And, in relation to this, they also pointed out the articulation of a less Eurocentric and ethnocentric history and a more self-critical history taught by some teachers regarding the colonial past than that which predominates for textbooks and social imaginaries (Maroto 2017). One of them stood out, for example, that the teacher started the unit on Imperialism and the First World War "by asking the students why the first world powers of our world are mainly concentrated in the West," with the aim that they "reflect on the consequences of the current looting of other continents by Western countries" (S12). However, some problems were also pointed out in this task of making "the others" visible, particularly the risks of stereotyping and othering.

For example, this is noticed regarding an annual project on Africa in a state high school with a large number of Afro-descendant students. These students "had been excluded from the project since the beginning and, at least, should have been asked for advice on certain controversial decisions." "Although the idea was good" in trying to make the African continent visible, the activities, such as a carnival in which they had to disguise as "African tribes," conveyed a deeply reductionist and "excessively negative perception about Africa (hunger, wars, epidemics, and so on), without any criticism of neo-colonialism." This caused discomfort among many of the students (S15). Faced with the inhibition of the head teacher's team and the tutor, this and other prospective GH teachers in training reacted holding a "Workshop on Africa" to problematize and socio-historically contextualize the stereotypes that had come to light. Likewise, this is a magnificent example of how critical observation can translate into critical action.

Alternative narratives in history of Spain

Also in relation to the expansion of subjects, we are interested in highlighting a final issue, as many students underlined how some teachers challenged the hegemonic narrative on the history of Spain. In the reports, inside this narrative, it could be found two key historical stages that are worth mentioning. First, with regard to Middle Ages, it was pointed out the challenging of deeply rooted clichés about the ancient and Catholic Spanish nation in the Spanish nationalist narrative, which continues to generate debate on national identity, religion and multiculturalism (Sáiz/López-Facal 2016). Thus, facing the clear predominance in textbooks and teaching programs of the medieval Catholic Kingdoms, understood as the spark of the Spanish nation, some teachers try to delve into Al-Andalus, the contemporary Iberian Islamic society so frequently invisible and stereotyped in classrooms. At the same time, it is contested the simplified and teleological idea of the long and complex Middle Ages as a coherent and programmed process of "Reconquest" by the "Christians": "in order for the students to understand the time dimension of the so-called Reconquest and cast doubt on the name of this historical process, the teacher made the students calculate approximately how many generations there have been in eight centuries" (S1).

Second, regarding the modern history of Spain, there were some teachers who developed a deeper understanding of some issues, opening up a major split in the time canon of history in schools, which has traditionally prioritized a remote past (González 2014). It stands out the way in which they challenged the – still very influential – "Francoist" and "equidistant" narratives, instead promoting a critical vision of the dictatorship and restoring the visibility and dignity of the victims of Francoism and of those who fought against it, usually not shown in textbooks (Valls/López-Facal 2019).

Several students presented it as a "critical" feature – "his passion was noticeable when he exposed some key issues of the twentieth century" (S27) –, which contrasts with the avoidance or superficial treatment of other observed teachers who seemed to be trying to avoid conflict with students or families when facing a still disruptive issue in today's Spanish society. The special focus on the Spanish Civil War (1936–1939), the Franco dictatorship (1936/1939–1976) and the transition to democracy (1976–1982) is observed both in the classes of the two courses that have these topics as major curriculum units and in the project implementation in school centers, minor research projects or other types of extracurricular activities involving students from other courses.

A last example illustrates particularly well all this, but also the reflections of trainee teachers. There was this employed teacher who "we could think" he would be "eminently technical," if we take into account he was performing lectures and following the textbook in the last year of post-compulsory secondary school. However, he was more multifaceted, because "he always introduced problematic concepts, notions or situations in order to make the students reflect." Similarly, the student highlighted that, despite the pressure of the external university entrance exam, his tutor prioritized activities with great critical potential, such as the organization of "a talk on democratic memory with some local residents with relatives murdered by the Franco regime during the post-war period."

Also the endless preparation of "improvised debates" on "problematic situations" that arose in the classroom, in which the student perceived a teacher adopting an attitude away from the illusion of neutrality and partiality, and approaching the "committed impartiality" (Kelly 1986), the most consistent position with a critical model. His tutor "encouraged the students to participate and to have an orderly and respectful dialogue with the ideologies of others," but at the same time he "seemed very aware of the need of self-disclose in the debates, either from a moral or political perspective." A good example of this was given during a class on the transition to democracy in Spain in which a student defended the State terrorism practiced against the members of ETA, the Basque terrorist organization: "he explained that it seemed fine to him that the State had 'murdered terrorists for the common good'." Before this comment, the trainee teacher emphasized that the tutor, "although not arrogantly or humiliatingly, but convincingly, pointed out that he was not going to allow any apology for (counter) terrorism" and "specifically encouraged this student to reflect on the consequences of his remark" (S9).

Conclusions

The analysis carried out allows drawing several conclusions and open new questions about the representations of prospective GH teachers. The adoption among half of the sample of a "methodological" approach shows the limits of the theoretical learning and points to the strength of previous representations about GH teachers, perhaps reinforced by the "methodological" emphasis of both theoretical subjects of the MAES and informal speeches of some tutors. However, the influence of theoretical learning on the critical curricular paradigm is traceable in a general way in the different tools analysed, considering its contribution for a richer and more nuanced observation and reflection on processes of GH teaching, sometimes pointed out by students.

In relation to this, it should be noted that the other half of the sample does internalize and apply a "critical" approach in the analysis of the tutors' teaching practice. Thus, it is observed how the graduates in history mostly assume the critical perspective since they tend to pay greater attention to content selection and approach and purposes. Generally, these students articulate a complex discourse in which they also focus on a limited and insufficient diffusion of critical practice. In short, it would be desirable to deepen the coherence between the analysis of their tutors and the implementation and self-analysis of their own classes during the internship period. Further research will be done on this issue.

References

Adler, Susan: "The Education of Social Studies Teachers". In: Levstik Linda S./ Tyson Cynthia A. (eds.): *Handbook of Research in Social Studies Education*. Routledge: New York 2008, pp. 329–350.

Apple, Michael: *Ideology and Curriculum*. Routledge and Kegan Paul: Boston 1979.

Bagley, Carl/Beach, Dennis/Marques da Silva, Sofia: *The Wiley Handbook of Ethnography of Education*. Wiley-Blackwell: United States 2018.

Boyle-Baise, Marilynne: "Doing Democracy in Social Studies Methods". *Theory & Research in Social Education* 31(1), 2003, pp. 51–71.

Burke, Peter: *Eyewitnessing. The Uses of Images as Historical Evidence*. Reaktion Books: London 2001.

Díez, Mª Consuelo/Fernández, Antonia: "Enseñanza de las Ciencias Sociales con perspectiva de genero". *Clío. History and History Teaching* 45, 2019, pp. 2–10.

Fernández, Carlos/García, Olga/Galindo, Enrique: *Escuela o barbarie. Entre el neoliberalismo salvaje y el delirio de la izquierda*. Madrid: Akal 2017.

Giroux, Henry A.: *Schooling and the Struggle for Public Life: Democracy's Promise and Education's Challenge*. Taylor & Francis: United States 2015.

González, María Paula: *La historia reciente en la escuela: saberes y prácticas docentes en torno a la última dictadura*. Ediciones Universidad Nacional de General Sarmiento: Los Polvorines (Argentina) 2014.

González, María Paula: *La enseñanza de la historia en el siglo XXI: saberes y prácticas*. Ediciones Universidad Nacional de General Sarmiento: Los Polvorines (Argentina) 2018.

Harvey, David: *Spaces of Capital: Towards a Critical Geography*. Edinburgh University Press: Edinburgh 2001.

Iggers, George: *Historiography in the Twentieth Century. From Scientific Objectivity to the Postmodern Challenge*. Wesleyan University Press: Hanover 1997.

Jenkins, Keith: *Rethinking History*. Taylor & Francis: London 1991.

Kelly, Thomas E.: "Discussing Controversial Issues: Four Perspectives on the Teacher's Role". *Theory and Research in Social Education* 14, 1986, pp. 113–138.

Legardez, Alain/Simonneaux, Laurence: *L'école a l'épreuve de l'actualité. Enseigner les questions vives*. ESF éditeur: Issy les Moulineaux 2006.

Maroto, José Manuel: "Prejuicios sobre África en los libros escolares de Educación Secundaria". *Historia Actual Online* 43, 2017, pp. 169–186.

Merchán, Francisco J.: "El examen en la enseñanza de la Historia". *Didáctica de las Ciencias Experimentales y Sociales* 15, 2001, pp. 3–21.

Moon, Jennifer. A.: *A Handbook of Reflective and Experiental Learning. Theory and Practice*. Routldege Farmer. New York 2004.

Parra, David: "Representación de la historia escolar y crítica del consenso del sentido común". In David Parra/Carlos Fuertes. (eds.): *Reinterpretar la tradición, transformar las prácticas*. Tirant: València 2019, pp. 73–97.

Pinheiro dos Santos, María F.: "El Practicum en Geografía: un Locus para problematizar la formación inicial del docente". In David Parra/Carlos Fuertes. (eds.): *Reinterpretar la tradición, transformar las prácticas*. Tirant: València 2019, *pp*. 327–345.

Rockwell, Elsie: *La experiencia etnográfica: Historia y cultura en los procesos educativos*. Paidós: Buenos Aires 2009.

Rozada, José María: *Formarse como profesor. Ciencias Sociales, Primaria y Secundaria Obligatoria*. Akal: Madrid 1997.

Sáiz, Jorge/López-Facal, Ramón: "The National Historical Narratives of Students and Trainee Teachers". *Revista de Educación* 374, 2016, pp. 118–141.

Seixas, Peter/Morton, Tom: *The Big Six Historical Thinking Concepts*. Nelson: Toronto 2013.

Souto, Xosé M.: "¿Qué escuelas de Geografía para educar en ciudadanía?". *Didáctica de las Ciencias Experimentales y Sociales* 24, 2010, pp. 25–44.

Stenhouse, Lawrence: *An Introduction to Curriculum Research and Development*. Pearson Education: London 1975.

Tadeu da Silva, Tomaz: *Documentos de identidade: uma introdução às teorias do currículo*. Autêntica: Belo Horizonte 1999.

Valls, Rafael/López-Facal, Ramón: "Spain". In: Luigi Cajani/Simone Lässig/ Maria Repoussi (eds.): *The Palgrave Handbook of Conflict and History Education in the Post-Cold War Era*. PalgraveMcMillan: London 2019, pp. 605–618.

Zabalza Beraza, Miguel Ángel: *Diarios de clase: Un instrumento de investigación y desarrollo profesional*. Narcea: Madrid 2004.

José A. Armas Castro, Jorge Conde Miguélez, Luis Alberto Marques Alves, and Cristina María Maia

Chapter 12 Learning to teach history in secondary education: Preservice teachers' attitudes when faced with emotional and controversial issues[1]

Abstract: This chapter offers the results of a quantitative, descriptive and non-probabilistic study on the attitudes of students training to be secondary education history teachers when faced with teaching sensitive and controversial topics. 646 trainee teachers from 23 Spanish universities participated in the research. Data was collected using a quantitative questionnaire created ad hoc. The acceptance, rejection and doubts of trainee teachers with regard to the options of *exclusive neutrality*, *exclusive partiality*, *neutral impartiality* and *committed impartiality* adapted from Kelly (1986) were analysed. The results indicate that the trainee teachers are, on the whole, in favour of the options of neutral impartiality and committed impartiality. It also appears that the prior training of the participants and the training programme received have more influence than other sociodemographic variables.

Keywords: Teacher education, History education, Secondary education, Controversial issues, Emotions and teaching

Introduction

Reviews on teacher education have dealt with aspects such as the professional knowledge of teachers, training programmes, the relationships between theory and practice, spaces in which training is carried out, the identity of teachers, the emotional dimensions of teaching practice and the role of place and community in teacher education (Cochran-Smith & Zeichner, 2005; Darling-Hamond & Bransford, 2005; Loughran & Hamilton, 2016).

Studies which have analysed the evolution of the approaches and paradigms which guided teacher education from the end of the 19th century to the present

---

1 This work is part of the research funded by the National R&D&I Plan of the Spanish Ministry of Economy (PGC2018-094491-B-C31) and co-funded with ERDF funds of the EU.

day have identified three moments (or "waves") on a global level, with particularities for each nation (Alexander, Fox, & Aspland, 2020; Cochran-Smith & Demers, 2008; Placier, Letseka, Seroto, Loh, Montecinos, Vásquez, &Tirri, 2016). The first period, dating from the end of the 19th to the middle of the 20th century, was marked by the objective of providing the number of teachers necessary for ensuring the schooling of the population. The training model focused on the mastery of the contents along with some teaching techniques. The second period, corresponding to the second half of the 20th century, was characterised by the quest for quality teacher training. A constructivist orientation of teaching and learning inspired training models oriented towards the development of teachers' professional knowledge aimed at promoting their pupils' learning. In the third period (extending from the end of the 20th century to the present day), training processes have been immersed in the tension between two trends. On the one hand, "accountability" (Cochran-Smith & Demers, 2008; Zeichner, 2010) is reflected in learning standards and the measurement of teaching effectiveness by way of pupils' results, while on the other hand, the "critical resistance" trend orients teacher training towards the promotion of education for social justice and democratic citizenship (Zeichner, 2016).

In terms of the initial training of history and social science teachers, three lines of research are of particular interest for the object of this study: the professional knowledge of history teachers, the teaching of difficult and controversial historical issues and the emotional component of the teaching of history (Adler, 2008; Crowe & Cuenca, 2016; van Hover & Hicks, 2018).

The concept of Pedagogical Content Knowledge (PCK) has occupied a dominant role in research on the professional knowledge of teachers. Introduced by Shulman (1986, 1987, 2015) in order to describe the knowledge used by teachers to transform the knowledge of a discipline into learning in a specific school context, it has been re-elaborated to include a broad range of knowledge, skills and dispositions: the knowledge of the discipline, the characteristics and the interests of the pupils, learning theories, the curriculum and its controversies, the specific teaching context and a wide range of pedagogical strategies and classroom management, not to mention the aims of education in democratic societies or the role of the emotions in teaching and learning. Gudmundsdottir and Shulman (1987) studied the cases of two social studies teachers (one a novice and the other a veteran) in the United States revealing notable differences in their mastery of PCK. Both teachers demonstrated expert knowledge in the contents of social studies. However, there were differences in their knowledge of the pupils' characteristics and of the teaching methods employed in the classroom. Seixas (1999) has demonstrated that the use of primary and secondary sources enables the construction of a shared discourse between historians and history teachers regarding "learning history via research." Harris and Bain (2011) achieved significant progress in overcoming the traditional separation between the study of contents and teaching methods integrating the study of global history with the analysis of teaching strategies and methods employed by teacher trainers. Monte-Sano (2011) analysed the progress of three trainee history teachers in relation to their understanding of the acquisition of aspects of historical thinking, such as interpretation and the use of evidence by their secondary pupils. In our case, we attempt to analyse the way in which preservice teachers incorporate pedagogical knowledge with regard to controversial or conflictive contents in history.

Dealing with sensitive and controversial social issues in the classroom, understood as a fundamental aspect of the education of young people for a deliberative democracy, has a long tradition in critical currents of education (Apple, 1979; Evans & Saxe, 1996; Giroux, 1988). Studies on the teaching and learning of difficult or conflictive history have increased significantly in recent years (Epstein & Peck, 2018; Gross & Terra, 2019; van Boxtel, Grever & Klein, 2016). Barton (2019) identifies three areas of study related with difficult and controversial history: the place it occupies in curricula and school textbooks, how pupils understand and position themselves when faced with controversial issues, and the approaches and attitudes of teachers towards these difficult and controversial issues. It is the latter aspect which is of particular interest in this chapter.

Kelly (1986) analyses four attitudes of teachers when working on controversial topics: *exclusive neutrality*, *exclusive partiality*, *neutral impartiality* and *committed impartiality*, depending on the teacher's degree of neutrality or commitment when faced with controversial issues. Along similar lines, Kello (2016), based on interviews of 53 history teachers from Estonia and Latvia, distinguished five strategies employed in teaching sensitive pasts: *hiding or avoiding*, *smoothing edges*, *just doing the job*, *enhancing heterogeneity*, and *leaving the truth open*. On the other hand, Kitson and McCully (2005) identified three perspectives among teachers, which they define as *avoiding*, *containing* and *risk-taking* in the context of history teaching in a divided society such as that of Northern Ireland. Last of all, Barton and McCully (2007) made some recommendations for dealing with controversial history topics at schools in Northern Ireland. The first of these recommendations is "dealing with emotions." If, as teachers, we aim to involve our pupils in the discussion of difficult and controversial history, we must be prepared to find emotional responses.

The so called “affective turn” (Zembylas, 2019) has situated the role of the emotions in the foreground of educational research (Day & Lee, 2011; Schutz & Pekrun, 2007; Zembylas, 2005). In the field of the teaching of history and the social sciences, several authors have analysed the role of the emotions when dealing with difficult and controversial history in school contexts. Employing observational processes, Bekerman and Zembylas (2011) studied the attitudes and emotions of a group of 32 Palestinian and Jewish pupils in a bilingual Israeli primary school when dealing with the death of the Palestinian leader Yasser Arafat. Sheppard and Levy (2019) interviewed 17 secondary education social studies teachers in the cities of New York and Washington in order to analyse the teachers descriptions of the role of the emotions in their history classrooms and the way in which their conceptions of the emotions had an impact on decisions regarding the social studies curriculum and their teaching. Zembylas and Kambani (2012) analysed, via interviews, the perceptions and emotions of 18 Greek-Cypriot primary teachers when teaching controversial historical issues in the strongly divided, in social terms, Cypriot context. These and other studies demonstrate a certain degree of contrast between the relevance teachers attribute to the emotions in the teaching of socially conflictive issues and their deficit in training with regard to managing these emotions in their classrooms.

In this chapter, we attempt to respond to the following questions: (a) What are the attitudes and strategies of a large group of preservice history and social sciences teachers from 23 Spanish universities regarding the teaching of controversial historical issues? (b) How do variables such as sex, prior training and training programmes influence their understanding of the teaching of difficult historical issues? (c) How do emotions linked to individual or social identity influence their attitudes and teaching practice?

Method

Participants

646 preservice teachers participated in the study, of whom 358 (55.4 %) were men and 280 (43.3 %) women, with eight people (1.2 %) providing no answer to the question of gender. The age ranges of the participants are shown in Tab. 1.

Tab. 2 shows the university studies of the participants. The data is grouped into two levels: (a) undergraduate studies = 628 people (97.2 % of the total), (b) postgraduate studies including master’s degrees/master of advanced

Tab. 1: Ages of the participants. Authors' own work

Age (years)	n	%
18–24	334	51.703
25–34	245	37.926
35 and above	49	7.585
NA	18	2.786
Total	646	100.000

Tab. 2: Academic training of the participants. Authors' own work

Academic training	n	%
Doctoral studies	5	.7740
Master's degree/MAS	7	1.0836
Graduate degree	627	97.0588
Diploma	1	.1548
NA	6	.9288
Total	646	100.0000

studies/doctoral studies = 12 people (1.8 %) and (c) six people (.92 %) who did not respond in terms of their prior studies.

Of the 646 participants, 64 (9.9 %) claimed to have carried out teaching innovation courses, 577 (89.3 %) stated that they had not done any such training and five (.8 %) provided no answer.

The participants were contacted in 23 Spanish universities. Tab. 3 shows the total data of the participants according to their university of origin.

Design of the research

This is a quantitative, descriptive and non-probabilistic study with an intentional sample (Creswell & Creswell, 2018). Descriptive and comparative analyses were carried out among independent groups of participants from 23 Spanish universities by way of a Likert-type interval measurement scale of ascending values ranging from 1 to 5 for discontinuous quantitative variables. Due to the nature of the variables, the groups and the distribution, non-parametric statistical tests work carried out, such as the Kruskal-Wallis test and the Mann-Whitney U test (Roni, Merga, & Morris, 2020).

Tab. 3: Number of participants in the study according to their university of origin. Authors' own work

University of origin	n	%
Autonomous University of Madrid	22	3.406
Complutense University of Madrid	46	7.121
University of Alicante	24	3.715
University of Almería	35	5.418
University of Barcelona	23	3.560
University of Castilla-La Mancha	18	2.786
University of Córdoba	28	4.334
University of Extremadura	33	5.108
University of Granada	51	7.895
University of Huelva	15	2.322
University of Jaén	27	4.180
University of La Rioja	20	3.096
University of Málaga	37	5.728
University of Murcia	40	6.192
University of Oviedo	9	1.393
University of Santiago de Compostela	24	3.715
University of Seville	23	3.560
University of Valencia	70	10.836
University of Valladolid	21	3.251
University of Zaragoza	29	4.489
University of the Basque Country	21	3.251
Jaume I University	9	1.393
Rovira i Virgili University	21	3.251
Total	646	100.000

Data collection tool

The data was collected via a quantitative questionnaire which was designed ad hoc. For this research, block five of the questionnaire, entitled "Opinions and conceptions on the teaching of history and educational competences," was used as it contained items 75, 76, 77 and 78, which were used as the dependent variables of the study. Each item was evaluated from 1 to 5 on a Likert-type scale. Four independent sociodemographic variables were also used: the university of origin, sex, academic training and the completion of teaching innovation courses.

Tab. 4: Items from Part II-Block 5 of the questionnaire

75. The teacher should avoid socially conflictive issues in the classroom: the teacher should be neutral and not influence his/her pupils as these are issues which correspond to the families.	1	2	3	4	5
76. The teacher should be involved in socially conflictive issues and promote the position which he/she considers to be ethically correct.	1	2	3	4	5
77. The teacher should enable debate among pupils in order for them to learn how to argue opinions, but without expressing his/her own opinions so as to avoid influencing the pupils.	1	2	3	4	5
78. The teacher should propose debate between different perspectives on socially conflictive issues and manifest his/her opinion along with the others, albeit without imposing his/her point of view.	1	2	3	4	5

The present study provides, first of all, data for each item taking into account the total number of participants. The degree of agreement of their opinions with each of the options proposed on the attitude which should be adopted by a teacher in the classroom when dealing with a controversial issue is evaluated in a quantitative manner (see Tab. 4). In order to achieve this, the following question was posed:

In your opinion, what attitude should be adopted by a teacher when dealing with a conflictive issue in the classroom? The questions in the questionnaire should be answered by marking the box for the answer which best represents your opinion, bearing in mind that 1= extremely irrelevant; 2= not very relevant; 3= neither relevant nor irrelevant; 4= somewhat relevant; 5= extremely relevant.

The items of the questionnaire was submitted to the validation of seven experts in terms of sufficiency, clarity and relevance in a questionnaire with a range of values from 1 to 4 points. The four items of this study were adapted from Kelly (1986) and were accepted as relevant by the experts with no objections.

Procedure, ethics and data analysis

The data was collected during the 2019–2020 academic year in a classroom context by members of the teaching and research staff of the universities participating in the study. In order to collect the data, information was sent in which the objectives of the project were explained and a web link for participants in the survey was provided. On the first page of the questionnaire the participants were informed of: the anonymous treatment of the data, how the procedure was to be carried out, how the data was to be returned, the voluntary nature of their participation and other aspects relating to ethics. The study was approved by the Ethics Committee of the University of Santiago de Compostela.

The data was analysed using the R program version 3.6.3 (R Core Team). This tool enabled the creation of: (a) tables of descriptive statistics (mean, median) with percentages as the units of measurement; and (b) graphs scaling the degree of agreement in percentages in an ascending range of values from 1 (extremely irrelevant) to 5 (extremely relevant). The coloured graphs made it possible to observe the visual pattern of responses. The dark brown and light brown tones represented negative values (rejection), grey represented neutral values (doubt) and light green and dark green, positive values (acceptance). Significant differences were sought according to the different sociodemographic variables. In order to achieve this, non-parametric tests were used: the Mann-Whitney U test for factors of two levels, and the Kruskal-Wallis K test for factors of three or more levels (denoting statistics with W and K respectively). The Department of Statistical Support (SAE) of the Scientific Area of the University of Murcia (www.um.Web/acti) was responsible for the analysis of the data.

Results

In block 5 of the questionnaire, the participants were requested to express their degree of agreement with four attitudes of teachers when working on difficult or controversial pasts on a Likert-type scale (with ascending values from 1 to 5 points) (see Tab. 4). According to the order in which they appear in the questionnaire, the attitudes are as follows: (1) *exclusive neutrality*, by adopting this attitude the teacher avoids problematic issues; (2) *exclusive partiality*, the decision is taken to provide a partial interpretation of the issue, which is that considered ethical by the teacher; (3) *neutral impartiality*, different versions of the issue to be taught are promoted, although the teacher does not express his/her point of view; and (4) *committed impartiality*, different perspectives are promoted, but in this case emotions are shown and the teacher expresses, but does not impose, his/her point of view.

Data are provided in graphs grouping the five values of the Likert scale into three categories distinguished by colours: (a) the values in brown tones indicate *rejection*, objections or non-acceptance of the attitudes; (b) the values in grey represent the degree of *doubt*, intermediate values which do not indicate rejection or acceptance; and (c) the values shown in green tones indicate the frequency of agreement with the attitudes, thereby indicating *acceptance*.

Tab. 5: Descriptive results according to the total number of participants. Authors' own work

	N	Min	Max	Mean	Median	1	2	3	4	5	Sd
Exclusive neutrality	645.0	1	5	2.56	2.00	29.30	22.30	22.90	14.40	11.00	1.34
Exclusive partiality	644.0	1	5	3.26	3.00	14.00	12.10	26.90	28.10	18.90	1.29
Neutral impartiality	645.0	1	5	4.08	4.00	4.03	5.12	14.30	32.10	44.50	1.07
Committed impartiality	644.0	1	5	4.12	4.00	4.04	5.28	15.40	25.60	49.70	1.10
Total items	645.5	1	5	3.50	3.25	12.84	11.20	19.88	25.05	31.02	1.20

Descriptive results according to the total number of participants

Tab. 5 shows the evaluations of the four attitudes according to the total number of participants and reflects the frequency of total and relative participation, minimum and maximum values, mean, median and percentages of evaluation of the four dependent variables.

The results indicate that the options of *exclusive neutrality* and *exclusive partiality* are those which are least valued by the teachers, although the former (item 75) received more rejection than acceptance, whereas the latter (item 76) accumulated more acceptance than rejection. The attitudes of *neutral impartiality* and *committed impartiality* are those chosen by the majority to be adopted in class when dealing with conflictive issues. The opinions which expressed doubt (values associated to option 3 in Tab. 5) are higher with respect to items 75 and 76, which represent the attitudes of *exclusive neutrality* and *exclusive partiality.*

Differences according to university of origin of the participants

Below, the data from items 75, 76, 77 and 78 are analysed taking into account the universities of origin of the participants. The data are presented in five chromatic varieties which qualify the degree of agreement of the participants with each attitude. The percentages have been grouped into three categories: the negative evaluations of the degree of agreement (extremely irrelevant, not very relevant), which shall be named "disagreement" or "rejection," are in brown tones and situated on the left; in the centre and in grey are the neutral or average evaluations (neither relevant nor irrelevant), which shall be called "doubts"; and on the right in green can be found the positive evaluations of the attitudes (somewhat relevant, extremely relevant), which shall be denominated "acceptance."

In Fig. 1 (item 75 = exclusive neutrality) it can be observed that, of the 23 universities, only three demonstrated more acceptance of this attitude than rejection (Almería, Castilla la Mancha and Huelva); in one (Rovira i Virgili) the frequencies are balanced between rejection and acceptance and in all the other universities (19) rejection is higher. The participants from the universities of Santiago de Compostela and Valencia are those who expressed the greatest rejection of this attitude, whereas those from the universities of Almería, Castilla la Mancha and Huelva expressed the greatest acceptance. The participants who maintained a higher level of doubt or neutrality were those from the universities of Valladolid and Jaume I.

In Fig. 2 (item 76 = Exclusive partiality) it can be seen that only two of the 23 universities had more participants against this attitude than in favour of it (Autonomous University of Madrid and Castilla La Mancha). In 18 universities, there were more participants in favour (acceptance) than against (rejection), with the universities of Santiago de Compostela, Extremadura and Alicante being those which offered a better evaluation of this attitude. Lastly, two universities (Jaén and Valladolid) had the same percentages of acceptance as rejection, whereas the universities of Málaga and the Basque Country are those which offered the least defined positions.

Fig. 3 (item 77 = neutral impartiality) shows how the participants of the 23 universities clearly accept this attitude in the classroom, with percentages always higher than 65.2 %. The universities in which this position is accepted with the highest frequencies are those of Extremadura and Córdoba with it being accepted with somewhat lower frequencies by the participants of the universities of Seville, Santiago de Compostela, Valencia and Huelva. Higher percentages in terms of rejection were offered by the participants of Murcia and Valencia and those who would never reject it were the groups from Oviedo and Jaume I. Lastly, in terms of doubt, the highest frequencies corresponded to the universities of Seville and Jaume I and the lowest to Córdoba, Extremadura and La Rioja.

In Fig. 4 (item 78 = committed impartiality), it can be seen how almost all of the universities accept the use of this attitude in the classroom. The highest frequencies of acceptance were shown by the groups from the Basque country, Zaragoza, La Rioja and Huelva (the latter with 100 % of acceptance). The lowest frequencies of acceptance were shown by the universities of Rovira i Virgili and Valladolid. Although it has been observed that the majority of the participants

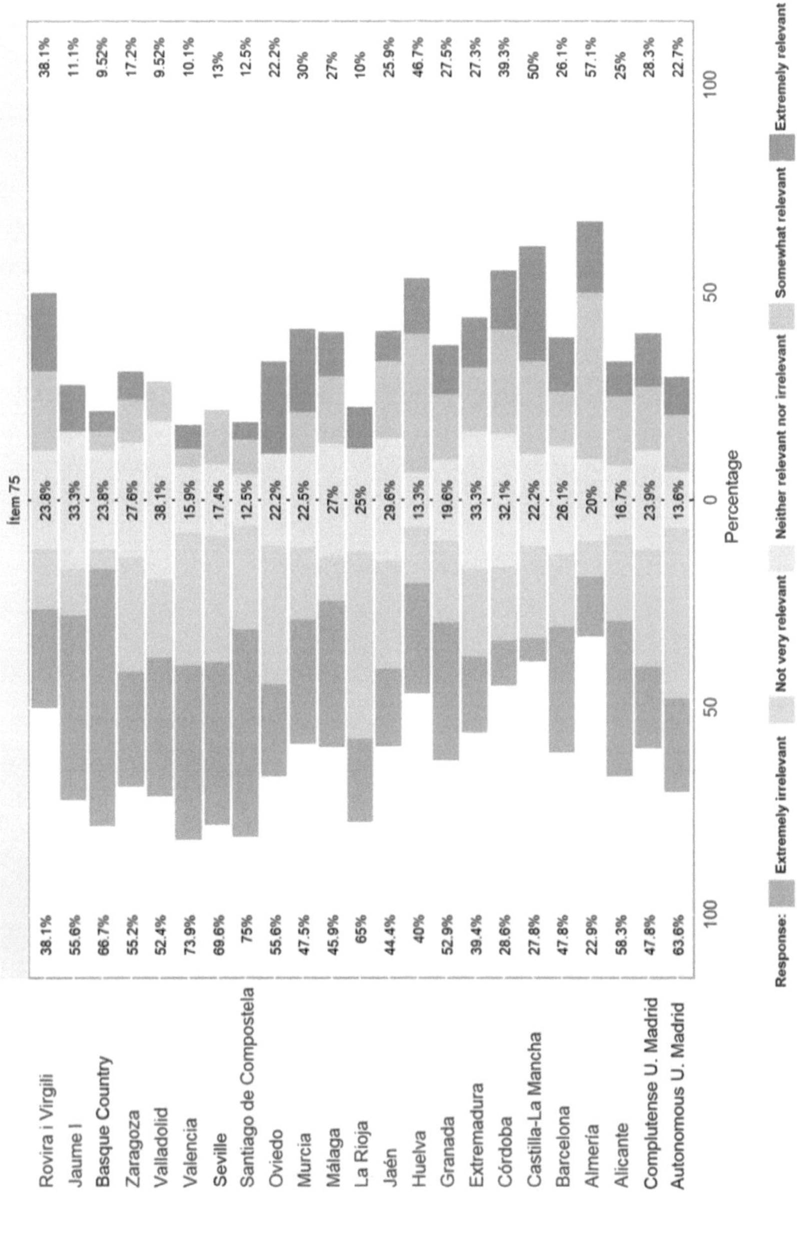

Fig. 1: Data for Item 75. Exclusive neutrality. Choice according to university of the participant

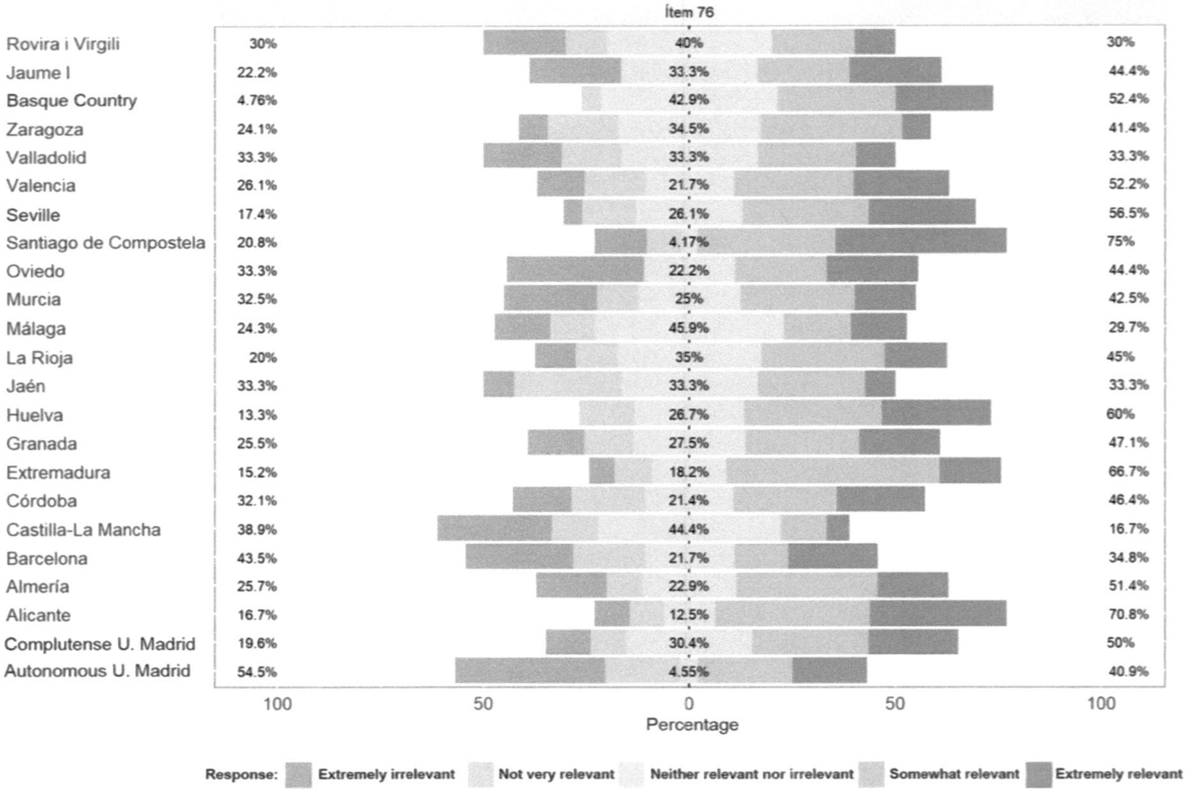

Fig. 2: Data for Item 76. Exclusive partiality. Choice according to university of the participant

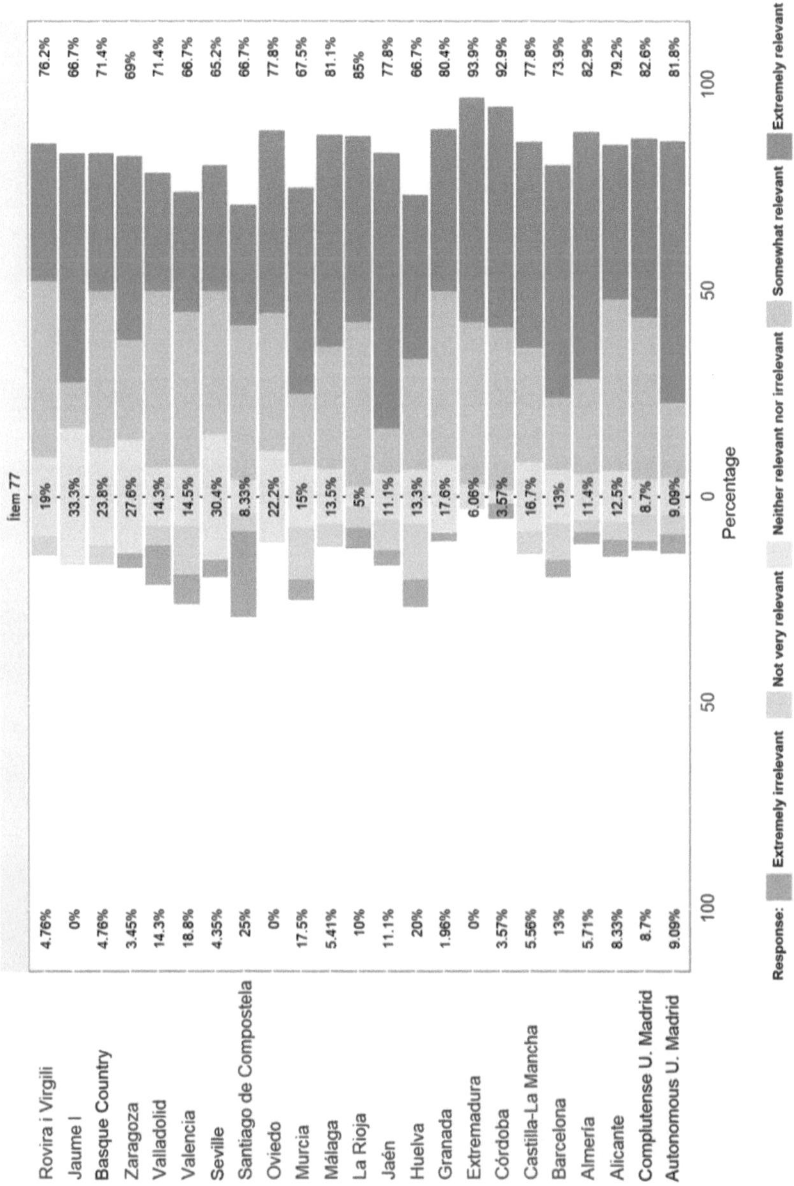

Fig. 3: Data for Item 77. Neutral impartiality. Choice according to university of the participant

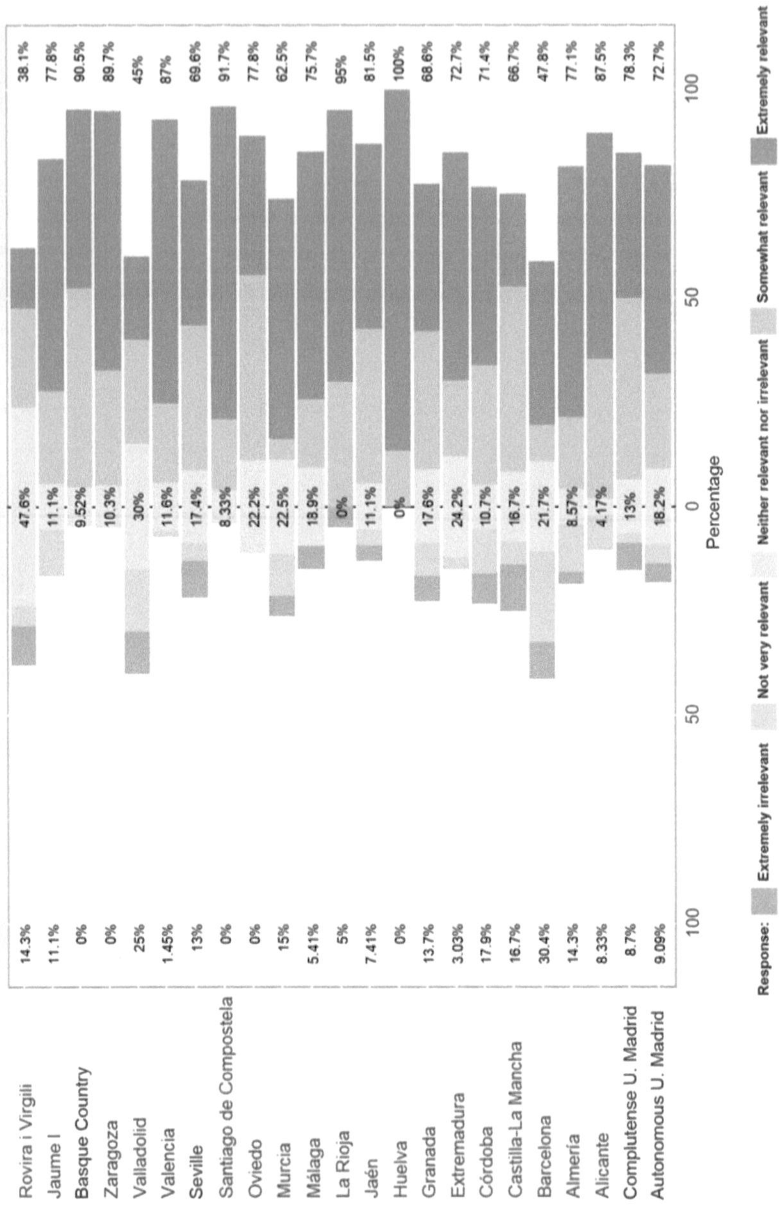

Fig. 4: Data for Item 78. Committed impartiality. Choice according to university of the participant

accept this point of view, there are opinions against this attitude in the universities of Córdoba, Castilla La Mancha and Barcelona. On the other hand, the groups from Santiago de Compostela, Oviedo, the Basque Country, Zaragoza and Huelva did not present any objections or veto to this attitude. The highest frequencies of doubt with regards to this attitude were shown by the universities of Rovira i Virgili and Extremadura, while the lowest frequencies corresponded to Alicante, La Rioja and Huelva.

Results according to sex of the participants

In Fig. 5, it can be observed that the perspectives of men and women do not vary much, either in the figure of their frequencies or in the homogeneity of the visual pattern. Taking these data into account, it can be stated that the sex of the participants does not seem to significantly influence the choice of attitude which teachers should assume in their work on conflictive issues in history classes.

Results according to the academic training of the participants

Fig. 6 shows the results (percentage units) of the acceptance, doubt or rejection of the four attitudes proposed, taking into account the variable of academic training (undergraduate or postgraduate) of the participants. Both the figures of frequencies and the visual pattern seem to indicate that the academic training of the participants does indeed have an influence on the choice and evaluation of attitude when faced with controversial issues.

For item 75 (exclusive neutrality), the hegemonic position with regard to this attitude is that of rejection in the two groups established according to training. Even so, having completed postgraduate studies implies a 25 % higher frequency of rejection. As for responses of acceptance, the data are almost uniform. However, the data corresponding to doubts again offers a discrepancy: participants with undergraduate studies demonstrated 23.4 % more doubts than their classmates with postgraduate studies.

For item 76 (exclusive partiality), the posture of both groups was to accept it as valid for use in the classroom. Again, dissonant data appeared in the category of doubts: those with postgraduate studies have 15 % more doubts regarding exclusive partiality than their classmates with undergraduate studies.

For item 77 (neutral impartiality), the groups established according to their academic training have a similar posture of acceptance and consider this attitude to be valid for use in the classroom. Participants with postgraduate studies

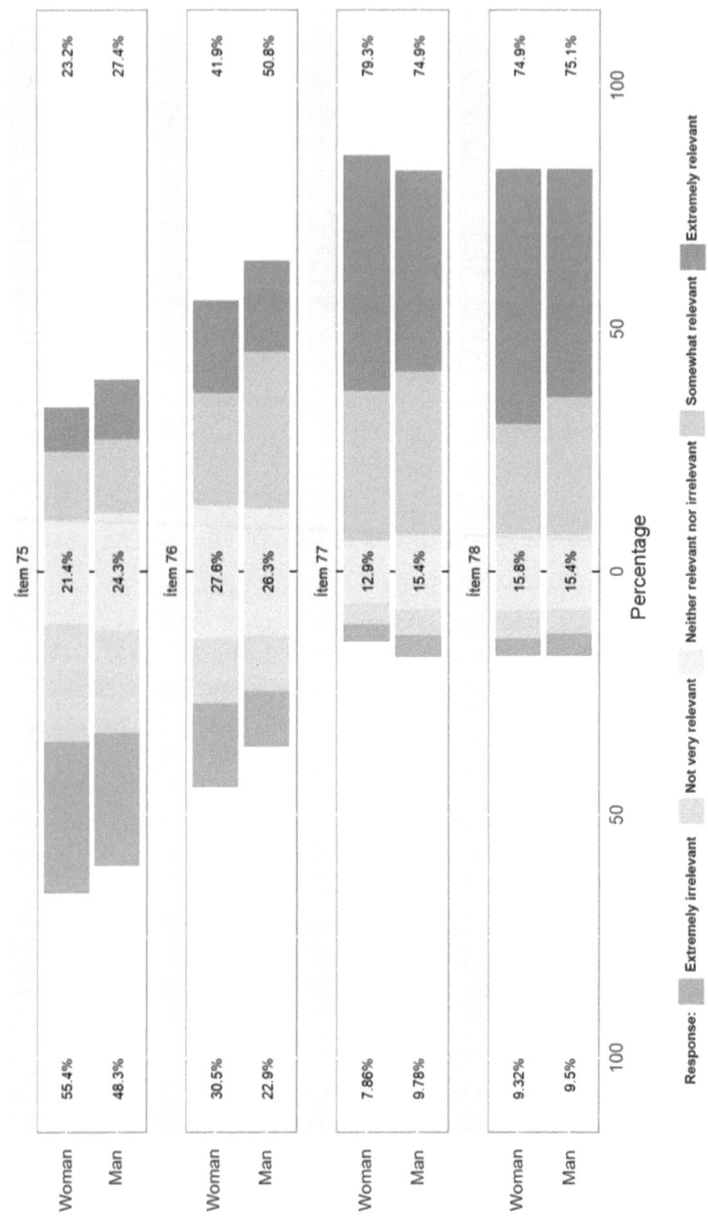

Fig. 5: Choice of attitudes to be assumed by a teacher when dealing with conflictive issues in the classroom according to the sex of the participants

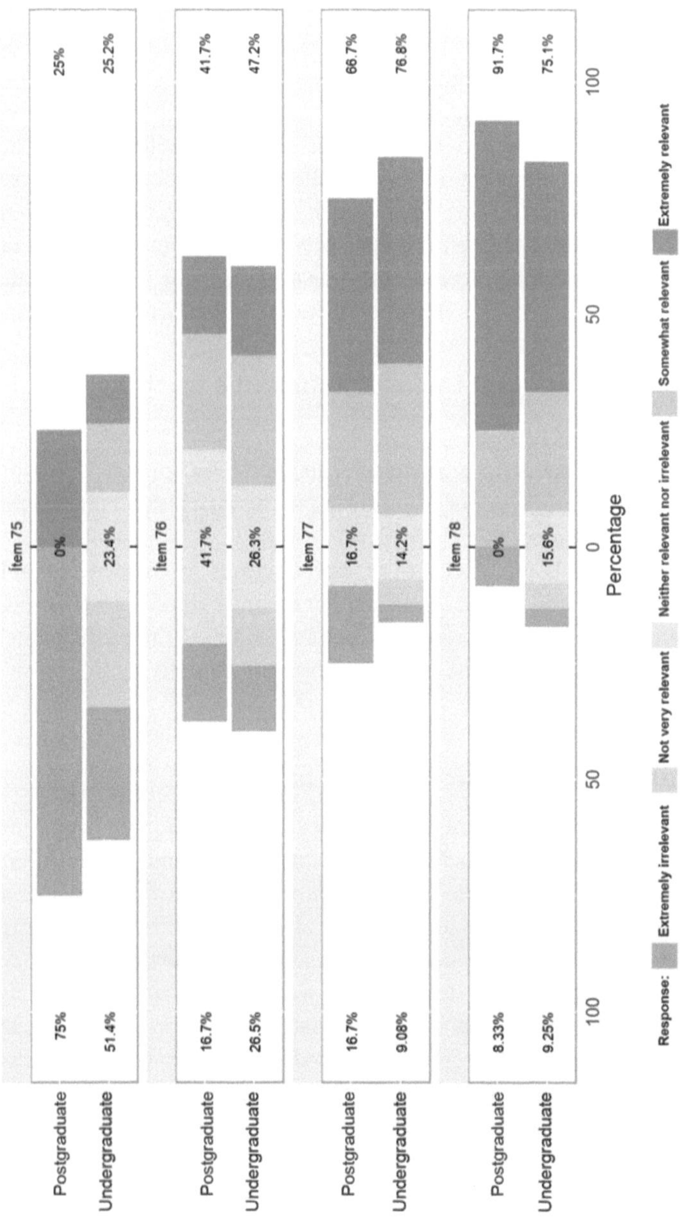

Fig. 6: Choice of attitudes to be assumed by a teacher when dealing with conflictive issues in the classroom according to the academic training of the participant

demonstrated higher frequencies in terms of rejection than their classmates with undergraduate studies. As far as acceptance is concerned, the scenario is inverted: participants with undergraduate studies expressed a higher degree of acceptance than those with postgraduate studies.

For item 78 (committed impartiality), the shared posture indicates a high level of acceptance of this attitude, with it being the best valued of the four. While the values of rejection show only slight differences, there are significant differences when it is valued positively: the participants with postgraduate studies showed a higher positive evaluation with a difference of almost 16 points. In terms of doubts, another differentiating piece of data was obtained: the undergraduate group had 15.4 % more doubts than those of the postgraduate group.

Results taking into account the completion of a teaching innovation course

Fig. 7 summarises the results (in percentage units) of the acceptance, doubt or rejection of the four attitudes proposed according to the variable of "completing a teaching innovation course."

Fig. 7 shows relative homogeneity in terms of figures of frequencies and visual pattern, which would seem to indicate that having completed a teaching innovation course does not have a significance influence on the participants' choice and evaluation regarding the attitude to be adopted with regard to controversial issues in the classroom.

Discussion and conclusions

This research is based on the assumption that dealing with conflictive or controversial issues in history and social science classes in secondary education is necessary and useful in order to educate young people in the values of a deliberative democracy and in the development of attitudes of respect and the acceptance of differences. Furthermore, prior research has shown that history teachers often show resistance to dealing with sensitive and controversial issues in the classroom for various reasons: the relative lack of presence of these contents in official curriculums focused on closed disciplinary contents, difficulties in controlling the times and emotions of pupils when dealing with issues which provoke divisions and controversies in society, doubts regarding how to evaluate, according to dominant models, certain historical contents with no definitive solutions which are susceptible to being evaluated from different perspectives and, last but not least, the resistance of some family and school

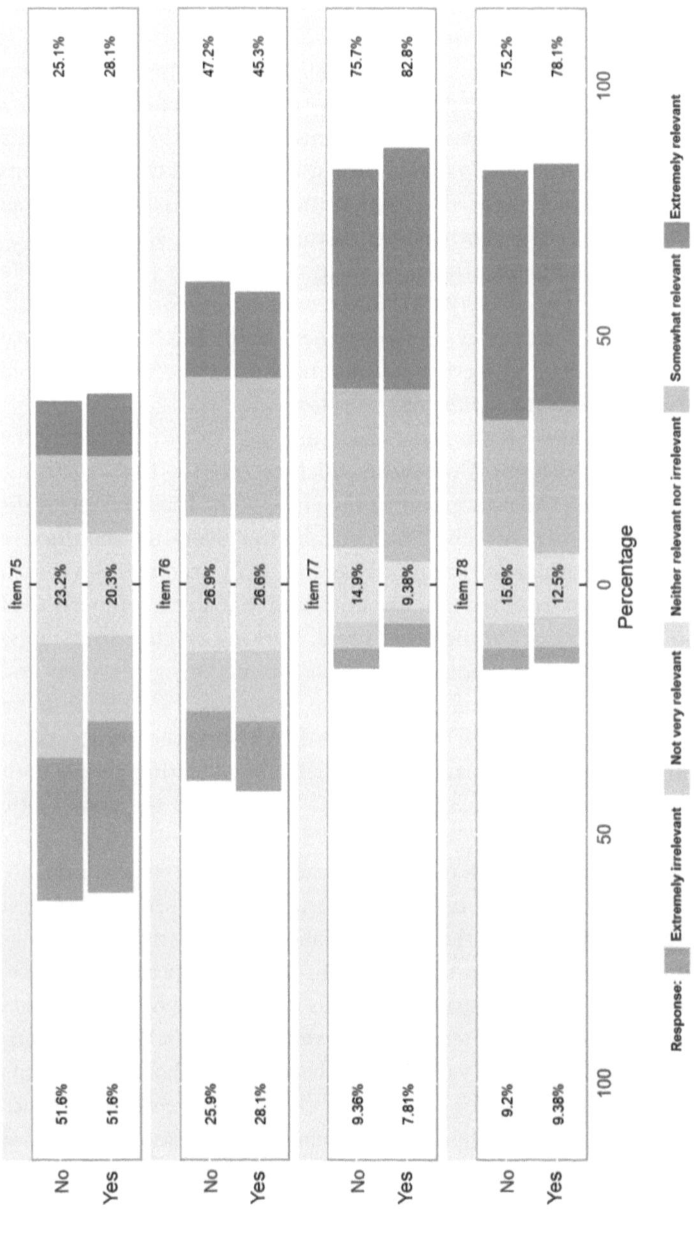

Fig. 7: Choice of attitudes to be assumed by a teacher when dealing with conflictive issues in the classroom according to whether the participant has completed a teaching innovation course

contexts regarding the teaching of these controversial issues. For these reasons, analysing the attitudes and mentalities of trainee history and social science teachers towards the teaching of sensitive and difficult historical contents is necessary in order to progress towards a training model which provides the necessary pedagogical content knowledge for overcoming these types of obstacles.

From this perspective, this quantitative research, with the participation of 646 preservice teachers from 23 Spanish universities, required the participants to express their position with regard to four attitudes which teachers can adopt in classroom when teaching controversial historical issues. These attitudes, adapted from Kelly (1986), are defined as *exclusive neutrality* (EN), *exclusive partiality* (EP), *neutral impartiality* (NI) and *committed impartiality* (CI) according to the degree of neutrality or commitment of the teachers. Each item was evaluated from 1 to 5 on a Likert-type scale and the responses were grouped into three categories: *rejection*, *doubt* and *acceptance*.

The attitudes of *EN* and *EP* are those which are least valued overall by the participants, whereas *NI* and *CI* are those which receive the highest degree of acceptance regarding their adoption in the classroom. These results indicate that preservice teachers on the whole consider that avoiding a conflictive issue (*EN*) or offering a partial or biased version (*EP*) are attitudes which can be epistemologically rejected and are undesirable if a critical method of educating pupils is the aim. On the other hand, working on controversial issues by promoting different versions (*NI*) and accommodating emotions and demonstrating one's own point of view without imposing it (*CI*) seem to be attitudes which are epistemologically in accordance with the construction of historical knowledge and educationally effective in educating people with the capacity for actively participating in social debates and the democratic process.

In this research, four sociodemographic variables were taken into account: sex, academic training, the university of origin and the prior completion of teaching innovation courses. The data indicate that the variable *sex of the participant* does not introduce any significant variation and, as a consequence, does not influence the choice of attitude towards the teaching of controversial issues. The variable *completion of teaching innovation courses* behaves in a similar way and only offers anecdotal variations in the sense that those participants who have completed such activities express slightly higher levels of acceptance in the *NI* attitude. On the other hand, the variables of academic training and university of origin do, indeed, offer significant variations, thereby demonstrating their influence in the participants' evaluation regarding attitudes towards controversial issues.

The variable *academic training* (undergraduate or postgraduate studies) seems to significantly influence the evaluation of the different attitudes towards teaching. All of the participants clearly rejected the *EN* attitude, but those with undergraduate studies showed a higher level of indecision, whereas those with postgraduate studies demonstrated a much higher level of rejection of this attitude. At the other extreme, the *CI* attitude is that which is most accepted by all of the participants, although those with postgraduate studies evaluate it 16 % higher whereas those with undergraduate studies present a higher level of doubt (up to 15 %). In summary, a broader and deeper training in history facilitates a better understanding of the disciplinary and educational approaches to controversial historical issues.

Last of all, the variable *university of origin* also seems to significantly affect the evaluation of the different attitudes. The majority of the 23 universities offer similar patterns of acceptance/doubt/rejection of the different attitudes towards the teaching of controversial issues. The *EN* and *EP* attitudes are those which concentrate most rejection, while *NI* and *CI* are those which accumulate most acceptance. However, variations from one university to another can be perceived, which is surely related with the training programme received by the students from these universities and the attention paid to the teaching and learning of difficult and controversial historical issues.

If we wish to have history and social science teachers who are able to deal with controversial historical issues in the classroom with the aim of educating young people in the democratic values of equality, diversity and respect for differences, teacher training programmes will have to create opportunities for trainee teachers to review their points of view regarding the disciplinary and educational aspects which intervene in the teaching and learning of difficult and conflictive histories. They must also facilitate the design and experimentation of teaching proposals which take into account the teachers' own involvement and emotions, as well as those of their pupils.

References

Adler, Susan: "The Education of Social Studies Teachers". In: Linda S. Levstik/Cynthia A. Tyson (Eds.): *Handbook of Research in Social Studies Education*. Routledge: New York 2008, pp. 329–351.

Alexander, Colette/Fox, Jillian/Aspland, Tania: "Third Wave Politics in Teacher Education: Moving Beyond Binaries". In: Jillian Fox/Colette Alexander/Tania Aspland (Eds.): *Teacher Education in Globalised Times. Local Responses in Action*. Springer: London/Singapore 2020, pp. 1–21.

Apple, Michael W.: *Ideology and Curriculum*. Routledge: New York/London 1979.

Barton, Keith C.: "Teaching Difficult Histories. The Need for a Dynamic Research Tradition". In: Magdalena H. Gross,/Luke Terra (Eds.): *Teaching and Learning the Difficult Pasts. Comparative Perspectives*. Routledge: New York/London 2019, pp. 11–25.

Barton, Keith C./McCully, Alan: "Teaching Controversial Issues... Where Controversial Issues Really Matter". *Teaching History*, 127, 2007, pp. 13–19.

Bekerman, Zvi/Zembylas, Michalinos: "The Emotional Complexities of Teaching Conflictual Historical Narratives: The Case of Integrated Palestinian-Jewish Schools in Israel". *Teachers College Record*, 113(5), 2011, pp. 1004–1030, retrieved on 20.10.2020, from http://www.tcrecord.org.

Cochran-Smith, Marilyn/Demers, Kelly E.: "How Do We Know What We Know? Research and Teacher Education". In: Marilyn Cochran-Smith/Sharon Feiman/D. John McIntyre (Eds.): *Handbook of Research on Teacher Education. Enduring Questions in Changing Contexts* (pp. 134–156). Routledge/Tylor & Francis Group: New York/London 2008.

Cochran-Smith, Marilyn/Zeichner, Kenneth M. (Eds.): *Studying Teacher Education: The Report of the AERA. Panel on Research and Teacher Education*. Lawrence Erlbaum Associates Publishers/AERA: Mahwah/Washington 2005.

Creswell, John W./Creswell, J. David (2018). *Research Design: Qualitative, Quantitative, and Mixed Methods Approaches*. Sage: Los Angeles/London/Singapore 2018.

Crowe, Alicia R./Cuenca, Alexander (Eds.): *Rethinking Social Studies Teacher Education in the Twenty-First Century*. Springer: New York/London 2016.

Darling-Hamond, Linda/Bransford, John (Eds.). *Preparing Teachers for a Changing World. What Teachers Should Learn and Be Able to Do*. Jossey-Bass: San Francisco 2005.

Day, Christopher/Lee, John Chi-Kin (Eds.): *New Understandings of Teacher's Work. Emotions and Educational Change*. Springer: Dordrecht/London/New York 2011.

Epstein, Terrie/Peck, Carla L. (Eds.): *Teaching and Learning Difficult Histories in International Contexts. A Critical Sociocultural Approach*. Routledge: New York/London 2018.

Evans, Ronald W./Saxe, David Warren (Eds.): *Handbook on Teaching Social Issues*. National Council for the Social Studies: Washington 1996.

Giroux, Henry A.: *Schooling and the Struggle for Public Life: Critical Pedagogy in the Modern Age*. University of Minnesota Press: Minnesota 1988.

Gross, Magdalena H./Terra, Luke (Eds.): *Teaching and Learning Difficult Pasts: Comparative Perspectives*. Routledge: New York/London 2019.

Gudmundsdottir, Sigrun/Shulman, Lee: "Pedagogical Content Knowledge in Social Studies". *Scandinavian Journal of Educational Research*, 31(2) 1987, pp. 59–70, retrieved on 20.10.2020, from DOI 10.1080/0031383870310201.

Harris, Lauren McArthur/Bain, Robert, B.: "Pedagogical Content Knowledge for World History Teachers: What Is It? How Might Prospective Teachers Develop It?". *The Social Studies*, 102, 2011, pp. 9–17, retrieved on 20.10.2020, from DOI 10.1080/00377996.2011.532724.

Kello, Katrin: "Sensitive and Controversial Issues in the Classroom: Teaching History in a Divided Society". *Teachers and Teaching: Theory and Practice*, 22(1), 2016, pp. 35–53, retrieved on 20.10.2020, from DOI 10.1080/13540602.2015.1023027.

Kelly, Thomas E.: "Discussing Controversial Issues: Four Perspectives on the Teacher's Role". *Theory and Research in Social Education*, XIV(2), 1986, pp. 113–138, retrieved on 20.10.2020, from DOI 10.1080/00933104.1986.10505516.

Kitson, Alison/McCully, Alan: " 'You Hear about It for Real in School'. Avoiding, Containing and Risk-Taking in the History Classroom". *Teaching History*, 120, 2005, pp. 32–37.

Loughran, John/Hamilton, Mary Lynn (Eds.): *International Handbook of Teacher Education* (2 vols.). Springer: London 2016.

Monte-Sano, Chauncey: "Learning to Open Up History for Students: Preservice Teachers' Emerging Pedagogical Content Knowledge". *Journal of Teacher Education* 62(3), 2011, pp. 260–272, retrieved on 20.10.2020, from DOI 10.1177/0022487110397842.

Placier, Peggy L./Letseka, Moeketsi/Seroto, Johannes/Loh, Jason/Montecinos, Carmen/Vásquez, Nelson/Tirri, Kirsi: "The History of Initial Teacher Preparation in International Contexts" In: John Loughran/Mary Lynn Hamilton (Eds.): *International Handbook of Teacher Education*. Springer: London 2016, vol. 1, pp. 23–68.

Roni, Saiyidi M./Merga, Margaret K./Morris, Julia E.: *Conducting Quantitative Research in Education*. Springer: Singapore 2020.

Schutz, Paul A/Pekrun, Reinhard (Eds.): *Emotion in Education*. Elsevier: Burlington/London 2007.

Seixas, Peter: "Beyond 'Content' and 'Pedagogy': In Search of a Way to Talk about History Education". *Journal of Curriculum Studies*, 31(3), 1999, pp. 317–337, retrieved on 20.10.2020, from DOI 10.1080/002202799183151.

Sheppard, Maia/Levy, Sara A.: "Emotions and Teacher Decision-Making: An Analysis of Social Studies Teachers' Perspectives". *Teaching and Teacher Education*, 77, 2019, pp. 193–203, retrieved on 20.10.2020, from DOI 10.1016/j.tate.2018.09.010.

Shulman, Lee S.: "Those Who Understand: Knowledge Growth in Teaching". *Educational Researcher*, 15(2), 1986, pp. 4–14, retrieved on 20.10.2020, from DOI 10.2307/1175860.

Shulman, Lee S.: "Knowledge and Teaching: Foundations of the New Reform". *Harvard Educational Review*, 57(1), 1987, pp. 1–23, retrieved on 20.10.2020, from DOI 10.17763/haer.57.1.j463w79r56455411.

Shulman, Lee S.: "PCK. Its Genesis and Exodus". In: Amanda Berry/Patricia Friedrichsen/John Loughran (Eds.): *Reexamining Pedagogical Content Knowledge in Science Education*. Routledge: New York/London 2015, pp. 3–13.

Van Boxtel, Carla/Grever, Maria/Klein, Stephan (Eds.): *Sensitive Pasts. Questioning Heritage in Education*. Berghahn Books: New York/Oxford 2016.

Van Hover, Stephanie/Hicks, David: "History Teacher Preparation and Professional Development". In: Scott Alan Metzger/Lauren McArthur Harris (Eds.): *The Wiley International Handbook of History Teaching and Learning*. John Wiley & Sons, Inc.: New Jersey 2018, pp. 391–418.

Zeichner, Kenneth: "Competition, Economic Rationalization, Increased Surveillance, and Attacks on Diversity: Neo-Liberalism and the Transformation of Teacher Education in the U.S.". *Teaching and Teacher Education*, 26(8), 2010, pp. 1544–1552, retrieved on 20.10.2020, from DOI 10.1016/j.tate.2010.06.004

Zeichner, Kenneth: "Advancing Social Justice and Democracy in Teacher Education: Teacher Preparation 1.0, 2.0, and 3.0". *Kappa Delta Pi Record*, 52(4), 2016, pp. 150–155, retrieved on 20.10.200, from DOI 10.1080/00228958.2016.1223986.

Zembylas, Michalinos: *Teaching with Emotion. A Postmodern Enactment*. Information Age Publishing: Greenwich/Connecticut 2005.

Zembylas, Michalinos: "Intentos por discernir la compleja imbricación entre emoción y pedagogía: contribuciones del giro afectivo". *Propuesta Educativa*, 1(51), 2019, pp. 15–29, retrieved on 20.10.2020, from http://propuestaeducativa.flacso.org.ar/wp-content/uploads/2019/11/PropuestaEducativa51-dossier-ZEMBYLAS.pdf.

Zembylas, Michalinos/Kambani, Froso: "The Teaching of Controversial Issues during Elementary-Level History Instruction: Greek-Cypriot Teachers' Perceptions and Emotions". *Theory & Research in Social Education*, 40(2), 2012, pp. 107–133, retrieved on 20.10.2020, from DOI 10.1080/00933104.2012.670591.

Belén Castro-Fernández, Ramón López Facal, Guadalupe Jiménez-Esquinas, and Daniel Schugurensky

Chapter 13 What can be learned from decontextualised heritage?[1]

Abstract: This paper analyses the perceptions of trainee primary education teachers on the inclusion of relevant and socially conflictive issues in heritage education. In university classes, a strategy based on collaborative learning and research methodology oriented towards the development of critical thinking is taught. Places which have been flooded by the construction of dams are studied due to the conflict which this causes between local interests and social benefit. This research employs mixed methods in order to analyse the perceptions of trainee teachers regarding conflictive heritage and its use in the classroom. The learning is evaluated in order to understand social problems and to analyse to what extent they consider that such issues can contribute towards constructing a socially committed citizenship. The results show that trainee teachers modify their perceptions regarding how to approach heritage education when research methodology is incorporated into their training process.

Keywords: Teacher training, Heritage education, Social problems, Conflict, Critical thinking

Introduction

The conflict which surrounds certain types of heritage is frequently attributed to situations of political violence in which oppression or social confrontation have led to pain and trauma. In such situations, heritage has been the object of attack during armed conflicts, opening up debate on the morality of the destruction of heritage (Frowe and Matravers, 2019). In Spain, the semantic attribute "conflict" has been linked with historical memory and the Francoist dictatorship (1936–1975). In the present day, uncomfortable heritages which have a symbolic legitimation but which nobody wants to recognise (Prats, 1997) coexist with others which the perspective of time and the desire for conciliation have sought to forget or to conceal.

---

1 This research is part of the coordinated projects PGC2018-094491-B-C31 with funding from the Spanish State Research Agency (AEI) and the European Regional Development Fund (ERDF).

So-called low intensity or low impact conflicts, in which heritage becomes a scenario of confrontation (Sánchez and Jiménez, 2016) are linked with everyday life. The decision to build dams which flood inhabited places gives rise to situations which are difficult to resolve if they are not handled democratically. The construction of a dam and the loss of a local heritage generate a social conflict with conflicting views regarding who has the legitimacy to take the decision to flood a territory or not. Furthermore, the mechanisms of restitution to be put in place for the area in question and how the possibility of moving buildings to a new location is evaluated are also conflictive issues.

Communities which create, protect and transmit heritage do not generally participate in this debate, giving way to a narrative which excludes the most vulnerable. In order to manage heritage conflicts and to promote peaceful coexistence, it is necessary for communities to regain the leading role. The ethical principles for the safeguarding of intangible cultural heritage drawn up by UNESCO state that the community must play a key role in the identification of the main threats to its heritage, among which is its decontextualization: "Communities, groups and, where applicable, individuals should play a significant role in determining what constitutes threats to their intangible cultural heritage including the decontextualization, commodification and misrepresentation of it and in deciding how to prevent and mitigate such threats"(UNESCO, 2015).

The term "decontextualised heritage" alludes to the process of the forced relocation of a community and, with it, the loss or alteration of the meaning and physical or intangible nature of its heritage. This process causes a conflict of interests and an exchange of opinions regarding the relocation of the group and the resignification of its heritage. It makes the recovery of local memory and the reconstruction of a shared identity difficult. Decontextualised heritage becomes a controversial issue and dealing with it in the classroom may favour the formation of a citizenship with a critical conscience and the capacity for acting accordingly.

The main objective of the present study is to identify the perceptions of trainee teachers regarding how to include relevant and socially conflictive issues when designing heritage education activities for primary schoolchildren. In their classes at university, they analyse certain places which have been flooded by the construction of dams. The learning sequence is based on collaborative learning and research methodology. Several dilemmas are proposed in order to encourage their investigative, ethical, emotional, reflective and argumentation skills. Situations are approached in which the loss of local heritage and the spheres of everyday coexistence have been imposed. The trainee teachers must evaluate the heritage and its local and cultural context.

The following specific objectives have been formulated:

- Specific objective 1 (SG1). To analyse the perceptions of trainee teachers regarding heritage and its use in the classroom.
- Specific objective 2 (SG2). To evaluate the learning of trainee teachers in the understanding of problems via the handling of sources, historical contextualisation, comprehension and critical argumentation and ethical commitment.
- Specific objective 3 (SG3). To analyse to what degree they assume the usefulness of heritage education in dealing with conflictive and socially relevant issues in the primary classroom.

The training of teachers in dealing with heritage conflicts

Heimberg (2010: 49) defines education for democratic citizenship as the "reasoned and critical study of social problems based on the disciplinary knowledge of the social sciences." Dealing with relevant social problems makes it possible to promote democratic habits, such as respect for conflicting opinions in a plural society. It requires the teacher to have developed civic skills and methodological strategies which facilitate interaction among pupils.

Conflicts are inherent to human relationships and studying real problems helps pupils to learn how to manage them (Soley, 1996). Working on controversial issues "stimulates critical conscience, helps to build opinions and reasoned judgments, to accept positions different to our own, promotes participation and connects the reality of the world, its conflicts and its complexity with the knowledge acquired at school" (Santisteban, 2019: 63).

The selection of the controversial issue is as important as the attitude of the teacher when approaching it in the classroom. Heimberg (2010) states that teachers should avoid false neutrality: they must show their point of view in order to demonstrate to their pupils that several legitimate opinions can coexist. López and Santidrián (2011) classify the way in which teachers approach conflictive issues: (a) the teacher avoids discussion: exclusive neutrality; (b) the teacher indoctrinates: exclusive partiality; (c), the teacher stays on the side-lines: neutral impartiality; (d) the teacher expresses his/her own opinion but does not impose it and encourages debate: committed impartiality. This research investigates whether future teachers assume the latter attitude when approaching social conflicts.

Heritage stirs up contradictory attitudes. At times, economic profit and political instrumentalisation interrupt or modify the process of social identification. Heritage tourism has played a significant role in the construction of national identity (Light, 2015). However, it often leads to discrepancies: a source of income or a sense of property – "for them" or "ours"; memory or amnesia – natural narrative or indifference due to lack of knowledge; desiccation or renovation – anachronic preservation or contemporary social: "The making of heritage may give rise to two opposing impacts simultaneously – increased social cohesion and place pride, on the one hand, and envy and competition (and, thus, social atomisation), on the other hand – and residents are totally cognizant of the tension between the two" (Silva, 2014, 14). Managing these disputes entails an exercise in seeking consensus among the community and other agents involved in the guardianship of heritage.

The practice of historical thinking makes it possible to understand interaction between the past and the present in order to take grounded decisions looking to the future. In teacher training, principles regarding the value of cultural heritage to society must be present in order to "establish processes for conciliation to deal equitably with situations where contradictory values are placed on the same cultural heritage by different communities;" (Council of Europe, 2005, Art. 7). Heritage is not something which has passed and is closed. Rather, it is a reviewable link from the present among the members of a community in order to understand the meanings of their union.

Method

The research hypothesis is that investigating a relevant and socially conflictive heritage problem employing case studies as a strategy facilitates the formation of critical thinking.

Tools

Four tools have been employed for measuring students' perceptions regarding how to approach relevant and socially conflictive issues from the perspective of heritage education, the improvement of knowledge and the semantic contextualisation of heritage and of its educational potential.

- Tools 1 and 2 (SG1). Pretest and posttest questionnaires on heritage and its use in the classroom.
- Tool 3 (SG2). Group work: A case study on decontextualised heritage.

- Tool 4 (SG3). A semi-structured questionnaire for individual critical reflection.

Tools 1 and 2 had previously been employed in a study with a group of trainee teachers in order to measure the transformative capacity of heritage education from an experience-based approach. On this occasion, the focus was on the traumatic past of a community which was forcibly removed to another location due to the construction of a dam which flooded a town of medieval origin (Castro and López, 2019). The same tools were used to confirm its validity. Changes in the conception of heritage and in discursive associations were measured.

Tool 3 collected narratives which were contextualised and argued in small groups on the study of a conflictive case and reflections after pooling information with the group as a whole. The aim was to show that learning based on the resolution of social problems contributes to forming social and critical thinking.

Tool 4 was submitted to validation by an expert in history and heritage education, which made it possible to adapt it in terms of structure and content. The changes introduced contributed to reducing the number of questions from 32 to 18, refining the statements to facilitate understanding and avoiding responses which were too open and dichotomic. The questionnaire was designed to analyse the students' reflections regarding: (a) the social conflict generated by the case study, (b) the role of education in developing the capacity for critical thinking when faced with social problems related with heritage; and (c) the strategy of critical training based on social problems dealt with in the classroom.

Characterisation of the sample

An intentional and deliberate sample of 29 students was selected for the pretest, 24 of whom were maintained for the rest of the tools. The 24 people who participated in all of the tools are taken as a reference point: 17 women and 7 men. Twenty of them were less than 25 years of age; three were between 25 and 30 and only one was older than 30.

Selection criteria: students from the 3rd and 4th years of the undergraduate primary education degree at the University of Santiago de Compostela during the 2019–2020 academic year who took the optional subject "History, heritage and education in Galicia." The group had some prior knowledge of heritage education. More than half (51.7 %) stated that they had obtained this knowledge during their degree, while a minority (10.3 %) had done so in an extracurricular manner.

Procedure

1st Phase. Starting point: pretest questionnaire at the beginning of the course

Twenty-nine pretest questionnaires containing 13 questions (tool 1) were distributed in order to discover the students' perception of heritage and its educational potential in accordance with objective 1 of this study.

2nd Phase. The contextualisation of the learning sequence

Work was carried out in class on the specific contents of the subject, such as heritage education for the construction of a democratic citizenship, the role of heritage in schools, the legal context of heritage education and the process of heritagisation and the construction of identities. At the same time, the decontextualised heritage of Portomarín, a small town in the interior of Galicia flooded in 1963 by the construction of a dam, was examined. A new town was built at a nearby location and some buildings were moved brick by brick. The trauma of the conflict remains latent but concealed among the town's population due to a lack of intergenerational dialogue (Castro and López, 2018, 2019). This is a controversial issue which seeks to promote the capacity of reflection among the students by way of research. In other words, they are encouraged to learn to apply concepts, to analyse social phenomena, to take decisions and to contribute alternatives in a reasoned manner.

3rd Phase. Development: research and case study

In a context of cooperative learning (Johnson, Johnson and Holubec, 1994) the students were required to investigate the flooding of Portomarín in small groups. This collaborative process of analysis of the case promotes responsibility and the students' control over their own learning, enabling them to plan, supervise and evaluate it (Coll, Mauri and Onrubia, 2006). It also contributes to developing the competence of learning to learn with a critical sense (Zahonero and Martín, 2012).

Working in small groups responds to objective 2, which seeks to evaluate the capacity for integral understanding of conflicts associated to heritage via the analysis of problems, the correct handling of sources, historical contextualisation, critical argumentation and ethical commitment. The students employ the above-mentioned skills and must also situate and characterise the educational potential of the case, and heritage contents in general, in primary education.

Four group assignments (tool 3) were made on the problems caused by the construction of dams in Aswan (Egypt), Capel Celyn (Wales), Uglich (Russia), and Ilisu (Turkey). These are all places in which the exploitation of hydroelectric energy generated conflict directly affecting the heritage of the local area. The forced removal to which the inhabitants of these places were subjected led to a process of the multifaceted revision of identity, similar to the case of Portomarín explained in class. When examining these cases, the students were required to identify and define the heritage elements in conflict, to situate the poles of the confrontation and, above all, to propose a possible adaptation for primary education.

Each group worked on a conflictive situation, employing a structured process of research: (1) presenting the case connected with the problem of conflictive heritage and arguing the social and educational value of the topic; (2) identifying and explaining the causes and intentions which led to the destruction of the heritage site in question; (3) constructing their own interpretation of the problem by way of contrasting data and relevant materials; (4) selecting and justifying four representative images of the dimensions analysed in the case study; (5) evaluating to what extent the problem could be adapted to the primary classroom in this context. This learning sequence aims to teach the students how to elaborate a historical explanation of a conflictive social problem.

The students presented their projects to their classmates and a pooling of information was organised regarding the application of the knowledge learned to nearby social conflicts. A virtual chat room was also made available in order to gather the students' opinions. In this way, it was hoped that the development of communicative competences could be encouraged by way of dialogue, informed opinion and the critical interpretation of the problems analysed.

4th Phase. Evaluation of the learning sequence and prospects for a desirable future: semi-structured questionnaire

Objective 3 of the study seeks to discover to what degree trainee teachers consider that heritage education can contribute towards building desirable future action based on socially relevant problems. A semi-structured questionnaire (tool 4) was employed, making it possible to reflect individually on the corresponding case study and to evaluate the learning sequence implemented.

Twenty-four questionnaires were completed which revealed the trainee teachers' conceptions on the research method for dealing with conflictive problems in the classroom, aimed at the socialisation and formation of their pupils as citizens.

5th Phase. Posttest questionnaire

In order to achieve the first specific objective and to contrast the data collected in the first phase, the 24 posttest questionnaires of six questions (tool 2) were distributed. This questionnaire evaluated the evolution of the students' knowledge and their associations regarding heritage and its educational use.

Data analysis

A mixed methodology, differentiated for each tool, was designed:

Quantitative

Quantitative methodology was applied for tools 1, 2 and 4. The data were processed with the SPSS program. A joint database for tools 1 and 2 was created, along with a specific one for the critical reflection of tool 4. The variables were categorised, and the open questions recoded according to a criterion of simplicity in order to facilitate the operability and comparability of the responses. Subsequently, a statistical analysis of the data was carried out. In the first stage, a frequency analysis was carried out in order to discover the direct results of the questions. Afterwards, a chi-square significance analysis was carried out in order to explore the relationships between the different variables. The values yielded were not conclusive due to the reduced size of the sample. In addition to the frequency analysis, descriptive analyses were carried out in the measurement or evaluation scale variables in order to obtain mean scores.

As far as the words solicited in the pretest and posttest questionnaires regarding heritage and Portomarín are concerned, a frequency analysis was carried out in order to dump these results in the Nvivo program, which then generated word clouds. In addition to this quantitative exploitation, the questions were also analysed qualitatively, taking into consideration the semantic field of the words obtained. This procedure is explained in the following section.

Qualitative

The research contained three elements which required qualitative analysis: the words solicited in tools 1 and 2, the work carried out by the students on a case study of decontextualised heritage, which constitutes tool 3, and a large proportion of the responses and reasonings contained in tool 4. These are sources with different qualities which have required individualised approaches.

A frequency analysis was carried out in order to discover the incidence of the different words in each question. These words were categorised in a deductive manner and were compared between their stages, paying particular attention to the beginning and the end of the learning sequence developed. In order to facilitate the visualisation of these results, the frequencies were dumped in the Nvivo program, by way of which word clouds were obtained.

A double analysis was carried out of the students' work. First of all, a critical reading was made in order to identify contents relating to the teaching of decontextualised heritage. The different components which proved to be representative of the topic of the research were placed in a table, from which the critical analysis was complemented by the visualisation of word clouds.

In order to analyse the critical reflection (tool 4), the students' responses were read and a frequency analysis of the recoding of their justifications and reflections was carried out.

Results

Specific objective 1. To analyse the perception of trainee teachers regarding heritage and its use in the classroom

The comparative analysis of tools 1 and 2 indicate an increase in the consideration that emotions have an influence on the preservation of heritage. The mean score increased from 4.62 in the pretest to 4.75 in the posttest. Between these two moments, the students matured their justifications and related them more with heritage education in the posttest (Tab. 1). The change of perspective stands out compared to emotions as conditioning factors of perception: from a majority of answers in the pretest to a minority in the posttest. The importance attributed to emotions in strengthening or generating a link with heritage remained stable. The value attached to the role of emotions in the development of feelings of identity in cultural terms increased. In other words, a change occurred from primary recognition of emotions as a factor in perception to an argumentation construction, which explains the transformation of the justifications focused on perception to those which point towards their effect on the link with heritage and its preservation.

The students evaluated the importance (on a scale of 1 to 5) of different educational activities relating to heritage (Fig. 1). The most highly-valued option in

Tab. 1: Evaluation regarding emotions

Justifications	Pretest		Posttest	
	N	%	N	%
They affect our perception and capacity of identifying heritage	12	41.4	4	16.7
They generate/strengthen the link with heritage	10	34.5	9	37.5
They foster a good attitude towards preservation	4	13.8	6	25
They strengthen feelings of ownership/identity	1	3.4	5	20.8
They need to be taught at school	1	3.4	-	-
They develop community links	1	3.4	-	-
Total	29	100	24	100

Fig. 1: Mean evaluation of educational activities

both questionnaires was direct contact with heritage elements through visits to sites or specific places (the score improved in the posttest). The development of the subject served to increase their evaluations of all the educational activities proposed. Talks by specialists was the element whose score improved the most, going from 2.9 to 3.5. The worst-valued element in both cases was explanation by the teacher, which did not exceed a score of 3 out of 5.

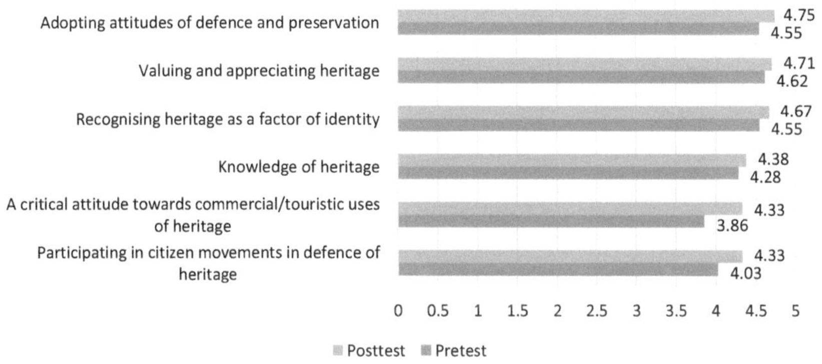

Fig. 2: Mean evaluation of educational contribution

As far as the contribution of heritage to primary education is concerned, in both cases, the change in attitude towards defence and preservation which can be generated among schoolchildren stands out, as it became the best-valued element between the pretest and the posttest (Fig. 2). The value and appreciation of heritage is the item with the next best score, although it went from being the best-valued item in the pretest to having the second-best score in the posttest. The perception of all of the educational contributions improved. However, the improvement in the critical attitude towards commercial and touristic uses particularly stands out, as its mean score changed from 3.86 to 4.33.

Furthermore, semantic associations were solicited with "heritage," in order to measure the dominant conception, and others with "Portomarín" which was the decontextualised heritage explained and analysed in the classroom (Tab. 2). In relation to the 10 words which the students had to indicate in relation to "heritage," 135 different words were collected in the pretest and 100 in the posttest. The most repeated words in both phases were "culture," "history" and "identity," although their specific weight varied. However, it cannot be stated that this change, in which "identity" became the word with most incidences, is due to a greater degree of importance being attributed to this issue (it only increased from 12 to 15 repetitions). Rather, it may be due to the fact that the incidences are more widely distributed and concentrated in other words which have a clearer relationship with the content worked on in the subject.

The students were required to indicate words related with Portomarín. To begin with, they were asked whether they knew anything about the history of the town, with only a minority (7 %) stating that they had some knowledge of

Tab. 2: Association of words with heritage: 10 words with the highest incidence

Pretest		**Posttest**	
Culture	23	Identity	15
History	17	Culture	14
Identity	12	History	12
Society	10	Inheritance	8
Art	8	Preservation	8
Monument	7	Respect	7
Surroundings	7	Education	7
Past	7	Feelings	6
Important	6	Link	6
Architecture	6	Protection	6

it. In the pretest, 10 words evoked spontaneously were solicited, followed by another 10 after having searched for information about the town. In the posttest, the students were only required to indicate 10 words in a spontaneous manner. Ninety-five different words were obtained in the spontaneous exploration of the pretest, 115 after searching for information and 118 in the posttest (Tab. 3). In addition to a greater semantic variety between the two tools, which reflects the evolution of knowledge, there are changes which extend beyond errors, such as the reference to the sea in the initial spontaneous association, which confirms the lack of knowledge of the history and location of Portomarín, which is located in the interior of Galicia. Their knowledge evolved towards the educational content with regard to heritage. Between the spontaneous and informed responses a semantic evolution took place towards the specific, and correct, geographical location and more specific heritage elements linked to tourism were highlighted. However, the spontaneous responses in the posttest emphasise the heritage conflict. It is also significant that one of the five most used words in the spontaneous responses of the posttest related to a feeling. Indeed, semantic contents related with feelings can only be found in the responses to the second questionnaire. Ultimately, the semantic associations solicited from the students show an evolution in the capacity of analysis regarding heritage. The participants went from evaluating a geographical location in its most superficial aspects to a resignification in terms of heritage.

The semantic associations solicited with "heritage" and "Portomarín" yielded relatively similar results to those obtained in a prior study via the data collection tools 1 and 2 (Castro and López, 2019).

Tab. 3: Association of words with Portomarín: 10 words with the highest incidence

Pretest Spontaneous		Pretest Search		Posttest Spontaneous	
Sea	14	Lugo	14	Dam	11
Fishing	14	Camino de Santiago	13	Heritage	10
Galicia	14	River Miño	10	Removal	8
Port	12	Church	9	Identity	7
Town	9	Bridge	8	Loss	7

Specific objective 2. To evaluate the learning of trainee teachers in the comprehension of problems

Tool 3 sought to measure the scope of the research by carrying out group work. The data on questions which were expected to be found are summarised in Tab. 4.

All of the assignments contained definitions of heritage offered by UNESCO or by experts, and heritage typologies affected by the construction of dams were identified, albeit with reflections of differing levels of depth. Those students who examined in more depth the effects of the conflict were also those who formulated a more complete educational approach, particularly as far as the development of competences relating to empathy and favourable attitudes towards the preservation and care of heritage are concerned. The students understood and applied their analytical capabilities, above all towards aspects linked with material elements and the population and demonstrated more limitations in evaluating dimensions such as landscape and intangible heritage.

Difficulties were identified in searching for information, not about the historical event in itself, but rather concerning the conflict of interests which it provoked. A lack of the use of empirical data for the contextualisation of the problems studied seems to indicate a low level of ability in the use of sources and critical resources on the part of the students. Thus, they are unable to analyse the problem by confronting different points of view. Only one of the assignments made use of maps and only one provided data in order to empirically evaluate the impact of these conflicts on a social and economic level. However, none of them made mention of the evolution of the situation once the dam had been built.

The groups unanimously considered it of value to include this type of problem relating to heritage in the contents of the primary curriculum. Three of the groups even believed that, with the corresponding adaptation to a similar event

Tab. 4: Comparison of the assignments on decontextualised heritage

Analysis	Dimension	Aswan	Capel Celyn	Kalyazin	Ilisu
Heritage education	Definition of heritage	Yes	Yes	Yes	Yes
	Characterises and distinguishes typologies of heritage affected	Yes	Yes	Yes	Yes
	Argues for the educational application of the case study	Yes	Yes	Yes	Yes
Identifies different levels of impact	Population	Yes	Yes	Yes	Yes
	Cultural	Yes	Yes	Yes	Yes
	Material heritage	Yes	Yes	Yes	Yes
	Intangible heritage	Yes	No	No	No
	Natural	Yes	Yes	No	Yes
Contextualises	Historically	Yes	Yes	Yes	Yes
	Geopolitically	Yes	Yes	Yes	Yes
	Sociologically	Yes	Yes	Yes	Yes
	Contrasts benefits and losses	Yes	No	No	No
	Identifies and argues the condition of victims	Yes	Yes	Yes	Yes
Provides empirical data	On economic impact	No	No	No	No
	On social impact	Yes	No	No	No
	Maps	No	No	No	Yes

of more geographical or cultural proximity, the same case could be worked on in the 6th year of primary education (11–12 years of age). The fourth group considered that working on the Uglich dam in Kalyazin was not very relevant due to its distance in terms of time and geography and because dealing with nearby cases implies a greater availability of resources and visits or talks by experts.

Specific objective 3: To analyse to what degree they assume the usefulness of heritage education in dealing with conflictive and socially relevant issues in the primary classroom

In order to measure the third objective of this study, the students were required to respond to a questionnaire (tool 4), which sought to analyse: (a) their critical capacity regarding the social conflict generated by the decision to build a dam and flood the town, (b) their evaluation of a heritage education model committed to the resolution of socially conflictive problems, and (c) their evaluation of the learning sequence developed in the subject based on the research strategy of the case study.

Heritage and social conflict

The assignments were required to discuss the appropriacy or not of the decision to build the dams and the questionnaire returned to this issue in order to explore in greater detail the students' opinions. All of them, without exception, considered that the decision to build the dams was not correct as the opinions of the local population were not taken into account and due to the fact that the benefit obtained did not compensate for the loss of heritage. 37.5 %justified the dams as the search for the lesser evil (Fig. 3), although the majority considered that there were other alternatives or that a greater social consensus should have been sought. 29.2 % of the students made reference to the lack of participation of the local population or, rather, the complete inexistence of consideration for the affected population. On the other hand, almost 21 % stated that if other options were in existence, other geographical alternatives should have been considered to the eventual location of the dam and that more prospective studies should have been carried out in order to exhaust all possibilities before proceeding to flood a town and eliminate its heritage. 12.5 % of the students considered that the heritage was of more value than the hydroelectrical infrastructure. Only one of the 24 students, in reference to the case of Kalyazin, considered that there was no other option than to build the dam.

The objective of this study is not to discover the detailed results of each of the cases studied, given their different geopolitical or procedural situations. However, it seems interesting to compare these justifications in order to

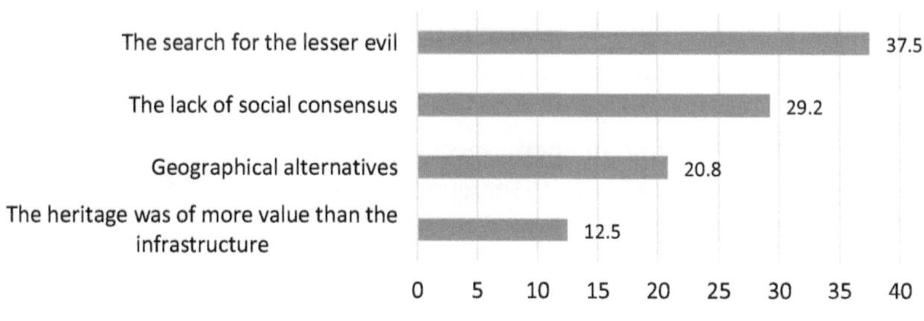

Fig. 3: Justification of the alternatives to the construction of the dam (%)

Tab. 5: Justification for the alternatives to the construction of the dam according to the case analysed

Justification	Aswan (Egypt)		Capel Celyn (Wales)		Kalyazin (Russia)		Ilisu (Turkey)		Total	
	N	%	N	%	N	%	N	%	N	%
Geographical alternatives	0	0	1	16.7	2	28.6	2	33.3	5	20.8
The search for a lesser evil	3	60	0	0	3	42.9	3	50	9	37.5
The heritage was of more value than the infrastructure	1	20	0	0	1	14.3	1	16.7	3	12.5
A lack of social consensus	1	20	5	83.3	1	14.3	0	0	7	29.2
Total	5	100	6	100	7	100	6	100	24	100

visualise the capacity of analysis and the identification of the conflictive elements (Tab. 5). The case of Capel Celyn has been characterised as a conflict in which social consensus was one of the main factors. The remaining cases have been identified with a greater general conflict and the justifications have been divided between the need to search for geographical alternatives or to search for the lesser of two evils. It is also worthy of note that the issue of heritage did not have greater prominence in cases such as that of Aswan, which are usually more clearly associated with this issue. Consensus or the search for social agreement with regard to the construction of the dam has proved to be a crucial element in the definition of the conflict.

Imagining the occurrence of a similar situation to the cases analysed today, the students clearly considered that legitimacy in the decision to flood a town should fall upon the community via its local government, with a score of 4.04 out of 5. The role of the national government, taking into account the opinion of society as a whole, received a score of 2.46. This score seems to indicate that the students clearly identified that the community should participate in the resolution of the conflict as the main agent affected by the decision.

Compensation is a fundamental aspect of the social evaluation which these types of conflicts end up having. Therefore, it was also desirable to know the students' perception regarding how the loss of heritage could be compensated in a similar situation which would lead to the loss and modification of local heritage (evaluated from 1 to 5). In general terms, none of the solutions for

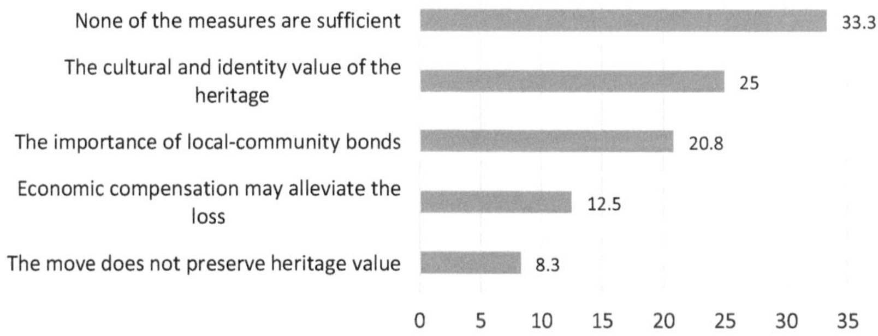

Fig. 4: Justification of compensation for the construction of the dam (%)

compensation proposed received very positive evaluations: economic compensation to continue living in another place (2.29); the removal brick by brick of certain buildings and important constructions to the new site of the community (2.67); living in proximity to the flooded site so that the community can maintain the links between its members (2.29). When asked for a justification in this response, the majority stated the insufficiency of these possible measures (33.3 %). Value in terms of the culture and identity of heritage is another of the most frequently expressed arguments, along with the importance of the component of space in the construction of community links (Fig. 4).

In the majority of the cases analysed, when the dam was built, heritage elements in the area to be flooded were transferred to a new location, and our aim was to look in depth at the potentialities of this fact. In general, the students showed a negative evaluation in terms of the relationship between community and identity in the local population. 83 % considered that this type of move converts heritage elements into a tourist attraction for outsiders, thereby limiting or reducing the community's sense of identification with them. When justifying this answer, the weight of the argumentation, in the majority of the cases (79 %), fell on the situation of the decontextualisation of the heritage. However, 21 % of the participants believed that these elements can be taken advantage of for tourism and that this can, to a certain extent, repair part of the loss derived from the move. The most significant point, in any case, is the extreme importance given to the concept of contextualisation, one of the key aspects of the subject, thereby showing the greater semantic wealth with regard to heritage that dealing with these case studies and the contents of the subject has brought to the students.

Heritage education and critical thinking

The students were asked whether the cases of decontextualised heritage analysed, with arguments both in favour and against, would be suitable for teaching to children from the 4th to the 6th years of primary education (9–12 years of age). The great majority stated that these issues could be taught in order to learn to take decisions when faced with possible social problems which may arise in the schoolchildren's environment (91.7 %). The remaining students considered that this is not a very relevant issue for children of these ages. As far as the appropriacy of dealing with these social problems in schools located in the area where the conflict arose is concerned, in general, it was considered necessary mainly so that the memory and history of the place should not be lost, and also to contribute towards conciliation. The possibility of not approaching these issues due to the risk of opening up old wounds was valued extremely negatively (Fig. 5).

Based on the reading of the paper "Los conflictos sociales candentes en el aula" (Controversial social conflicts in the classroom) (López and Santidrián, 2011), the majority (92 %) considered that committed impartiality is the most appropriate attitude for the teacher to adopt in order to work on conflictive heritage in the primary classroom, followed equally by neutral impartiality (4 %) and exclusive neutrality (4 %). No students selected the fourth option proposed of exclusive partiality. On the other hand, all of the competences proposed for the improvement of learning associated to dealing with this problem in primary schoolchildren's surroundings were valued favourably with scores of above 4 out of 5 (Fig. 6). The fostering of critical questioning was the best valued (4.75), followed by the argumentation of critical and informed opinions (4.71) and the fostering of civic competence (4.67).

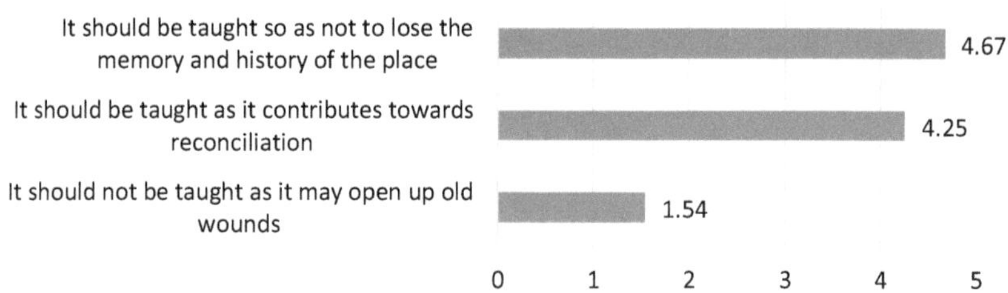

Fig. 5: Mean evaluation of the reasons for teaching, or not, the decontextualised heritage in the schools of the places affected

Fig. 6: Mean evaluation of the educational usefulness of the decontextualised heritage. Note: Each student was able to select more than one option

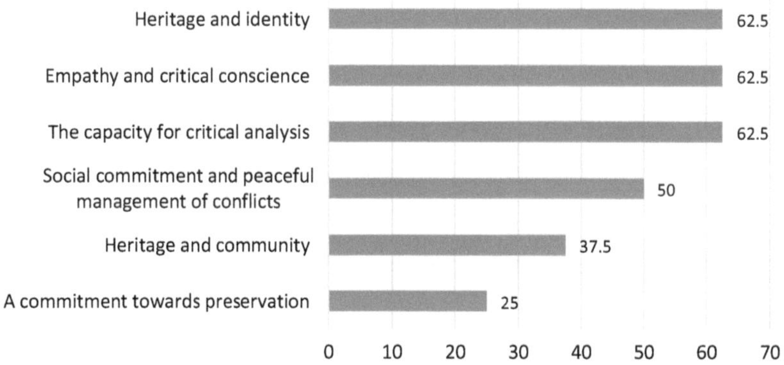

Fig. 7: Desired learning competences (%)

Last of all, the students were asked to state the three things which they would like primary schoolchildren to learn when dealing with a social problem such as their case study in the classroom. Given that this was an open question, the responses were recoded in order to be able to operate with them (Fig. 7). The relationship between heritage and identity, empathy and critical conscience and the capacity for critical analysis are the three elements mentioned the most, present in 62.5 % of the responses. 50 % also mentioned learning how to manage conflicts peacefully and social commitment

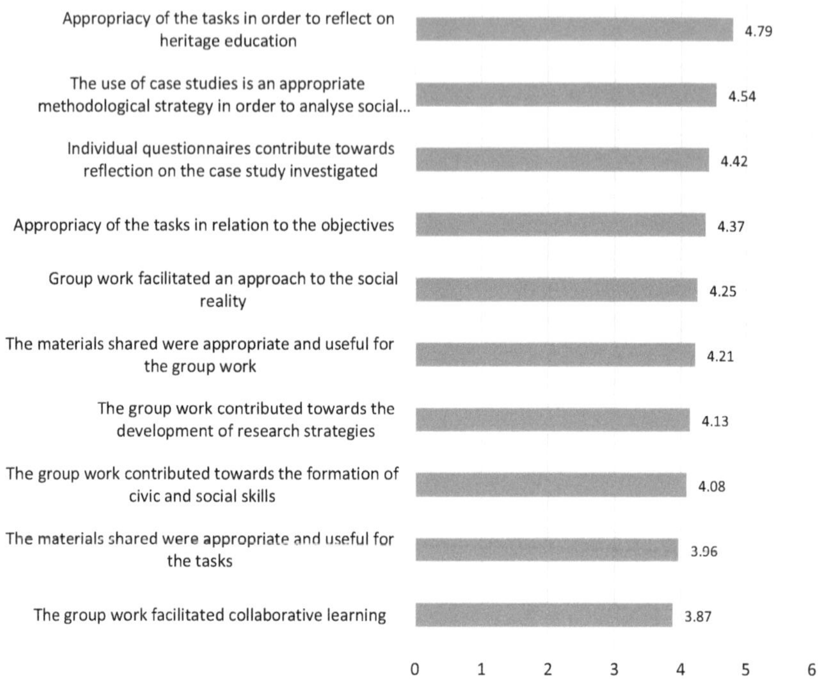

Fig. 8: Evaluation of the learning sequence developed in the subject

Social problems and teacher training

The learning sequence developed in the classroom obtained a positive evaluation (Fig. 8). The two most highly-valued aspects were the suitability of the tasks carried out in order to reflect on the role of heritage education and the use of case studies as a methodological strategy for analysing social reality.

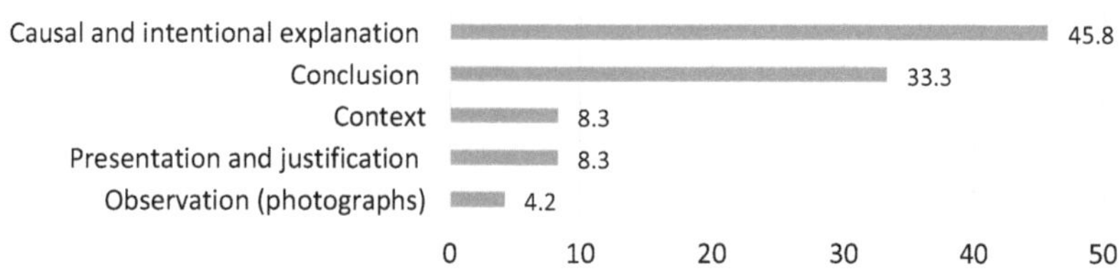

Fig. 9: Evaluation of the case study (%)

Tab. 6: To what extent can I incorporate the contents of this subject into a teaching proposal on socially conflictive problems?

Aims of heritage education	*Heritage*	■ "What I liked most about the subject was learning to give importance to heritage. Beforehand, I was not aware of everything that this concept encompassed or of the need to know how to look after it and how to work on it in the classroom with our pupils" (A10).
	Historical thinking	■ "I did not know how important it would be for me, as a future teacher, to work on these cases [. . .]. However, as I contextualised, analysed and reflected on my case study, I realised the great benefits it presented" (A5).
	Critical thinking	■ "In primary classes, it is unlikely that we would work on these cases due to a lack of preparation [. . .] these cases allow us to consolidate all of the competences contemplated in the curriculum and to develop that critical and reflective thinking in the pupils, based on their own interests, with the aim of them becoming autonomous beings capable of taking their own decisions in everyday life, learning to value what is theirs in order to be able to do so with the rest, in order to create a better future" (A5).
	Social thinking	■ "(I would like to emphasise) the importance of fostering a critical attitude towards all of the cases approached, to involve the children in all of the realities and to encourage them to establish links with society and to be aware of the importance of heritage in their lives, their history and their identity" (A1).
Research strategy	*Emotional trigger*	■ "In tackling an issue such as heritage, I believe that the emotional aspect is extremely important along with involving the children so that they feel part of the reality which we wish to show them" (A1).
	Case study	■ "I must say that not only did the contents of the subject help me to reflect, but also the methodology which we carried out, as a concept so ambiguous as heritage education would be extremely difficult to understand in its totality only by making use of educational materials" (A7). ■ "Working by seeking for information implies a process of exploration of the world which implies asking many questions and making different discoveries. Furthermore, it implies that pupils take the leading role and become autonomous . . . the fact of working on real cases leads to an increase in the interest and curiosity of the pupil" (A9).
	Perspective of analysis	■ "A good way of speaking about socially conflictive places could be [studying] cities or towns which the children know nothing about, but in such a way they will not feel identified and will not experience the topic with the same enthusiasm and interest as with a place which they can visit someday" (A8).

As far as the research process carried out by way of group work is concerned, the sections which were considered to be of most use in applying the contents of the subject were: causal and intentional explanation of each case (45 %) and the prospective educational outlook solicited in the conclusion (33.3 %). This result is particularly significant in terms of the perception that the students built on their development of critical thinking. Some distance behind were both the historical and geopolitical context and the social educational presentation and justification of the case. The least-valued aspect was the use of photographic documentation as an iconographic source for measuring the scope of the conflict (Fig. 9).

The degree of satisfaction of the students with the functional learning acquired is high; they appreciate the potentiality of working on the analysis of social problems in primary education by way of cooperative work. Their opinions clearly show that they attribute sense to their learning and that they value the transformative capacity of heritage education in order to acquire useful knowledge which can be transferred to everyday life within the framework of a democratic education. In this regard, they considered that heritage education contributes towards the development of key competences, such as conflict resolution, critical capacity and civic commitment (Tab. 6).

Conclusion

The hypothesis has been confirmed that researching about a socially conflictive problem by way of heritage education contributes towards the formation of critical thinking if case studies are employed as a knowledge strategy. The results obtained seemed to confirm that the future primary teachers have learned the educational potentialities of dealing with cases of decontextualised heritage in the classroom. A positive evolution of their social thinking and capacity for analysis has been observed, particularly in terms of how their semantic universe around "heritage" and the particular conflict of Portomarín explained in the class, has gone from being merely descriptive to incorporating elements of heritage evaluation in material, intangible and emotional terms. The students have demonstrated great accomplishment in dealing with their case studies. Not only have they adequately developed the capacities required in the handling of information (reasoned use of sources, critical and committed thinking), but they have also argued about the educational use of their case studies, searching for a place for them in the primary curriculum and proposing alternatives for adapting the content to different year groups and to the local area of their future pupils. Finally, the students have also recognised the contribution of heritage education in the development of their critical capacities for dealing with social problems (identification, contextualisation, explanation, argumentation and discussion), and in the development of key skills for primary schoolchildren.

References

Castro-Fernández, Belén/López-Facal, Ramón. "Patrimonio cultural y competencias sociales: bases para una propuesta de intervención didáctica en Portomarín". *Educatio Siglo XXI*, 36(1), 2018, pp. 129–148, from DOI: 10.6018/j/324201.

Castro-Fernández, Belén/López-Facal, Ramón. "Portomarín: la memoria herida de un desarraigo". *Revista Electrónica Interuniversitaria de Formación del Profesorado*, 22(2), 2019, pp. 95–110, from DOI: 10.6018/reifop.22.2.363841.

Coll, César/Mauri, Teresa/Onrubia, Javier. "Análisis y resolucié Luis/ Gutiérrez, Sué: "Evaluación de programas educativos en museos: Una nueva perspectiva". *Bordón 69* (1), 2016, pp. 45–65.

Council of Europe. (2005). *Framework Convention on the Value of Cultural Heritage for Society.*

Frowe, Helen/Matravers, Derek. "Conflict and Cultural heritage: A Moral Analysis of the Challenges of Heritage Protection". *J. Paul Getty Trust Occasional Papers in Cultural Heritage Policy*, 3. Getty Publications: Los Angeles 2019.

Heimberg, Charles. "¿Cómo puede orientarse la educación para la ciudadanía hacia la libertad, la responsabilidad y la capacidad de discernimiento de las nuevas generaciones?". *Íber Didáctica de las Ciencias Sociales, Geografía e Historia*, 64, 2010, pp. 48–57.

Johnson, David/Johnson, Roger/Holubec, Edythe. *Cooperative Learning in the Classroom*. Association for Supervision and Curriculum Development, Alexandria: Virginia 1994.

Light, Duncan. "Heritage and Tourism". In: Waterton Emma/Watson Steve (eds.): *The Palgrave Handbook of Contemporary Heritage Research*. Palgrave Macmillan: London 2015, pp. 144–158.

López-Facal, Ramón/Santidrián, Víctor. "Los «conflictos sociales candentes» en el aula". *Íber Didáctica de las Ciencias Sociales, Geografía e Historia*, 69, 2011, pp. 8–20.

Prats, Llorenç. *Antropología y patrimonio*. Ariel: Barcelona 1997.

Sánchez-Carretero, Cristina/Jiménez-Esquinas, Guadalupe. "Relaciones entre actores patrimoniales: gobernanza patrimonial, modelos neoliberales y procesos participativos". *Revista PH*, 90, 2016, pp. 190–197, from DOI: 10.33349/2016.0.3827.

Santisteban Fernández, Antoni. "La enseñanza de las Ciencias Sociales a partir de problemas sociales o temas controvertidos: estado de la cuestión y resultados de una investigación". *El Futuro del Pasado*, 10, 2019, pp. 57–79, from DOI: 10.14516/fdp.2019.010.001.002.

Silva, Luis. "The Two Opposing Impacts of Heritage Making on Local Communities: Residents' Perceptions: A Portuguese Case". *International Journal of Heritage Studies*, 20(6), 2014, pp. 616–633.

Soley, Mary. "If It's Controversial, Why Teach It?". *Social Education*, 60(1), 1996, pp. 9–14.

UNESCO. (2015). *Ethics and Intangible Cultural Heritage.*

Zahonero, Adela/Martín, Mario. "Formación integral del profesorado: hacia el desarrollo de competencias personales y de valores en los docentes". *Tendencias pedagógicas*, 20, 2015, pp. 51–70.

Part III Heritage and ICT in the training of geography and history teachers

Olaia Fontal Merillas and Silvia García-Ceballos

Chapter 14 Heritage education in teacher education and training[1]

Abstract: The necessary presence of heritage content in teacher's formation makes us propose, in the courses for future teachers, more globalized subjects that integrate the values and social commitment to heritage. In practice, we propose a sequencing of heritage teaching through methodologies that combine research, collaboration and awareness in a continuous and significant way. This chapter reflects on some of the issues related to the implementation of the discipline and the necessary search for common criteria in the comprehensive preparation of future teachers. In this sense, it proposes a definition of competencies to be developed in heritage education, a proposal shaped from the practice in the classroom with a sample of n=132 students that, after an exploratory study, show the existing disconnection between the knowledge that they must transfer to their future students and the knowledge that they have acquired during their formative stage. At the same time, the transformation of their reductionist knowledge towards an integral and humanist definition of heritage is evident after the implementation of an educational program based on awareness and connection with their context. An education that poses challenges and questions, that provides tools and proposes an experiential learning for future practices.

Keywords: Heritage education, Teacher training, Competencies, Awareness, Social Science

Introduction

The latent need to implement heritage education within teacher training programs in a continuous and meaningful way has in recent years, become one of the discipline's main challenges. This challenge leads us to insistently propose educational strategies that explicitly integrate the teaching of heritage-related contents into the social sciences education curriculum. In addition, there is a commitment to the integration in the classroom of the relational method (Fontal, 2003; 2013), since in the wake of several studies that have

---

1 This work is part of the research funded by the National R&D&I Plan of the Spanish Ministry of Economy (PID2019-106539RB-100) and co-funded with ERDF funds of the EU.

implemented such an approach (Fontal, 2013; Fontal & Marín, 2018; Fontal & García-Ceballos, 2019b), we now have the evidence that educational proposals inspired by this framework are effective. The purpose of introducing this method in the classroom is to provide relations and experiences that bridge the educational gap between our students and heritage assets. The ultimate goal is to raise the awareness of trainee-teachers around an open-minded, respectful and appreciative perspective towards their own legacy as well as that of others, whether it be individual or collective; past, present or future. Eventually, our students should develop a commitment derived from that experience, and from the assessment and identification processes involved in it (Gómez-Redondo, Fontal, & Ibáñez-Etxeberria, 2016); they should understand the importance of employing a teaching style based on social and civic values where heritage and its manifestations are key to understanding the former (Domínguez & López-Facal, 2014). Only if this chain of events is activated through an enriching experience that elicits an emotion or an echo in the prospective teachers, will the latter resonate with them until it is transmitted to future students.

In the present chapter new approaches will be highlighted that require the unification of criteria on educational standards in heritage issues and the mandatory compliance with these standards in formal education classrooms. What leads students to think that heritage is worth preserving? Who does heritage belong to and how do we guide them so that they make heritage into something of their own? What reasons underpin the assignment of value to heritage assets? What social or individual function does this assignment of value serve? What measures can we propose to make heritage sustainability possible and promote its care and conservation? Are heritages plural and multi-factor? Do they have multiple meanings and are they interchangeable? These are some of the issues around which our practice revolves as it pursues the development of teaching competencies in heritage matters, which means that our initial steps must be based on thoughtful consideration. Proposals based on active and comprehensive learning rely on inquiry, problematization, cooperation and heritage awareness.

In this way, we intend to forestall the common tendency to perpetuate insulated subjects that produce uncritical or purely mimetic teachers estranged from contextual reality and disconnected from society's interests. This is why, inspired by experiential practice, we support a participatory community that becomes aware of its own heritage, questions its value and challenges its function and its role in today's society: one that addresses issues regarding heritage conservation and dissemination and that provides teachers with teaching methods, instruments and procedures for their future practice.

The links along the chain: expectations and reality

It has long been suggested that teacher training in heritage education is scarce (Cuenca, Estepa, & Martín-Cáceres, 2017; López-Facal, 2015; Martínez &Fontal, 2020) and that, therefore, we are in front of a top priority need in the formative degrees linked to education. Despite the fact that more and more recognition has been demanded (Estepa, Ávila, & Ferreras, 2008), expectations have not been met. Only by increasing the presence of heritage education in this process, will we be able to train competent professionals who inspire the engagement of the present and future society with the inherited legacy, the legacy we are currently building and the one to come. The continual struggle to make this discipline visible (Fontal & Ibáñez-Etxeberria, 2017) has meant that for those closely involved in this line of research the term has become familiar, straightforward and clear; however, this understanding does not match the reality of the classroom (Fontal, García-Ceballos, & Aso, 2020). This is why we would like to pose two key questions, the first one being whether all education professionals understand what we mean by this term, since authors such as Calbó, Juanola, and Vallés (2011) or Cantón (2009) point out that the idea of a heritage by and for people has already been internalized and even surpassed by educators and institutions. The second, no less important question is whether society at large, the main agent along this chain of education and awareness, understands the breadth, uses and plurality inherent to the meaning of the term heritage (Smith, 2006).

In a recent study on the presence of heritage education throughout the whole of compulsory education, the results show that 80 % of respondents claim that they received superficial training, while 50 % state that they do not remember or did not participate in any heritage-specific project or activity. These data underscore the fact that there are broken links along the chain that prevent the latter from being completed. Whether or not there was a heritage-based educational approach, it can't have been significant enough. In the area of teacher training, it is observed that without a greater presence of heritage education, there is no correlation between the knowledge prospective teachers must transfer and the one they currently possess (Fontal & García-Ceballos, 2019b). For a large number of education students, the more formalistic and traditional sense of the term still survives (Fontal et al., 2020). Comparisons of the heritage categories referred to by students before and after receiving specific training yield very significant data. Following that specific training, they provide much richer definitions which expand and transform the concept of heritage towards a symbolic, identity-driven and humanistic perspective. They are able to identify the uses, values and effects that heritage contributes to them on a personal level and as members of society, and to define it in a way that expresses a responsible commitment (Fontal et al., 2020). So far, the actual knowledge of these teachers-in-training is far from reaching the desirable educational standards in matters of heritage (BOE, 2015; Domínguez & López-Facal, 2017b; Martín-Cáceres & Cuenca, 2015), which otherwise have as yet no clear definition or substantial development (Luna, Vicent, Reyes, & Quiñonez, 2019).

An incipient yet incessant discipline

More and more research hubs at the national level are committed to heritage education, as it is a global discipline present across several areas of compulsory education. This leads us to look back on the discipline's origin. While it has experienced a marked growth, we must not forget that in Spain heritage education began to develop no more than 25 years ago from the so-called didactics of heritage (Prats, 1997). Hardly 20 years have elapsed since the latter discipline morphed into the most common current conception of heritage education within the framework of the bonds-based relational educational approach proposed by Fontal (2003). This approach, which turned into a fundamental pillar of the discipline, has endured due to its effectiveness while being enriched by nuances such as the identization processes (Gómez-Redondo et al., 2016), the archaeology of bonds (Fontal & Marín, 2018), the quality standards in educational planning (Fontal & García-Ceballos, 2019a), or the latest contribution, included in this chapter: the teaching competencies in heritage education (Fontal et al., 2020).

This relational dimension of heritage, defined as the relationship of a subject with an asset that awakens feelings of cohesion, belonging or commitment, enters a conversation with our very genesis due to its being perceived as belonging to ourselves or to our collectivity; it causes and/or awakens in the subject a valuable experience that makes it meaningful. However, despite recognizing the richness of its implementation and the fact that it possesses a remarkable theoretical-practical consistency, its true transmission and integration in formal education requires a slow and gradual process, like any educational change whose overall significance does not become fully apparent until after a few years. This process has not yet completed its first cycle, since a significant part of the professionally active educators have not received training including the presence of heritage. This fact shows that despite ongoing efforts, there continues to be a large portion of professionals – and a large part of society – who have not received any heritage-related education in their classrooms or in their training programs. For this reason, we still continue to delve into the discipline's foundations and to promote its inclusion in teacher education and training along a common path that we must follow according to a set of rules, i.e. the competencies and skills to be attained.

Competencies in heritage education

The Council of Europe incorporated competencies into the education system in 2006, thus introducing a significant change in the concept of training. The term "teaching competencies" has been discussed by multiple authors and has been the subject of several ratings. We will be focusing, however, on the research by Perrenoud (2005) and the study by Tejada (2009), who compiled various classifications (Ayala, 2008; Pérez, 2005; Zabalza, 2003) in order to analyze them and propose his own taxonomy. While there are other contributions around this term, if what we want is to propose specific teaching competencies for heritage education which can provide the discipline with a roadmap, these classifications lie way off our educational practice.

In this sense, several studies have raised questions about teaching competencies in heritage education. Some start from the insights gained by students and teachers (Agrusti, Poce, & Re, 2017; Domínguez & López-Facal, 2014; 2017a; Kostel, Brekelmans, Korthagen, & Wubbels, 2005). Others analyze the presence of heritage as content in official formal education (Cuenca, Estepa, & Martín-Cáceres, 2011; Fontal, Ibáñez-Etxeberria, Martínez-Rodríguez, & Rivero, 2017). From a more practical perspective, the presence of heritage in classrooms has likewise been the subject of analysis (Castro & López-Facal, 2017; Cuenca, 2014; Fontal et al., 2020). Closer to a global vision of the competencies to be developed in teacher training, the study by Castro and López-Facal, (2019) attempts to define the existing needs in teacher training, while that by Domínguez and López-Facal (2017a) inquires into the skills already acquired by teachers.

The development of competencies in heritage education is not only necessary, but should moreover become one of the transversal educational axes that guide our teaching practice. Its integration is therefore essential, regardless of which knowledge areas should implement such an integration. In our own classroom-based research, 132 students were asked in which subjects they remembered working on the concept of heritage. Without losing sight of the devastating figure that indicates that 80 % of respondents claim to have received little training on heritage, the students referred to a total of 23 different subjects, including primary and secondary education. Since the questionnaire used the multiple-choice format, each subject was assigned an individualized frequency.

History was the subject most often mentioned by the students (87 %), followed by Language and Geography (both with 47.25 %), Music (41.25 %), Plastic Arts and Art History (both with 35 %). Less frequently, the presence of heritage-related contents was identified in the Social Sciences subjects in primary education (27.5 %), Philosophy (26.25 %), Religion (23.75 %), Economy (20 %), Classical Culture (16, 25 %), Natural Sciences (13.75 %), and Biology (12.5 %). Finally, more experimental or elective subjects were clearly under-represented in the students' answers, including Physical Education, Mathematics, Technology, English, French, Physics and Chemistry, Scientific Culture, Latin and Photography.

In order to put forward a proposal for teaching competencies in heritage education – following from the one already outlined in Fontal et al., (2020) – a common approach to all areas of knowledge was set up starting from the sequence established by Fontal (2003), the validated instruments for the evaluation of heritage education programs (Calaf, San Fabián, & Gutiérrez, 2017; Fontal, García-Ceballos, Arias, & Arias, 2019), the guidelines in the National Plan for Heritage Education (Domingo, Fontal, & Ballesteros, 2013), the notion of bonds posited by Fontal and Marín (2018), the heritage educommunication theory developed by Martín and Cuenca (2015) and the quality standards for the design of educational proposals collected in Fontal and García-Ceballos (2019a).

Tab. 1: List of teaching procedures and competencies (source: Fontal et al., 2020, pp. 3–5)

TEACHING PROCEDURES **As sourced from the oricedural sequence (Fontal, 2003)**	**TEACHING COMPETENCIES** **As sourced from Ayala (2008), Pérez (2005) and Tejada (2009)** **[TEACHING PERFORMANCE IN HERITAGE EDUCATION]**
KNOWING	**THEORETICAL COMPETENCY**
• In-depth knowledge about heritage is imparted: its categories, its legislation, the competent bodies, its epistemological evolution and its curricular coverage. • Through dialogue, experimentation and appropriation processes, the key elements of heritage are advanced, while a comprehensive knowledge of all its categories is encouraged, including factors like ownership, belonging and identity.	• Demonstrates familiarity with and use of current legislation on heritage, curricular and organizational matters, as well as various sources and bibliographic resources. • Possesses a command of the discipline and a broad, holistic and comprehensive understanding of heritage from a symbolical, identity-driven perspective that transcends historical, economic, or aesthetic values.

Tab. 1: Continued

TEACHING PROCEDURES **As sourced from the oricedural sequence (Fontal, 2003)**	**TEACHING COMPETENCIES** **As sourced from Ayala (2008), Pérez (2005) and Tejada (2009) [TEACHING PERFORMANCE IN HERITAGE EDUCATION]**
• The presence of heritage in curricular subjects is worked upon in a transversal way. • Connections are established between heritage and: a) the immediate environment; b) other territorial areas. • Existing projects, articles and lines of research are explored, so that target students become familiar with different approaches and work and planning styles in heritage education.	• Is familiar with the several methods of heritage teaching and puts them into practice in the theoretical design of educational programs by defining the groundings and criteria on which the program is established.
UNDERSTANDING	DIDACTIC COMPETENCY
• Through analysis, reflective processes, the problematization of heritage and its investigation, the understanding of heritage itself as an element in constant growth and transformation is promoted. • Experiential work leads to the understanding of heritage as the sediment of culture that shapes past and present societies and continues to be built with a view to the future. • Work is conducted not only on global heritage, but also on the local and specific; the meanings and the experiences of everyday life; and the exceptional, which is set in relation to the interests of the students themselves. • Resources are provided and contents selected in relation to heritage history and curricular subject-matter across the several knowledge areas.	• Designs processes and experiences that are meaningful, experiential, binding, relational, symbolic and identity-driven, reflective, creative, inquiring, critical and debate-oriented. In this way, the students put into practice the work done during their training using several methodologies, strategies, styles, resources and materials, etc. • Plans in a clear, precise, orderly and coherent way complete didactic structures, integrating in their processes the development of values, competencies, concepts, skills and attitudes. • Facilitates and guides learning processes seeking the integral development of students and providing keys and resources necessary for the proper knowledge and interpretation of such processes.

(*continued on next page*)

Tab. 1: Continued

TEACHING PROCEDURES **As sourced from the oricedural sequence (Fontal, 2003)**	**TEACHING COMPETENCIES** **As sourced from Ayala (2008), Pérez (2005) and Tejada (2009)** **[TEACHING PERFORMANCE IN HERITAGE EDUCATION]**
• Understanding is promoted through the dialogic relationship with human values in a significant way. • Proposals are put forward that challenge students in their work environment or in their practical implementation; premises are discussed, knowledge pills are created to advance heritage-related knowledge, new forms and spaces are designed to enhance its value or make it known, etc.	• Adapts to the educational level of the corresponding educational stages, as well as to the students' interests, concerns, diversity of learning patterns, contexts and subjects. • Relates contents with the immediate reality of students. • Uses Information and Communication Technologies in order to reach out to students, motivate and energize them, as well as to complement contents and activities or conduct heritage-related inquiries.
RESPECTING, VALUING AND CARING	SOCIAL COMPETENCY
• It involves acquiring attitudes of respect, appreciation and care towards heritage manifestations of diverse nature and origin. • Work is conducted in a multiple and flexible way so as to secure significance, awareness and acquaintance with the several manifestations resulting from cultures and social diversity. • An awareness of social responsibility and heritage commitment is promoted through the uses and values of heritage itself, fostering reflective, identity-driven, experiential, relational, binding, and critical processes. • Work is devoted to the value-enhancement of heritage in the framework of a constructive and problematizing paradigm, which allows debating and exchanging critical and value judgments, normally developed from conflicting heritage assets in the students' closest environment.	• Plans keeping in mind the social component associated with heritage, identity, meaning and memory, while promoting the process of symbolic appropriation. • Proposes activities aimed at acquiring knowledge of different cultures, contexts and manifestations, promoting attitudes of appreciation, respect and care for the diverse heritages. • Cooperates and collaborates in networks, and participates in or develops interdisciplinary, inter-territorial, intergenerational, intercultural projects, etc. • Makes visible and shares his/her practices as a key instrumental axis for collaboration and social exchange in order to continue enriching the discipline.

Tab. 1: Continued

TEACHING PROCEDURES **As sourced from the oricedural sequence (Fontal, 2003)**	**TEACHING COMPETENCIES** **As sourced from Ayala (2008), Pérez (2005) and Tejada (2009)** **[TEACHING PERFORMANCE IN HERITAGE EDUCATION]**
• There is room for free interpretation and the construction of significance, as well as the creation and intervention of new or existing collectively owned heritages, starting from the interests of the students and their closest surroundings.	• Engages in team work with fellow professionals from his/her institution or from external institutions proposing new projects that contribute to the students' experiential learning and further their contact with heritage. • Demonstrates and applies a social approach to heritage, showing attitudes of commitment, appreciation, respect and care towards it.
ENJOYING	COMPETENCY IN INNOVATION
• Work adopts a humanized approach in which the learner is creator, expert, mediator, recipient, agent of value enhancement and transmitter of heritage. • Discovery is encouraged through educational proposals; and so is the creation of new heritages, communities, heritage networks; and the intervention in our heritage assets or their dissemination. • Dialogues are established with one's own heritage and with the heritage of others through a relational process, which in turn enables the student to be placed at the center of learning. • Inquiry and research processes are encouraged around heritage assets and their context. • Work is conducted on historical narratives and the value of assets and traditions is delved into. Their meanings are explored, and so are the ways in which we relate to our legacy.	• Seeks to implement new dynamics, methodologies and innovative strategies in educational planning so as to apply them to heritage and promote new experiences of learning and inquiry among students. • Seeks to be a conscious and active teacher in the constant process of educational change and in an intercultural context. • Is open to active participation in educational and disciplinary research and innovation processes. • Promotes and participates in team work. • Promotes and participates in educational research and innovation projects. • Retrains and is continuously updating knowledge and strengthening weak points. • Promotes the exchange of projects, experiences, students, teaching staff and knowledge with other settings.

(*continued on next page*)

Tab. 1: Continued

TEACHING PROCEDURES As sourced from the oricedural sequence (Fontal, 2003)	**TEACHING COMPETENCIES As sourced from Ayala (2008), Pérez (2005) and Tejada (2009) [TEACHING PERFORMANCE IN HERITAGE EDUCATION]**
• Experience is sought from direct contact, the feeling of the didactic impact of objects, their sources and in situ locations (museums, sites, landscapes, historical places, etc.). The senses and emotions are appealed to through binding processes.	• Designs innovative materials, resources and tools to expand or promote the teaching and learning of heritage.
TRANSMIT	COMMUNICATIVE COMPETENCE
• Priority is given to teaching, communication and the dissemination of heritage in its broadest and most comprehensive range of possibilities through different media, mostly technological, in order to obtain greater visibility and social outreach. • Creative processes of heritage transmission or documentation are encouraged by appealing to emotional and sensory narratives that stimulate target audiences or make them actively participate in history and in the awareness-raising and learning process. • Work is implemented on the educommunication of heritage. This is achieved through the organization of ideas, the combination of images, sounds and texts, the creation of microvideos, the production and deployment of photographs, the use of storytelling, the gamification of heritage, advertising campaigns, awareness, sensitization; creation of virtual spaces such as blogs or websites to share heritages, encourage dialogues and debates or simply to include them and make them known.	• Adapts both language, (being accurate and understandable) and textual sources and resources used in the classroom, taking into account the latter's diversity so as to favor the understanding of heritage categories and the interpretation of multiple cultural, social, artistic, economic contributions, etc. • Stimulates multidirectional communication, dialogue, debate, argumentation, etc., in educational proposals implemented in the classroom. • Knows how to be, and how to act; is able to relate to others and interact cooperatively/collaboratively with other people in the professional context. • His/her communication is inspired by a socio-critical, open-minded and open-to-dialogue spirit which they use to train citizens committed to sustainable development in the field of heritage. • Integrates a holistic approach to heritage in his/her discourse, a meaningful and comprehensive perspective of culture including socio-emotional factors.

Tab. 1: Continued

TEACHING PROCEDURES **As sourced from the oricedural sequence (Fontal, 2003)**	**TEACHING COMPETENCIES** **As sourced from Ayala (2008), Pérez (2005) and Tejada (2009) [TEACHING PERFORMANCE IN HERITAGE EDUCATION]**
• The transmission of heritage is addressed as a basis for informal learning by taking advantage of current digital infrastructures and media, including the use of apps or social networks in order to stimulate and multimediatize heritage.	• Encourages the appreciation of the internal and external heritage assets of culture and the commitment to their care, appreciation and respect.
SENSITIZE	EVALUATIVE COMPETENCY
• This is the ultimate goal of the procedural sequence. It seeks to sensitize students around heritage and to ensure that through their work and creative outputs they also raise the awareness of others. • Work is carried out by following all of the above procedures, and this provides students with new concepts, views and meanings around heritage. New teaching models, resources and heritage policies are thus facilitated. • Unique practices are shared that reflect work based on the values, the knowledge and the stories that heritage affords us; and are based on identity, authenticity, fragility and respect for our heritage. Only in this way can a global awareness be achieved that allows learning to resonate and be significant and, therefore, applicable in future practice, while comprising a global approach and being implemented in a transversal way.	• Evaluates the students' learning process and is responsible for the continuous improvement of his/her group, providing feedback on its development. • Designs evaluation tools and systems, tables of values, rubrics, scales, observation matrices, in any work context. • Takes into account the quantitative and qualitative evolution of the students, individually and collectively. • Integrates different evaluation systems (hetero-evaluation, co-evaluation, self-evaluation). • Shows self-critical ability concerning his/her own practice and teaching and training work through self-assessment, analysis, revision and improvement. • Actively contributes to the improvement of educational quality.

Heritage education in in teacher training classrooms

By implementing these teaching procedures, the socialization of heritage takes place in classrooms through various binding, emotional and relational strategies between individuals and heritage itself. Sensitizing and making students responsible for heritage through a sociological perspective (Cuenca, 2014; Labrador & Silberman, 2018; Vienni, 2014) implies awakening values of awareness and sustainability in students. To this purpose, it is necessary to promote/provoke a stimulus that triggers off feelings of ownership, belonging and identity (Fontal & Gómez-Redondo, 2016; Gómez-Redondo, 2014). This experience should be as valuable as it is revealing so that later, in the actual practice of their profession, the now prospective teachers will be able to transfer the values of heritage to their students with the same nuanced conception: comprehensive, multiple, first-hand and diverse. The aim is to create an active, playful and dynamic methodology that turns students into the main subjects of learning (Miralles, Gómez-Carrasco, & Rodríguez, 2017). It is about creating a cognitive resonance in students that causes, among them and also in bodies or institutions, social synergies and initiatives driven by social cooperation (Fontal et al., 2020).

We may even speak of co-construction supported by a dialogic base: the creation of inter-heritage spaces arising from the sum of individualities and the search for commonalities. This will bring about a change from a personal SELF into a collective US; and from a single meaning to a multiple meaning made up of the various narratives and experiences that students exchange, which in turn involves attitudinal changes towards heritage resulting from the awareness process in which they are immersed. Now more than ever this discipline is critical: heritage education is the fundamental instrument in understanding cultural diversity. It is the vehicle needed to promote respect as a fundamental pillar in the construction of the personality of individuals committed to the distinctive and particular traits of multiculturalism. Its role for today's society is more than necessary.

From the point of view of the consistency between the current educational regulations concerning heritage-related contents that must be implemented and their actual representation in teacher training programs, authors like Fontal et al. (2017) already pointed out that there was a significant disconnection in this regard. Such a gap indeed requires considered thinking in order to integrate the new approaches. According to the above paper, only "6.78 % of our own state-funded universities teach compulsory/basic subjects involving heritage, and 18.64 %teach these contents in elective subjects" (Fontal et al., 2017, p. 85). These figures have not changed substantially since this study was published.

At present, some studies such as the one by Lucas and Delgado-Algarra (2020) already propose a model of good practices embodied in a teacher profile, dubbed as the "postmodern teacher," that contributes to this continuous reflection. Studies like the one authored by Castro and López-Facal (2019) highlight certain needs detected by the educational practitioners themselves, including a strong demand for training in heritage education. Finally, some publications refer to heritagization-driven educational proposals that have already been implemented in teacher training classrooms involving a service-learning approach in order to solve social problems through culture. Such is the case of the contributions by Vicent and Luna (2019) or Triviño and Rico (2015), while Gómez-Redondo (2014) presents an extracurricular proposal based on the processes of heritagization and "identization" in artistic education, and more particularly on the ARTEfacto project. At the international level, mention must be made of the publication by Poce, Agrusti and Re (2018) around a project that promotes digital innovation in cultural heritage education.

Our work in the classroom consists in raising learning challenges and awakening the motivation and interest of the students so that they delve deeper into this subject matter and become aware of its significance in a creative and collaborative way. We should not forget that in order for this rationale to materialize itself in primary education classrooms, initial teacher training is essential (Juanas et al., 2009; Perrenoud, 2010). Through these strategies, students achieve an awareness as well as a reconceptualization of heritage, evolving towards a more comprehensive (Cuenca, 2014) and humanized, bonds-centered understanding of the term (Fontal, 2003; 2013). This transformation highlights the need to design effective proposals whose implementation goes beyond research hubs on heritage education or specific curricular subjects.

Heritage education in the teaching of the social sciences

As a discipline, heritage education combines all the fundamental pillars that make up our tangible and intangible culture, and includes all the elements that we consider essential for our knowledge of the past and our understanding of the present. It brings us closer to a plural vision of the world in cultural, geographical, artistic or historical terms. It comprises a wide range of aspects like beliefs, manifestations, celebrations, rituals, gastronomy or folklore; and it enables us to approach history in its multiple dimensions: its context, imprint, settings, characteristics and evolution. Working with heritage implies transferring into the classroom the sources, documents and narratives that constitute footprints for our knowledge of history and culture; for the understanding of social complexity; and for the development of a democratic, critical, open and flexible thinking based on dialogue, argumentation, ethics and the validity and reliability of sources.

The integration in the classroom of heritage education allows us to take a further step in the teaching and understanding of the social sciences; it surpasses the scientific method (Feliu & Sallés, 2011) of historical thinking (Lévesque, 2009; Seixas, 2017) and establishes meaningful connections with events, places or assets. It works from the present, from the particular, from our own local meanings to then reach a global dimension. Heritage education makes it possible to work with the transversal contents related to citizenship and the diversity of its cultures and traditions; promotes respect, education for peace and the sustainable development of heritage resulting from its appreciation, care and conservation/transmission to future generations. It allows us to integrate factors related to coexistence, emotional competence and self-concepts based on identity and personal assets; it helps us value our legacy and that of other civilizations and cultures so that we enrich ourselves by taking up a multiple perspective. Heritage education also enables us to problematize conflict, vandalism, destruction, ideological or religious confrontation; and to raise awareness about the issues that affect a globalized and changing world. It allows us to work on meanings and resignification, to address controversial issues and conflicting heritages; to handle major concepts that sustain the world, such as identity, culture, ownership, law, equality, freedom (of thought, expression, interpretation). Finally, heritage education promotes our awareness of values and pluralism, the development of key social and civic competencies and of cultural manifestations.

A practical view: heritage as the active principle of culture

In our teaching practice, we try to combine all the elements that should be involved in an effective work process in order to achieve a human and comprehensive understanding of the term (Fontal & García-Ceballos, 2019b) and, ultimately, the development of teaching competencies (Fontal et al., 2020). One of the most enriching practices in our case consisted in placing in the classroom different objects that we used as a catalyst of our proposal: one based on observation, reflection and investigation. Students immersed themselves in a continuous methodological active principle, wondering, analyzing, arguing, hypothesizing and receiving new information as a result of a gamification strategy based on the reception of information clues. In this way, they were able to check and build new proposals related to the history of the object, later

Tab. 2: Example of heritage typologies defined during a classroom practice and frequency of responses

Typology	Description	Frequency
Categorical heritage	Generic, "container" heritage: that which contains various manifestations and includes the great heritage categories — natural, artistic, cultural, archaeological, historical, etc.	16.8 %
Deeply-rooted or family heritage	Social heritage that is ingrained in our family and affective circle, our home and traditions. It refers to house, family, homes in the town or village, coexistence with friends and neighbors. It is generally anchored in a sense of individual ownership.	21 %
Personal-relational heritage	Social, affective and particular (unique) heritage that refers to material goods linked to people and is purely relational. Gifts or inherited property, generally tangible, resulting from family ties or from ties with the closest social environment.	6.3 %
Personal-experiential heritage	Tangible heritage related to frequently used personal belongings that represent fundamental pillars for the individuals and color their daily lives. It involves an element of pleasurable enjoyment linked to experiences like music, books, instruments, a team's shirt or private collections.	14.7 %
Personal-intangible heritage	Personal heritage acquired or owned, which individuals care for and appreciate as something essential in their lives. It normally refers to intangible terms: identity-based and intellectual heritage, the education received, personal relationships or health.	5.3 %
Local heritage	Heritage referred to the places that people see as the backbone of their experiences: their town, their province, the geographic area they feel most attached to. It can also be a specific place, the Gràcia district in Barcelona, Zaragoza's Basilica of our Lady of the Pillar, the Pyrenees, Valladolid. . . It relies on a feeling of collective property to which a factor of personal appropriation has been assigned.	15.8 %
Faith heritages	Heritage linked to beliefs, indicative of religious creeds and faith, like places of worship, religious traditions such as Holy Week, symbols, sacred books, or iconographic representations of saints or deities, among others.	5.3 %

(*continued on next page*)

Tab. 2: Continued

Typology	Description	Frequency
Economic heritage	Heritage that is based on monetary value, but that may as well have emotional ties such as a house, a car, a laptop or an iPad.	5.3 %
Collective intangible heritage	Heritage that consists of non-tangible identifying aspects of a culture or group: the set of traditions of each place, its customs, popular songs, languages, festivities, autochthonous crafts, gastronomy. It often refers to memory and identity.	9.5 %

to share them with the whole group so as to enrich the latter's experience and encourage a diversity of meanings. Through group dynamics we reached into the students' motivations and got them to develop an attitude open to learning.

At this point the key question arose: what is your most precious heritage? Students pondered over this question and shared their answers. These were then arranged on a surface, and connections were mapped and drawn; possible interpretations were also shared through dialogic dynamics and defined in containing clusters. In this way, students identified the values attributed to their belongings and observed the prevalence of their responses so as to become aware of the conceptual change that had taken place in the course of this practical exercise (See Tab. 2).

Subsequently, these categories helped participants to think about new heritages contained within them and to reflect on whether the chosen object was indeed the most valuable. They asked themselves once again about their heritage and they co-built new ones, understanding the breadth of the term, experiencing their own heritage and that of others. This is an "affective-active" process in which prospective teachers discovered the narratives of each asset and established synergies. Once the practice was over, a discussion was started in order to promote linguistic competency and critical thinking; they understood the particular and the global, the personal and the collective value of heritage. The multiple readings and meanings, the various factors that make up heritage and its various manifestations were analyzed. Different lines of debate were raised dealing with controversial issues or heritages in conflict. Finally, through project-based learning, participants in the experience became members of a UNESCO committee of experts whose objective was to discuss and define a number of keys for the safeguarding of heritage, which resulted in three fundamental axes derived from this practice: the wealth of multicultural convergence, the testimonial value of conservation and its sustainability.

Conclusions

Throughout this chapter, we have delved into the contribution afforded by heritage education to teacher training and shown how the relational method and the affective, social and binding strategies can enrich the teaching of social sciences. The sum of historical education and heritage education complements the objective factor that characterizes historical thinking (Lévesque, 2009; Seixas, 2017), as well as the emotional component (Bisquerra, 2000) that is typical of the relational method (Fontal, 2013). This combination provides the teacher training experience with a significant nuance that proves critical. The use of our own heritage as a starting point provides an axis for educational practice and predisposes students to a constant search for connections, values and questions to be resolved. The integration of active methodologies makes feeds their curiosity and makes them rethink new ways of learning through heritage. Throughout history, sources have helped us know the past, analyze causes and consequences and understand change and continuity, historical time or multiperspective, among other metaconcepts (Seixas & Morton, 2013). However, heritage education additionally makes it possible to assign values, since all assets can be turned into heritage (Fracaso, 2016). This contribution means that not only does experience carry along with it a work process that enables knowing and understanding, but that it furthermore results from an awareness that attaches a greater significance to the construction of relations and the enjoyment of the lived experience. Emotions, together with a grounding on identity (Fontal & Gómez-Redondo, 2016) should not be excluded from the teaching-learning process, since they ultimately enrich it.

To conclude, we return to one of the key questions mentioned at the beginning of the chapter: what leads students to think that heritage is worth conserving? It is important to answer this question because the answer will provide the key to our concern: what do I want to get from the students? What challenge am I raising in the classroom? In order to educate competent teachers to train a citizenship that is socioculturally committed and respectful of cultural diversity (Martín-Cáceres & Cuenca, 2011), we must choose the appropriate strategies and the most suitable mediation in order to sympathize with the heritage perceptions experienced by others, awaken the interest and motivation over this subject and make students enjoy it (Fontal et al., 2020).

References

Agrusti, Francesco/Poce, Antonella/Re, Maria Rosaria: "Mooc design and heritage education. Developing soft and work-based skills in higher education students". *Journal of E-Learning and Knowledge Society 13* (3), 2017, pp. 97–107.

Ayala, Francisco: *El modelo de formación por competencias*, 2008, last accessed on 6 September, 2020, at http://practicadocente.weebly.com/uploads/5/2/5/7/5257083/evaluacion_en_la_formacion_por_competencias_tipo_de_evidenci.pdf.

Bisquerra, Rafael: *Educación emocional y bienestar*. Praxis: Barcelona. 2000.

Calaf, Roser/San Fabián, José Luis/Gutiérrez, Sué: "Evaluación de programas educativos en museos: Una nueva perspectiva". *Bordón 69* (1), 2016, pp. 45–65.

Calbó, Muntsa/Juanola, Roser/Vallés, Joan: *Educación del patrimonio: visiones interdisciplinares*. Universidad de Gerona: Gerona. 2011.

Cantón, Valentina: "La Educación Patrimonial como Formación para la Educación Ciudadana". *Revista Correo del Maestro 154*, 2009.

Castro, Belén M./López-Facal, Ramón: "O patrimonio alén as aulas". *Revista galega de Educación 69*, 2017, pp. 38–41.

Castro, Leticia/López-Facal, Ramón: "Educación patrimonial: necesidades sentidas por el profesorado de infantil, primaria y secundaria". *Revista Interuniversitaria de Formación del Profesorado 94* (33.1), 2019, pp. 97–114.

Council of Europe: "Recommendation of the European Parliament and of the Council of 18 December 2006 on key competences for lifelong learning". *Official Journal of the European Union*, (12), 2006, pp. 10–18.

Cuenca, José María: "El papel del patrimonio en los centros educativos: hacia la socialización patrimonial". *Tejuelo 19*, 2014, pp. 76–96.

Cuenca, José María/Estepa, Jesús/Martín-Cáceres, Myriam J.: "El patrimonio cultural en la educación reglada". *Instituto de Patrimonio cultural de España* 5, 2011, pp. 45–57.

Cuenca, José María/Estepa, Jesús/Martín-Cáceres, Myriam J.: "Patrimonio, educación, identidad y ciudadanía. Profesorado y libros de texto en la enseñanza obligatoria". *Revista de Educación* (375), 2017, pp. 136–159.

Domingo, María/Fontal, Olaia/Ballesteros, Paloma (Coords.): *Plan Nacional de Educación y Patrimonio*. Ministerio de Educación, Cultura y Deporte, Secretaría de Estado de Cultura: Madrid. 2013.

Domínguez, Andrés/López-Facal, Ramón: "La formación inicial del profesorado y educación cívica del alumnado de primaria". *Clío: History and History Teaching 40*, 2014, pp. 1–26.

Domínguez, Andrés/López-Facal, Ramón: "Formación de maestros y educación patrimonial". *Estudios pedagógicos 43* (4), 2017a, pp. 49–68.

Domínguez, Andrés/López-Facal, Ramón: "Patrimonios en conflicto, competencias cívicas y formación profesional en educación primaria". *Educación* (375), 2017b, pp. 86–109.

Estepa, Jesús/Ávila, Rosa. M./Ferreras, Mario: "Primary and secondary teachers' conceptions about heritage and heritage education: A comparative analysis". *Teaching and Teacher Education 24* (8), 2008, pp. 2095–2107.

Feliu, María/Sallés, Neus: "El método científico para enseñar Historia: una experiencia en la formación de maestros". *Clío 37*, 2011.

Fontal, Olaia: *La Educación Patrimonial. Teoría y práctica en el aula, el museo e internet.* Trea: Gijón. 2003.

Fontal, Olaia: *La Educación Patrimonial: del patrimonio a las personas.* Trea: Gijón. 2013.

Fontal, Olaia/García-Ceballos, Silvia: "Evaluación de programas de Educación Patrimonial: estándares de calidad". *ENSAYOS, Revista de la Facultad de Educación de Albacete 34* (1), 2019a, pp. 1–15.

Fontal, Olaia/García-Ceballos, Silvia: "Las plataformas 2.0 como herramientas de aprendizaje y adquisición de competencias en educación patrimonial". *Revista interuniversitaria de formación del profesorado 94* (33.3), 2019b, pp. 285–306.

Fontal, Olaia/García-Ceballos, Silvia/Arias, Benito/Arias, Víctor. B.: "Evaluación de la calidad de programas de educación patrimonial: construcción y calibración de la escala Q-Edutage". *Revista de Psicodidáctica 24* (1), 2019, pp. 31–38.

Fontal, Olaia/García-Ceballos, Silvia/Aso, Borja: "Desarrollo de competencias docentes en educación patrimonial mediante plataformas 2.0 y entornos digitales como herramienta de aprendizaje". *Investigación en la escuela 101*, 2020, pp. 1–14.

Fontal, Olaia/Gómez-Redondo, Carmen: "Heritage Education and Heritagization Processes: SHEO Methodology for Educational Programs Evaluation". *Interchange 47* (1), 2016, pp. 65–90.

Fontal, Olaia/Ibáñez-Etxeberria, Alex: "La investigación en Educación Patrimonial. Evolución y estado actual a través del análisis de indicadores de alto impacto". *Educación* (375), 2017, pp. 184–214.

Fontal, Olaia/Ibáñez-Etxeberria, Alex/Martínez-Rodríguez, Marta/Rivero, Mª Pilar: "El patrimonio como contenido en la etapa de Primaria: del currículum a la formación de maestros". *Revista electrónica interuniversitaria de formación del profesorado 20* (2), 2017, pp. 79–94.

Fontal-Merillas, Olaia/Marín-Cepeda, Sofía: "Nudos Patrimoniales. Análisis de los vínculos de las personas con el patrimonio personal". *Arte, Individuo y Sociedad 30* (3), 2018, pp. 483–500.

Fracasso, Liliana: *Lo "patrimoniable": Utopías Concretas, Prácticas Artísticas y Hábitat Popular.* XIV Coloquio Internacional de Geocrítica. Las Utopías y la Construcción de la Sociedad del Futuro, Barcelona. http://www.ub.edu/geocrit/xiv_lilianafracas.pdf, 2016.

Gómez-Redondo, Carmen: "El origen de los procesos de patrimonialización: la efectividad como punto de partida". *Educación artística: revista de investigación (EARI)5*, 2014, pp. 66–80.

Gómez-Redondo, Carmen/Fontal, Olaia/Ibáñez-Etxeberria, Alex: "Procesos de patrimonialización e identización patrimonial con uso de TIC en torno al arte contemporáneo". *Artnodes: revista de arte, ciencia y tecnología* (17), 2016, pp. 52–63.

Juanas, Angel de./Fernández, Pilar/Martín, Rosa./González-Ballesteros, Mairena/Pesquero, Encarnación/Sánchez, María E.: "Comparative study of the evaluation of professional competencies by experienced and trainee Spanish primary teachers". *European Journal of Teacher Education 32* (4), 2009, pp. 437–454.

Koster, Bob/Brekelmans, Mieke/Korthagen, Fred/Wubbels, Theo: "Quality requirements for teacher educators". *Teaching and Teacher Education 21* (2), 2005, pp. 157–176.

Labrador, Angela M./Silberman, Neil A.: "Public heritage as social practice". *The Oxford Handbook of Public Heritage Theory and Practice*, 2018, pp. 1–17.

Lévesque, Stephane: *Thinking Historically. Educating Students for the Twenty-First Century.* University of Toronto Press: Toronto. 2009.

López-Facal, Ramón: "La LOMCE y la enseñanza de las ciencias sociales". *Iber: Didáctica de las ciencias sociales, geografía e historia 79*, 2015, pp. 5–7.

Lucas, Laura/Delgado-Algarra, Emilio. J.: "El Profesor Posmoderno de Ciencias Sociales: Un Modelo de Buenas Prácticas en Educación Patrimonial". *REICE. Revista Iberoamericana Sobre Calidad, Eficacia y Cambio En Educación 18* (1), 2019, pp. 27–45.

Martín-Cáceres, Myriam/Cuenca, José María: "La enseñanza y el aprendizaje del patrimonio en los museos: la perspectiva de los gestores". *Revista de Psicodidáctica 16* (1), 2011, pp. 99–122.

Martínez-Rodríguez, Marta/Fontal Merillas, Olaia. "Dealing with heritage as curricular content in Spain's primary education". *Curriculum Journal 31* (1), 2020, pp. 77–96.

Miralles, Pedro/Gómez Carrasco, Cosme. J./Rodríguez, Raimundo A.: "Patrimonio, competencias históricas y metodologías activas de aprendizaje: Un análisis de las opiniones de los docentes en formación en España e Inglaterra". *Estudios pedagógicos 43* (4), 2017, pp. 161–184.

Pérez, María J.: La formación permanente del profesorado ante los nuevos retos del sistema educativo universitario. *XI Congreso de Formación del profesorado*, Segovia 17–19 February, 2005.

Perrenoud, Philippe. *Diez nuevas competencias para enseñar.* 3rd. ed. Graó: Barcelona. 2005.

Perrenoud, Philippe: "La formación del profesorado: Un compromiso entre visiones inconciliables de la coherencia". *Revista interuniversitaria de Formación del Profesorado 68* (24), 2010, pp. 103–122.

Poce, Antonella/Agrusti, Francesco/Re, Maria Rosaria: "Heritage education and initial teacher training: An international experience". *Journal of e-Learning and Knowledge Society 14* (2), 2918, pp. 127–143.

Prats, Llorenç: *Antropología y patrimonio.* Ariel: Barcelona. 1997.

Seixas, P.: "A model of historical thinking". *Educational Philosophy and Theory 49*(6), 2017, 593–605. https://doi.org/10.1080/00131857.2015.1101363

Seixas, Peter/Morton, Tom: *The Big Six Historical Thinking Concepts.* Nelson College Indigenous: Toronto. 2013.

Smith, Laurajane: *Uses of Heritage.* Routledge: London. 2006.

Tejada, José: Competencias docentes. *Profesorado. Revista de currículum y formación del profesorado 13* (2), 2009, pp. 1–15.

Triviño, Laura/Rico, Lidia: "Propuestas desde la metodología aprendizaje-servicio para fomentar el interés por la educación patrimonial en la formación del profesorado". *International Journal of Educational Research and Innovation (IJERI) 5*, 2016, pp. 1–13.

Vienni, Bianca: "Interdisciplinary socialization of archaeological heritage in Uruguay". *Journal of Cultural Heritage Management and Sustainable Development 4* (1), 2014, pp. 95–106.

Vicent, Naiara/Luna, Ursula: "Heritagization of the (un)known. A heritage project for elementary education teacher training". *Ensayos. Revista de la Facultad de Educación de Albacete 31* (1), 2019, pp. 41–53.

Zabalza, Miguel. A.: *Competencias docentes del profesorado universitario. Calidad y desarrollo profesional.* Narcea: Madrid. 2003.

Domínguez-Almansa, Andrés, Costa-Casais, Manuela, and Riveiro-Rodríguez, Tania

Chapter 15 Cliffs with memory: Historical and social teacher training via climatic and environmental transformations[1]

Abstract: This paper presents a project relating to teaching innovation and research in the context of primary teacher training. It takes Place-Based Education theory as a point of reference. Over the course of several academic years (191 pupils), an experiment has been carried out by way of which schoolchildren learn about a series of geomorphologically unique formations on the coast of Galicia (NW Spain). These formations, currently active cliffs, are Quaternary deposits which provide a significant amount of information on climatic and environmental changes which took place during the last Ice Age and the Holocene. The research deriving from this intervention is mainly qualitative in nature, supported by documents created by the students, in which it is possible to find: a) the level of prior knowledge of the trainee teachers; b) their reflections and evaluations on these formations and their educational and civic dimension; c) their role as disseminators of the scientific and heritage value of these deposits by way of ideas and proposals for educational interventions.

Keywords: Climate change, Edapho-sedimentary deposits, Trainee teachers, The teaching of the social sciences, History teaching, Geography teaching

Introduction

This paper examines the possibility of offering trainee primary teachers an interdisciplinary vision in order to approach the teaching of geography and history. A fusion between physical geography and social knowledge is employed to address, from a critical thinking perspective, a problematic and present-day issue which has become a relevant topic in education, namely climate change (Gaudiano, Cartea, 2020). This approach makes it possible to incorporate knowledge from other disciplines while, at the same time, employing an educational

---

1 This research is part of the coordinated project COMPROP (PGC2018-094491-B-C31: *Emotional dimension and controversial issues in teacher training* with funding from the Spanish State Research Agency and European Regional Development Fund.

formula open to lines such as heritage education, Place-Based Education and socio-critical education.

This approach stresses the educational and social value brought by reflecting in the classroom on spaces with intrinsic values but with hardly any or no civic recognition, opening up ways for their possible heritage appropriation within the framework of education (Domínguez-Almansa, López-Facal, 2017). This intention is integrated in influential proposals in the field of heritage education, such as the relationship of heritage with ideological issues and the power of the elites (Cuenca López, 2016; Estepa, 2019) and the idea of changing the prominence of objects for that of people (Fontal, 2013). It can also be framed in Place-Based Education, focused on the connection between the classroom and the local community (Gruenewald, 2003; Sobel, 2004) and the recognition of everyday places which can acquire meaning for the people who live within them (McInerney, Smyth, Down, 2011).

Both points of reference, which encourage critical judgement (Herodotou et al. 2019), defend the necessary inclusion of natural spaces. In this case, a representation of spaces which, in the field of geomorphology or geology, fit into the consideration of pattern sites, best sites and representative sites (Carcavilla, 2014) is discussed. These are edapho-sedimentary Quaternary deposits located on the coast of Galicia (NW Spain), which, in the present-day, function as active cliffs (Costa-Casais, Caetano Alves, 2013). To their scientific value must be added their educational potential, both in terms of addressing cognitive aspects, making it possible to gain a better understanding of climate change, and civic characteristics steered towards the evaluation and preservation of these environmental records of the memory of the past. Although such deposits are present in other territories, they become unique on the Galician coast due to the abundance and wealth of the information they provide on a global, regional and local scale (Threnhaile et al., 1999).

A justified educational action

When dealing with issues such as climate change and its marks on the Earth, the dissociation between recent research trends and educational routines is shown. It is necessary to change these dynamics by carrying out educational actions directed at trainee primary teachers. Making such actions suitable implies the commitment of university teachers as a point of reference for a transformative and critical practice (Johnston, 2006). Some previous studies based on the development of this research have been carried out along the same thematic

Fig. 1: Location of the field trip and Fonforrón beach

lines of this study (Costa-Casais, Domínguez-Almansa, 2018; Domínguez-Almansa, Costa-Casais, López-Facal, 2019).

The idea of the importance of learning based on the discovery and investigation of the surrounding area is shared (Hmelo-Silver, Duncan, Chinn, 2006). This action begins with an excursion to a beach bordered by deposits (Fig. 1).

Pupils are encouraged to find evidence of climate change, and a tendency to focus exclusively on the present and to leave aside prior knowledge of what is a natural process of the planet is demonstrated. When reflecting upon this, the idea of the memory of the Earth and its possible records appears. Some of these records are extremely precise but are hidden at the bottom of lakes and seas or are under ice (Hodzic, Kennedy, 2019), while others, such as the edapho-sedimentary deposits located in this site, which contain more fragmented information, are present in the landscape although they are invisible to those who do not possess the ability to recognise them (Costa-Casais, Caetano Alves, 2012). Once they have been identified, the question is asked how information about environmental changes can be obtained. This makes it possible to bring meaning to the idea of interdisciplinarity due to the fact that invisible elements of a physical and chemical nature are analysed in the laboratory, for example pollen or the variability of bromine, which is associated to a greater or lesser maritime influence (Martínez et al., 2016). Visible elements are also observed, such as how the intensity of climatic variations is reflected in the different sedimentary levels, which can easily be identified (Fig. 2).

Visualising climate change as a constant in the life of the planet leads to the appreciation of the succession of interglacial-glacial-interglacial (Holocene)

Fig. 2: Identifying the core of the sampling to analyze the invisible proxies and making visible the sedimentary levels (Caamaño deposit-Porto do Son)

periods. It also assists in understanding that, in each period, there were significant climatic variations influenced by geographical location. In this case, these deposits were located in an area between permanent ice and less intense cold conditions with greater humidity, thus leading to the abundance of both deposits and their differentiated facies (Figs. 3 and 4).

However, new questions arise: What was the landscape like in the last Ice Age? It can be stated that the origin of these deposits is continental and that the sea level was much further away than it is today, leading to a significant evocative effect considering what are today the Rías Baixas and the valued islands at their mouths. Did the landscape change during the Holocene? In order to appreciate this aspect, the excursion then took in an archaeological context, the hillfort of Baroña (Porto do Son). This is a fortified pre-Roman settlement which receives many visitors due to its unique location on an elevation exposed to the sea (Fig. 5).

When asked to identify a noteworthy aspect of its location, some pupils point out that the sea attacks part of the settlement, deducing that its construction

Fig. 3: Reflecting on this cave (Fonforrón beach-Porto do Son). It existed in the previous interglacial. It was fossilized in the glacial stage and in the current (Holocene) has been newly discovered

took place in a colder environment in which the coastline would have had a different morphology. This makes it possible to understand the context of other archaeological spaces (Blanco-Chao et al., 2020). Noting the relationship between climate change and landscape makes it possible to transform a consolidated idea: non-industrialised societies of the past which did not pollute on a large scale were not affected by climate change. This can be refuted by discovering that this settlement experienced the Roman Warm Period and its effect on the sea level (Tallón-Armada et al., 2017), which, similar to the present level, would have made the hillfort difficult to live in.

A renewed perspective on this issue makes it possible to address it in the classroom from the parameters of a meaningful educational approach in order to understand social thinking which interweaves the past, the present and the future (Israel, 2012; Santisteban-Fernández, González-Monfort, Pagès-Blanch, 2020).

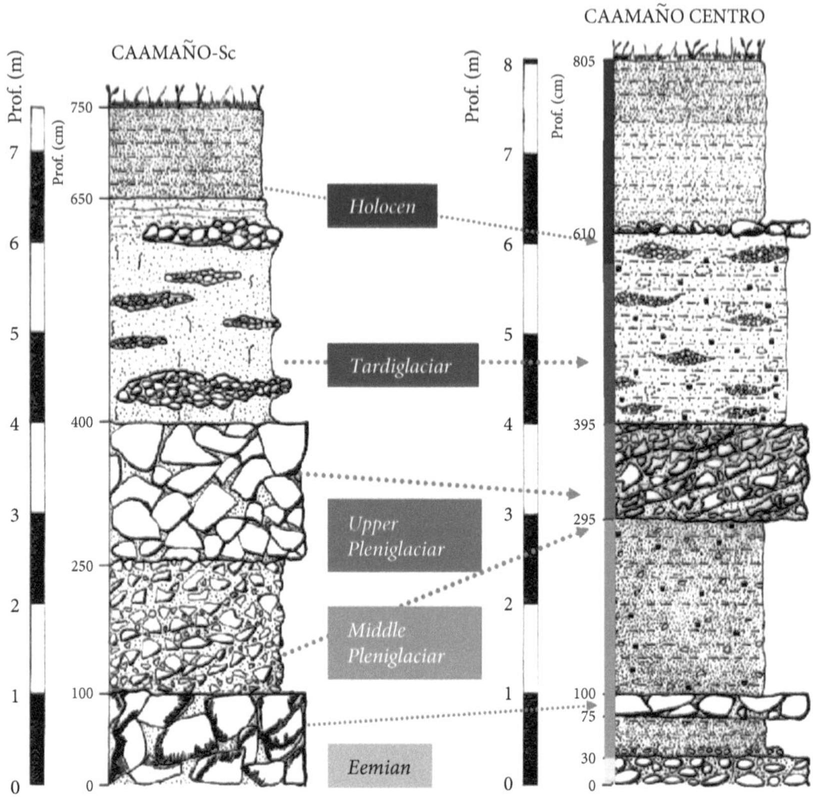

Fig. 4: Reconstruction of the levels identified in the edapho-sedimentary deposits. (Figure made by M. Costa-Casais)

In general, primary schoolchildren are introduced to historical knowledge by way of a graph of historical periods with no chronological proportionality. In a relaxed atmosphere (Hassinger-Das et al., 2017), and using 24 steps which represent the hours of one day, the temporal dimension of each period is visualised and compared, incorporating climate change as a historical determinant (Fig. 6). This issue is dealt with in research which combines documentation from the fields of history and the Earth sciences (Diamond, 2005; Brooke, 2014; Blom, 2017).

By combining these documents (Fig. 7) as a resource of great educational value (Bardsley, Bardsley, 2007), and the information recorded in the deposits,

Fig. 5: Hillfort of Baroña (Porto do Son)

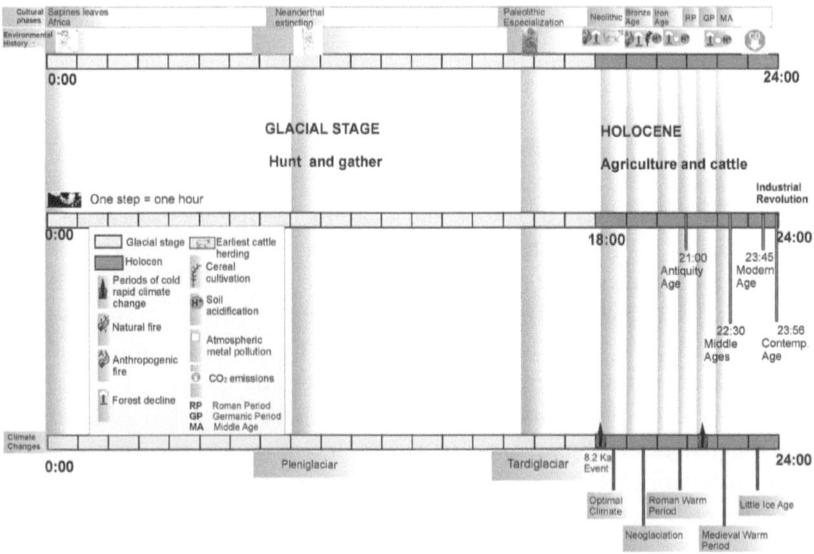

Fig. 6: Historical time and climate in history

a route is traced from the hunter-gatherers of the long Ice Age to the short warm time period of the contemporary world. This route passes through times of great climatic change, which generated the present-day climates and brought about agriculture, and a succession of climatic isolations which accompanied the

Fig. 7: Document to reflect on the Little Ice Age (approx. 1570–1700)

traditional stages of humanity. A long period dominated by agrarian activity is reflected upon, in which the need for climate stability brought about both joy and fear, giving rise to rituals and customs (Rodríguez Plasencia, 2013), transmitted by memory, which have come down to us today under the veil of Christianity. It is necessary to approach this issue in schools from the perspective of new intellectual and critical paradigms.

The contemporary world has developed up to the present moment during a growing warm period. Current research identifies this as a process of acceleration of global warming brought about by human influence. Indeed, scientists from different fields admit that we are already in a new stage, the Anthropocene (Crutzen, Stoermer, 2000). In order to discuss the didactics of climate change as a problem of the present, in addition to giving value to the past, it is necessary to:

- Be aware that we are arriving towards the end of an interglacial stage which is characterised by a natural process of accentuation of warming (Hodzic, Kennedy, 2019). This makes it possible to reflect on the true dimension of human influence, which has never before been present.
- Discover evidence (Veettil, Kamp, 2019). Without resorting to overused places, by comparing the same deposit twenty years ago and at the present time, it can be appreciated that it is being covered by vegetation, a

phenomenon associated to warming which can also be observed in dune systems (Jackson et al., 2019).

- Associate the problem to fundamental issues search as the culture of coal (Nordhaus, 2013) and economic growth. Regarding this aspect, the effectiveness of resorting to works which bring together rigour and dissemination is appreciated. In this case, Higgs (2015) shows how the investor of today does not make the most of the growth of yesterday. An individualism which preys upon resources and the planet itself imposes itself as an ideology.
- Propose the necessity of integrating climate change into education for environmental and human sustainability (Bangay, Blum, 2010; Collins, Genet, Christian, 2013). It is essential to integrate primary schoolchildren who can take in the aspects dealt with here and those who, at an early age, are educated in the responsibility of building a more sustainable society. It is necessary to exemplify this by way of real experiments carried out in schools.
- Return to the past in order to confer educational value upon the knowledge of cultures capable of generating sustainable spaces (UNESCO, 2016) and to seek educational paths which avoid the fictitious dissociation between the human race and nature.

From an educational perspective of a critical nature, it is of value to project these learnings towards the future (Estepa-Giménez, Martín-Cáceres, 2020). The memory stored both in edapho-sedimentary deposits and in other environmental records makes it possible to know that an interglacial period is followed by another glacial period. If we are at the end of an interglacial period, what hypotheses should teacher training consolidate? Two possible scenarios should be reflected upon, always debatable and controversial as scientific hypotheses and their social projection.

1) Induced global warming changes the natural dynamics of the climate (Mann, Kump, 2016). This would suppose an uncertain prolongation of the interglacial period and warming with consequences for the environment and humans. Thus, the thawing of the North Pole would reveal an infinity of resources which would ruin what, in recent decades, seems to have been the least violent period in the history of humanity (Pinker, 2011). It could also reveal viruses and bacteria of unpredictable consequences (Malavin et al., 2020).
2) The rise of the temperature would imply a rapid acceleration of thawing. That great quantity of freshwater would alter the marine currents, which regulate the climate, and it would accelerate the arrival of the new Ice Age. If these predictions, quite refuted (Waver, Hillaire-Marcel, 2004),

would finally occurred, they would cause problems of adaptation for a massified society, the inhabitants of which can no longer be described as hunter-gatherers

Working on climate change with trainee teachers is a way of conveying the need to approach a critical teaching method in the age in which they have to live. Inasmuch as they assume this method and put it into practice, natural climate change will not be avoided, but it could be avoided that a dominant ideology, based on growth and taken as unquestionable truth, defeates science, which would lead to the disappearance of critical thinking with the ability to reflect on the consequences of forcing natural dynamics or to prevent an adequate adaptation to these.

Objectives

To analyse:

1. To what extent trainee teachers understands the environmental records of the past and value those which exist in their surroundings.
2. The cognitive and civic competences acquired in the learning process, paying attention to the capacity of relating the past, the present and the future.
3. How this knowledge can be valued for their professional practice.
4. Their capacity for approaching educational designs so that primary schoolchildren can discover, know, give meaning to, and even disseminate the problem of climate change.

Method

This research is framed within the qualitative tradition (Flick, 2007) and interpretive research (Erickson, 1989). It is an inductive and bounded study which integrates the ethnographic method and the case study (Denzin & Lincoln, 2005). It is based on an educational action and its feedback carried out over three academic years (2016/2017; 2018/2019; 2019/2020) in the area of the teaching of the social sciences in a primary education teacher training degree (University of Santiago de Compostela – Lugo Campus).

191 students (n=191) took part in the study after receiving training in heritage and place-based education in the previous year. They also expressed their interests in dealing with the issue of climate change. At the end of the educational action, they were asked to produce empirical materials in order to approach this research: free individual reflections (n=191) and educational designs produced in groups (n=25) in order to put their learning into practice.

Tab. 1: Sections and key ideas

Individual reflections	Group designs
1. Evaluation of the deposits	1. Discovering the deposits
– Acquisition of knowledge and interdisciplinarity – Association of heritage, identity, preservation – Benefits for professional training	– Teaching the memory of the Earth – Promoting their value
2. A new understanding of climate change	2. Giving meaning and offering contents for working on climate change
– A different approach to geography and history – Connection between past, present, future – A critical and civic dimension	– Alternative procedures – Introducing time and space – A present-day problem
3. Teaching for climate change	3. The vocation of dissemination
– The importance of the surroundings – Cognitive learning – Critical education with social impact	– Pupils as agents of change

Authors' own work

The data were treated in an aggregate and holistic manner and were analysed by way of the method of constant comparisons (Glaser & Strauss, 1967), observing the ideas reiterated in the reflections and the group work. They were categorised into several sections and key ideas (Tab. 1) ascribing textual quotes to each one of them in order to exemplify them.

The quotes of the reflections are identified taking account the student and the academic year (e.g.: St1:2017/2018), the designs, work and academic year (e.g.: D1:2017/2018).

Results

Individual reflections

1. A high degree of consideration for the deposits.

   a) They are valued for being the source of appreciated knowledge and are considered as an example of interdisciplinarity:

[...] They are an extremely good tool for learning about the history of the Earth [...] (St4:2018/2019); The discovery of these deposits made me change my way of seeing the coast [...] (St3:2018/2019); Getting to know them close up [...] now I understand that they are archives of time [...] every climatic change undergone by our planet can be seen reflected in them [...] (St10:2016/2017); [...] they are of great interdisciplinary value [...] (St10:2019/2020); [...] They can be included in different subjects [...] from the social sciences it could be discovered that they are archives of time [...] in which climatic changes, variations of the sea, and even anthropic activity are reflected [...] from the natural sciences the composition of these natural elements could be studied (St10:2018/2019).

b) They are seen has being a heritage asset, a symbol of identity and a space to be preserved:

[...] due to the information which they provide us with [...] they should be declared as heritage (St10:2016/2017); [...] they are part of our natural heritage [...] (St45:2018/2019); They are more than a heritage asset [...] they should be declared as a world heritage site [...] (St3:2029/2020); They are unique and make [...] the Galician coast unique [...] (St10:2028/2019); Galicia is full of deposits [...] part of our heritage [...] of our identity [...] of our history [...] it is our obligation to preserve them, to take care of them and to value them as they deserve [...] (St7:2019/2020); They are an incredibly important part of our landscape [...] They confer both culture and beauty on our surroundings. They are a symbol of our identity which we must look after and value [...] (St5:2019/2020).

c) They are associated with the improvement of their professional training:

They are an important source of information which we must protect and study [...] (St10:2018/2019); [...] it has contributed to both my personal and professional training [...] (St2:2018/2019); Having knowledge of something so important gives meaning to the role of teachers (St5:2016/2017); [...] this should be a fundamental aspect in primary education [...] (St5:2028/2019); At no point in all of my education did I ever hear anything about them. Therefore, I believe that it is important to teach what they are in primary education. . .and what they hide inside them (St54:2019/2020); The more knowledge we accumulate about the deposits, the more reference points we will have in order to propose significant teaching [...] (St59:2019/2020).

2. A new understanding of climate change.

a) A different approach to focusing on geography and history is highlighted:

[...] the issue we dealt with last year about the hidden memory of the Spanish Civil War is related in that [...] both things are unknown by the population [...] (St9:2018/2019); [...] a fact of great relevance in the 21st century [...] we are talking about our own survival on the Earth [...] (St8:2016/2017); [...] knowing about the Earth throughout time [...] what climatic changes were like, the variations in the sea [...] (St45:2019/2020); it is important not to [...] forget issues like this [...] to go on maintaining the memory of everything that is alive on the planet (St50:2019/2020); [...] it is extremely important to work on these issues [...] on climate change [...] our ancestors [...] on different stages of history (St81:2019/2020); [...] it is related with geology [...] history [...] what climate there was, what resources there were [...] (St43:2018/2019).

b) Presentism is abandoned, and climate change is reflected upon paying attention to the past, the present and the future:

[...] how the landscape changed [...] the natural processes which took place [...] and those which will take place (St6:2019/2020); [...] civilisations evolve or disappear depending upon climate. Favourable climatic conditions facilitate the development of life and its prosperity [...] adverse climatic conditions often lead to wars, conflicts and other human catastrophes (St20:2019/2020); [...] on the one hand, it is a natural phenomenon [...] and, on the other, human [...] the impact of which began with the Industrial Revolution, accelerating the natural process [...] we worsened the consequences (St20:2019/2020); [...] the climate affects human beings a great deal [...] the rise in temperatures brought about droughts [...] a lack of food, conflicts [...] climatic migrations [...] in our days (St81:2019/2020).

c) A critical and civic dimension is assumed when approaching problematic issues:

[...] our industrialised society is worried more about economic profits than [...] the transformation of natural areas (St8:2018/2019); [...] we all need to be responsible, as it is we who are causing this climate change (St14:2019/2020); [...] a consumer society [...] helps to destroy the landscape [...] it is necessary to raise awareness and to transform this into habits (St37:2019/2020); [...] global warming is a contemporary issue [...] we could reflect on what is happening (St39:2018/2019); [...] instilling the importance of the natural world [...] the critical situation of the abandonment and mistreatment of the landscape [...] we must recover it and seek solutions. . .particularly among younger people [...] (St42:2018/2019).

3. The appropriation of climate change for teaching.

a) The idea of teaching linked to the surrounding area is defended:

[...] teaching them history by way of everyday items which they possibly do not know about [...] outside of the textbooks (St1:2019/2020); [...] seeking points of reference in the local environment (St6:2016/2017); it would be good [...] to visit historical places (St7:2019/2020); [...] educational excursions [...] we should be interested in what we often have around us [...] (St59:2019/2020); [...] we underestimate our own resources and possibilities in the surrounding area [...] (St61:2019/2020); I would organise excursions [...] a closer perspective [...] not studying merely from a textbook [...] giving value to what surrounds us [...] (St12:2018/2019).

b) The cognitive and emotional dimension is appreciated in learning about climate change:

[...] along with the science and the documentation, we can come to learn many things about the past, such as our evolution or that of our planet (St20:2019/2020);

The classroom [...] is the starting point for dealing with the variations in temperature which have taken place over time [...] human activity is bringing about an increase in the speed of the changes [...] (St34:2019/2020); [...] so that the pupils can see that nature belongs to them, to make them feel that the coasts, the cliffs and the seas belong to them, so that they are "moved" when they see an ecological catastrophe [...] (St15:2019/2020); [...] in order to achieve significant learning, to excite the pupils so that they want to learn more (St21:2018/2019).

c) They are committed to a critical model of teaching with a social impact:

[...] we should encourage our pupils to feel responsible for its care and to fight for it [...] (St5:2018/2019); [...] we should be active and critical subjects [...] in order to transmit these things to the children in the future (St8:2016/2017); [...] we attempt to go further by speaking about related issues, seeking to create a social consciousness in the children [...] (St15:2019/2020); [...] increasing awareness and participation among the children [...] to improve the quality of education (St13:2016/2017); Critical thinking is an essential skill [...] ask for solutions, moving away from the mere transmission of knowledge (St11:2019/2020).

Group designs

1. A commitment to the discovery and appreciation of the deposits.

   a) They are shown to be archives of the memory of the Earth:

All of the designs (n=25) present the edapho-sedimentary deposits as discovery. They relate the layers with geological and human history.

[...] a model which will represent layers of sediments in different colours [...] (D8:2018/2919); [...] they will be given a "puzzle" of a piece of cake [...] they can see the different layers [...] each layer has a climatic period written on it [...] (D1:2016/2017); [...] an urn with materials [...] salt, coloured chalk, stones of different sizes [...] (D11:2018/2019).

b) Educational activities are approached:

All of them do so (n=25). Excursions are employed (n=19) in order to generate interest. The participation of specialists is valued (n=8) and collaborative research is encouraged (n=6) along with comparative research with publicly recognised places (n=2).

"[...] we decided to take them to interpret [...] the Swabian deposits [...] in order to learn from a significant place (D10:2018/2019); on the excursion [...] a treasure hunt [...] which is the cliff itself, making them see its importance [...] (D15:2018/2019); we will need an expert who can explain [...] (D17:2018/2019); [...] a fictitious resolution [...] which says that all the deposits in the area will be eliminated [...] reaction [...] they should investigate" (D17:2018/2019); [...] one of the roads which passes next to a cliff is about to collapse [...] investigate [...] (D1:2016/2017); The Grand Canyon, The Eiffel

Tower, The Colosseum, the deposits [...] they are as unique as the rest [...] (D16:2018/2019).

2. Providing meaning and offering contents for working on climate change.

a) Alternative procedures are proposed:

All of the designs (n=25) move past the transmissive model. Debate and reflection are explicitly valued (n=18). Entertaining materials and resources are created (n=11). Films and audio-visual materials are employed (n=8) along with activities associated with information technology (n=7).

[...] we will give each group a series of questions to guide them [...] (D2:2016/2017); [...] we will read a story we have written [...] guided [...] to build knowledge (D3:2018/2019); [...] a story [...] about Mr Oli [...] who lives in an Ice Age (D19:2018/2019); an experiment for the greenhouse effect [...] (D13:2018/2019); a game [...] the timeline of the ages [...] if it is hot or cold [...] it will be explained (D4:2016/2017); [...] the first 6 minutes of Ice Age [...] the migration of animals and humans [...] climate change [...] survival (D10:2018/2019).

b) The dimensions of space and time are worked upon:

All of the groups proposed working on time and space. There is a commitment to approaching changes and continuities (n=7), reflections on glaciers (n=3), animal and vegetable species (n=8) and seasonal and climate changes in relation to history (n=3).

[...] regarding [...] changes in the climate which have occurred throughout history, we will jump to the present (D1:2016/2017); [...] some tribes set off on a journey in search of food [...] enormous fruit [...] if you ate it you fell asleep for thousands of years [...] they woke up [...] where before there was land, now there was sea [...] (D3:2018/2019); [...] the island of Sálvora [...] the hunter went there from where the port is located today [...] how is it possible that he went on foot? (D19:2018/2019); [...] a glacier [...] would you say that that is a typical landscape of Galicia? Could it have been in another time? (D13:2018/2019); [...] Galicia [...] ecosystems such as taiga and tundra [...] (D8:2018/2019); [...] to reflect on [...] changes in the climate [...] Galicia between the 17th and 18th centuries [...] in comparison with other parts of Europe [...] (D18:2018/2019).

c) It is approached as a problem of the present:

All of the groups approached it as a present-day problem (n=25) in three ways: pollution and global warming (n=11); the excessive consumption of resources and the destruction of landscapes; an attack against natural heritage and its necessary preservation (n=10). The future perspective is opened up by way of catastrophes associated to a climate which becomes more and more extreme (n=7).

> [...] reflecting on what is happening in the present day, changes brought about by human action [...] global warming and destruction [...] (D3:2016/2017); [...] looking for related news [...] in different parts of the world [...] (D1:2016/2017); [...] the Industrial Revolution [...] would bring with it the beginning of large-scale pollution. . .global warming (D21:2018/2019); [...] flooding on the Mediterranean coast [...] the thawing of the ice caps [...] droughts in Galicia [...] refugees (D1:2016/2017); [...] promoting the defence and respect of the environment, and understanding, from an objective, critical and scientific point of view, the concept of climate change (D13:2018/2019).

3. A concern for the civic dissemination of academic work.

    a) A commitment to educating pupils as agents of transformation:

All of the groups manifest this point (n=25) and the majority specify it with activities for disseminating the knowledge acquired (n=21); by way of converting the school into an exhibition centre (n=10); by producing reports and written assignments which involve the families (n=6); via civic initiatives which require action on the part of the community (n=5); and drama and exhibitions for younger schoolchildren (n=4).

> [...] opening the doors of the school to the surrounding area [...] (D2:2016/2017); [...] disseminating what has been learned [...] to the other children in the school and then to the community [...] (D19:2018/2019); [...] a report outside of the classroom, asking people what they think [...] (D3:2016/2017); Sending videos to the mayor, in which the deposits can be seen (D1:2016/2017); [...] asking their family members [...] to reflect on this together [...] (D19/2018/2019); [...] to transmit [...] the importance of the deposits, their history up to the present day [...] performing a play [...] actors [...] scriptwriters, directors [...] (D8:2018/2019); [...] gathering signatures [...] in change.org for climate change [...] going to the council [...] (D16:2018/2019).

Discussion

It is extremely relevant that from an action which proposes a new teaching approach regarding climate change, the trainee teachers have unanimously paid special attention to what has been presented to them as an environmental archive, the edapho-sedimentary deposits of the Galician coast. This is a confirmation of the educational importance of working with documents (Morgan, 2002; Prats, Santacana, 2011), particularly if they are places which single out everyday landscapes (Morón, 2013; Liceras, 2018). Furthermore, their high evaluation is rooted in different lines, which are, in many cases, interwoven: (1) Contributions on a cognitive level, making it possible to state that knowledge also has an emotional dimension (Domínguez-Almansa, Riveiro-Rodríguez, 2017), and understanding the value of interdisciplinarity (Jones,

2010), which is intellectually and socially necessary in order to scientifically address issues which occupy and preoccupy. (2) Without making mention of heritage in the educational actions, it has been proven how prior training in this field has generated a process of heritage appropriation, which is fundamental in the aspirations of this educational approach (Fontal, Ibáñez, 2015). Thus, the scientific value of these geomorphological formations is accompanied by a strong sense of identity (Cuenca-López, Estepa-Giménez, Martín-Cáceres, 2017),which has its origins in the reflection on their frequency and uniqueness on the Galician coast and is assumed as a differentiating element and with a certain degree of pride, thereby leading to a commitment to their preservation and dissemination. (3) In coherence, the trainee teachers attribute the deposits with a high degree of value for their future teaching work and they are able to transfer this from reflection to the design of educational actions. In these actions, it can be appreciated which aspects of what are today cliffs with memory are used, via a variety of procedures, as an educational route for discovering, providing meaning to and disseminating climate change. This demonstrates that what has been absorbed in their training process is assumed as important for proposing and designing their own educational actions.

Apart from the appropriation of the deposits, the reflections and designs show that: (1) the students have understood the scientific and educational dimension of climate change, recognising it to be a natural reality which can be integrated into social knowledge of a better quality; (2) they value the need for employing an alternative focus to geography and history in their training and teaching careers, as opposed to traditional routines (Domínguez-Almansa et al., 2020); (3) they do not deal with space and time with the desired depth in their reflections, in spite of indicating an interest in these aspects during the recent training experience. This lack demonstrates an urgent need for changes to be made the secondary education (Souto, 2011; Sáiz, Fuster, 2014). However, in the designs, the incorporation of geography and history is undertaken with more robustness, which indicates the benefits of cooperative work (Koles et al., 2010; Genç, 2016;), allowing for the skills of some to improve the competence of the group; (4) they have assumed the need to deal with present-day problems by way of the interaction between the past, the present and the future (Israel, 2012; Santisteban-Fernández, González-Monfort, Pagès-Blanch, 2020), valuing the importance of documentation in order to undertake a critical method of teaching of higher quality (Prieto, Gómez, Miralles, 2013); (5) they have demonstrated an interest in the civic and transformative dimension of education (DePalma, 2019; Schweisfurth, 2006). Thus, in addition to cognitive aspects, they reason and position themselves ethically and emotionally regarding a conflictive issue such as climate change; a problem which they have been able to transmit in their designs, which enable their pupils to understand, think and act.

Conclusions

The objectives of this research responded to a prior educational action, which explicitly approached climate change as a problem of society today. However, it carried a sufficient burden of implicit elements upon which it was also necessary to verify the attention which the trainee teachers could pay to them. Thus, the proposal was made to analyse aspects ranging from what type of evaluation they would make of the edapho-sedimentary Quaternary deposits, which were presented as environmental archives adorning the Galician coast, to the scope of the new contents and methodological approaches for dealing with the topic. The idea was for them to abandon the social representation, too predominant and presentist, that associates climate change with starving bears at the North Pole, the credibility and emotional burden of which can engulf what should be an optimum educational approach. With this starting point, the aim was to enable the trainee teachers to carry out a critical reflection, with the freedom granted by knowledge, on their teaching role and even to approach reasoned and socially active teaching actions.

The results presented here enable us to be prudently optimistic, as, although it is true that they have detected structural cognitive problems which may impact this and any other action focusing on geographical and historical knowledge, it is no less true that they have appreciated extremely relevant aspects regarding how to approach the issue. One extremely relevant example is the high degree of appreciation of the deposits, along with their understanding of climate change by way of new parameters, which has been well expressed in their reflections and designs.

Aside from the fulfilment of the objectives, it is necessary to pay attention to two relevant issues bearing future research in mind. One of them, concerning the need for better initial training of future primary education teachers, is difficult to solve. This would imply that secondary teachers should show a commitment to methodologies focusing on geographical and historical knowledge from perspectives which make it more meaningful. The other issue will depend exclusively on the authors of this study. It concerns the capacity for designing educational actions approaching climate change in primary education and to continue their real practice in schools. This will make it possible to research another level of teaching competence, which concerns educating pupils in a more competent manner, both in the cognitive sense and in the personal and social sense.

References

Bangay, Colin/Blum, Nicole: "Education responses to climate change and quality: Two parts of the same agenda?". *International Journal of Educational Development*, 30(4), 2010, pp. 359–368, from DOI 10.1016/j.ijedudev.2009.11.011.

Bardsley, Douglas. K./Bardsley, Annette. M.: "A constructivist approach to climate change teaching and learning". *Geographical Research*, 45(4), 2007, pp. 329–339, from DOI 10.1111/j.1745-5871.2007.00472.x.

Blanco-Chao, Ramón/Cajade-Pascual, Daniel/Costa-Casais, Manuela: "Rotation, sedimentary déficit and erosion of a trailing spit inside ria of Arousa (NW Spain)". *Science of the Total Environment*, 749, 2020, pp. 1–14. doi.org/10.1016/j.scitotenv.2020.141480.

Blom, Philipp: *Nature's Mutiny. How the Little Ice Age of the Seventeenth Century Transformed the West and Shaped the Present.* Norton & Company: London. 2017.

Brooke, John L.: *Cimate Change and the Course of Global History: A Rough Journey.* Cambridge University Press: Cambridge. 2014.

Carcavilla Urquí, Luis: "Guía práctica para entender el patrimonio geológico". *Enseñanza de las Ciencias de la Tierra*, 22(1), 2014, pp. 5–18.

Collins, Dwight E./Genet, Russell M./Christian, David G.: "Crear una nueva narrativa para promover la Sostenibilidad". In: Worldwatch Institute: *The State of the World 2013: Is Sustainability Still Possible?* W.W. Norton: New York. 2013.

Costa-Casais, Manuela/Caetano Alves, Isabel M.: "Geological heritage and geodiversity: Quaternary deposits as palaeoenvironmental records in the 'Southern Coast' (Galicia, NW Spain)". In: González Díez, A. (Coord.): *Avances de la Geomorfología en España 2010–2012*: actas de la XII Reunión Nacional de Geomorfología, Santander Publican Ediciones, Universidad de Cantabria. 2012, pp. 27–30.

Costa Casais, Manuela/Caetano Alves, Isabel M.: "Geological heritage at risk in NW Spain. Cuaternary deposits and landforms of 'Southern Coast' (Baiona-A Garda)". *Heritage*, 5, 2013, pp. 227–248, from DOI 10.1007/s12371-013-0083-7.

Costa-Casais, Manuela/Domínguez-Almansa, Andrés: "Posta en valor dos depósitos edafo-sedimentarios cuaternarios do litoral galego (NW da Península Ibérica): unha perspectiva dende a educación patrimonial".

In: Blanco-Chao, Ramón/Castillo, Francisco/Costa-Casais, Manuela/García, Jesús H./Valcárcel, Marcos (eds.): *Xeomorfoloxía e Paisaxes Xeográficas. Catro décadas de investigación e ensino. Homenaxe a Augusto Pérez Alberti.* Universidade de Santiago de Compostela: Santiago de Compostela. 2018, pp. 115–138.

Crutzen, Paul J./Stoermer, Eugene F.: "The 'anthropocene'". *Global Change Newsletter*, 41, 2000, pp. 17–18, retrieved from: http://www.igbp.net/download/18.316f18321323470177580001401/1376383088452/NL41.pdf.

Cuenca López, José M.: "Escuela, patrimonio y sociedad. La socialización del patrimonio". *UNES: Universidad, escuela y sociedad*, 1, 2016, pp. 22–41, retrieved from: https://1library.co/document/y96dnrdy-escuela-patrimonio-sociedad-socializacion-patrimonio.html.

Cuenca-López, José M./Estepa-Giménez, Jesús/Martín-Cáceres, Myriam: "Patrimonio, educación, identidad y ciudadanía. Profesorado y libros de texto en la enseñanza obligatoria". *Review of Education*, 375, 2017, 136–159, from DOI 10.4438/1988-592X-RE-2016-375-338.

Denzin, Norman K./Lincoln, Yvonna S.: "Introduction: The discipline and practica of qualitative research". In: Denzin, Norman/Lincoln, Yvonna (eds.): *The Sage Handbook of Qualitative Research* (Third ed.). Sage: Thousand Oaks, London. 2005, pp. 1–32.

DePalma, Renée: *La educación para el desarrollo y la ciudadanía global: una experiencia de investigación-acción participativa*. Graó: Barcelona. 2019.

Diamond, Jared M.: *Guns, Germs and Steel: The Fates of Human Societies.* Norton & Company: New York. 2005.

Domínguez-Almansa, Andrés/Costa-Casais, Manuela/López-Facal, Ramón: "Educar para reconocer: Apropiación patrimonial de los depósitos cuaternarios del litoral gallego por estudiantes de magisterio". *Revista Electrónica Interuniversitaria de Formación del Profesorado*, 22(1), 2019, pp. 57–70, from DOI 10.6018/reifop.22.1.357591.

Domínguez-Almansa, Andrés/López Facal, Ramón: "Conflictive heritages, civic competence and professional training in primary education". *Review of Education*, 375, 2017, pp. 86–109, from DOI 10.4438/1988-592X-RE-2016-375-336.

Domínguez-Almansa, Andrés/Riveiro-Rodríguez, Tania: "Lo emocional y lo racional en la enseñanza de las ciencias sociales". In: López Facal, Ramón/Miralles Martínez, Pedro/Prats Cuevas, Joaquim (Dirs.): *Enseñanza de la historia y competencias educativas*. Graó: Barcelona. 2017, pp. 49–65.

Domínguez-Almansa, Andrés/Riveiro-Rodríguez, Tania/Monteagudo-Fernández, Jose/López-Facal, Ramón: "Conflictive memory and heritage education in the initial training of primary teachers". In: Delgado Algarra, Emilio J./Cuenca López, José M. (eds.): *Handbook of Research on Citizenship and Heritage Education*. IGI Global: Hershey PA. 2020, pp. 471–499.

Erickson, Frederik: "Métodos cualitativos de investigación sobre la enseñanza". In: Wittrock, Merlin C. (Comp.): *La investigación en la enseñanza, II*. Barcelona: Paidós: Barcelona. 1989, pp. 195–301.

Estepa Giménez, Jesús: "Memoria, patrimonio y ciudadanía: una contribución desde una perspectiva didáctica". *PH: Boletín del Instituto Andaluz del Patrimonio Histórico*, 27(96), 2019, pp. 225–226, retrieved from: https://www.iaph.es/revistaph/index.php/revistaph/issue/view/111.

Estepa Giménez, Jesús/Martín Cáceres, Myriam: "Heritage in conflict: A way to educate in a critical and participative citizenship". In: Delgado Algarra, Emilio J./Cuenca-López, José M. (eds.): *Handbook of Research on Citizenship and Heritage Education*. IGI Global: Hershey PA. 2020, pp. 43–55.

Flick, Uwe: *Designing Qualitative Research*. Sage Publications: London. 2007.

Fontal Merillas, Olaia: *La educación patrimonial: del patrimonio a las personas*. Trea: Gijón (Spain). 2013.

Fontal Merillas, Olaia/Ibáñez Etxeberria, Alex: "Estrategias e instrumentos para la educación patrimonial en España". *Educatio Siglo XXI*, 33(1), 2015, pp. 15–32, from DOI 10.6018/j/222481.

Gaudiano, Edgar J.G./Cartea, Pablo A.M. "Educación para el cambio climático ¿Educar sobre el clima o para el cambio?". *Perfiles Educativos*, 42(168), 2020, pp. 157–174, from DOI 10.22201/iisue.24486167e.2020.168.59464.

Genç, Murat: "An evaluation of the cooperative learning process by sixth-grade students". *Research in Education*, 95(1), 2016, pp. 19–32, from DOI 10.7227%2FRIE.0018.

Glaser, Barney/Strauss Anselm: *The Discovery of Grounded Theory: Strategies for Qualitative Research*. Aldine Publishing Company: New York. 1967.

Gruenewald, David A.: "The best of both worlds: A critical pedagogy of place". *Educational Researcher*, 32(4), 2003, pp. 3–12, from DOI 10.3102%2F0013189X032004003.

Hassinger-Das, Brenna/Toub, Tamara S./Zosh, Jennifer M./Michnick, Jessica/Golinkoff, Roberta/Hirsh-Pasek, Kathy: "More than just fun: A place for games in playful learning". *Infancia y Aprendizaje*, 40(2), 2017, pp. 191–218, from DOI 10.1080/02103702.2017.1292684.

Herodotou, Christothea/Sharples, Mike/Gaved, Mark/Kukulska-Hulme, Agnes/Rienties, Bart/Scanlon, Eileen/Whitelock, Denise: "Innovative pedagogies of the future: An evidence-based selection". *Frontiers in Education*, 4(113), 2019, from DOI 10.3389/feduc.2019.00113.

Higgs, John: *Stranger Than We Can Imagine. An Alternative History of the 20th Century*. Soft Skull Press: Berkeley, CA. 2015.

Hmelo-Silver, Cindy E./Duncan, Ravit G./Chinn, Clark A.: "Scaffolding and achievement in problema-based and inquiry learning: a responce to Kirschner, Sweller, and Clark". *Educational Psychologist*, 42(2), 2006, pp. 99–107, from DOI 10.1080/00461520701263368.

Hodzic, Migdat/Kennedy, Ivan R.: "Time and frequency analysis of Vostok ice core climate date". *Periodicals of Engineering and Natural Sciences*, 7(2), 2019, pp. 907–923, retrieved from: http://pen.ius.edu.ba.

Israel, Andrei L.: "Putting geography into place: What geography educators can learn from place-based education and viceversa". *Journal of Geography*, 111(2), 2012, pp. 76–81, from DOI 10.1080/00221341.2011.583264.

Jackson, Derek W.T/Costas, Susana/González-Villanueva, Rita/Cooper, Andrew: "A global 'greening' of coastal dunes: An integrated consequence of climate change?". *Global and Planetary Change*, 182, 2019, pp. 1–9, from DOI 10.1016/j.gloplacha.2019.103026.

Johnston, Marylin: "The lamp and the mirror. Action Research and Self Studies in the Social Studies". In: Barton, Keith C. (ed.): *Research Methods in Social Studies Education. Contemporary Issues and Perspectives. A Volume in Research in Social Education*. Information Age Publishing: Greenwich. 2006, pp. 57–83.

Jones, Casey: "Interdisciplinary approach-advantages, disadvantages, and the future benefits of interdisciplinary studies". *ESSAI*, 7, Article 26, 2010. Retrieve from: https://dc.cod.edu/essai/vol7/iss1/26.

Koles, Paul H./Stolfi, Adrienne/Borges, Nicole, J./Nelson, Stuart/Parmelee, Dean X.: "The impact of team-based learning on medical student's academic performance". *Academic Medicine*, 85(1), 2010, pp. 1739–1745.

Liceras Ruiz, Ángel. "Notas para el estudio del paisaje". *UNES. Universidad, Escuela y Sociedad*, 4, 2018, pp. 26–39.

Malavin, Stas/Schakova Lyubov A./Claverie, Jean-Michel/Rivkina, Elizaveta: "Frozen Zoo: A collection of permafrost samples containing viable protists and their viruses". *Biodiversity*, 8(1), 2020, from DOI 10.3897/BDJ.8.e51586.

Mann, Michael E./Kump, Lee R.: *Dire Predictions. Understanding Climate Change 2nd Edition*. DK Publishing, Penguin Random House, Pearson: United States, 2016.

Martínez Cortizas, Antonio/Ferro Vázquez, Cruz/Kaal, Joeri/Biester, Harald/Costa-Casais, Manuela/Taboada Rodríguez, Teresa/Rodríguez Lado, Luis: "Bromine accumulation in acidic black colluvial soils". *Geochimnica et Cosmochimica Acta*, 174, 2016, pp. 143–155, from DOI 10.1016/j.gca.2015.11.013.

McInerney, Peter/Smyth, John/Down, Barry: "'Coming to a place near you?' The politics and possibilities of a critical pedagogy of place-based education". *Asia-Pacific Journal of Teacher Education*, 39(1), 2011, pp. 3–16, from DOI 10.1080/1359866X.2010.540894.

Morgan, Katherine R.: "Using primary sources to build a community of thinkers". *English Journal*, 4, 2002, 69–74, retrieved from: https://pdfs.semanticscholar.org/75a8/4f6d6e92fe2fd5bb02c9009e32d74e8188cb.pdf.

Morón Monge, Carmen: "Paisaje y Geografía: una oportunidad para educar en Patrimonio". In: Estepa Gimeno, Jesús (ed.): *La Educación Patrimonial en la Escuela y el Museo: Investigación y Experiencias*. Universidad de Huelva: Huelva. 2013, pp. 237–247.

Nordhaus, William: *The Climate Casino: Risk, Uncertainty, and Economics for a Warming World*. Yale University Press: USA. 2013.

Pinker, Steven: *The Better Angels of Our Nature. Why Violence Has Declined*. Viking Adult: New York. 2011.

Prats, Joaquim/Santacana, Joan: "Los restos arqueológicos, los monumentos y los museos como fuentes del pasado". In: Prats, Joaquim (Coord.): *Geografía e Historia. Investigación, Innovación y Buenas Prácticas*. Graó: Ministerio de Educación Cultura y Deporte, Secretaría General Técnica. 2011, pp. 39–68.

Prieto, José A./Gómez, Cosme J./Miralles, Pedro: "El uso de fuentes primarias en el aula y el desarrollo del pensamiento histórico y social. Una experiencia en Bachillerato". *Clío*, 39, 2013, retrieved from: http://clio.rediris.es/n39/articulos/historiasocial/PrietoGomezMiralles.pdf.

Ristinen, Robert: *Energy and Environment*. John Wiley & Sons: New York. 1999.

Rodríguez Plasencia, José L.: "De ritos y mitos agrarios". *Revista de folklore*, 382, 2013, pp. 19–29, retrieved from: https://dialnet.unirioja.es/ejemplar/353183.

Saíz Serrano, Jorge/Fuster, Carlos: "Memorizar historia sin aprender pensamiento histórico. Las PAU de historia de España". *Investigación en la Escuela*, 84, 2014, pp. 47–58.

Santisteban Fernández, Antoni/González-Monfort, Neus/Pagès Blanch, Joan: "Critical Citizenship Education and Heritage Education". In: Delgado Algarra, Emilio J./Cuenca-López, José M. (eds.): *Handbook of Research on Citizenship and Heritage Education*, IGI Global: Hershey PA, USA. 2020, pp. 26–42.

Schweisfurth, Michele: "Education for global citizenship: teacher agency and curricular structure in Ontario schools". *Educational Review*, 58(1), 2006, pp. 41–50, from DOI 10.1080/00131910500352648.

Sobel, David: "Place-based education: Connecting classroom and community". *Nature and Listening*, 4(1), 2004, pp. 1–7, retrieved from: http://www.kohalacenter.org/teachertraining/pdf/pbexcerpt.pdf.

Souto González, Xosé M.: "Geografía y otras ciencias sociales: la interdisciplinariedad y la selección de contenidos didácticos". In: Prats, Joaquim (Coord.). *Didáctica de la Geografía y la Historia*. Graó: Ministerio de Educación, Cultura y Deporte, 2011, pp. 131–144.

Tallón-Armada, Rebeca/Costa-Casais, Manuela/Blanco-Chao, Ramón/Taboada Rodríguez, Teresa/Martínez-Cortizas, Antonio: "Paleoenvironmental reconstruction of an urban archaeological site: The Roman Salt mines of Vigo, northwest Iberia". *Geoarchaeology*, 2017, pp. 1–15, from DOI 10.1002/gea.21648.

Threnhaile, Alan S./Pérez-Alberti, Augusto/Costa-Casais, Manuela/Blanco-Chao, Ramón: "Rock coast inheritance: an example from Galicia, North-western Spain". *Earth Surf Process Landforms*, 24, 1999, pp. 605–621, from DOI 10.1002/(SICI)1096-9837(199907)24:7%3C605::AID-ESP977%3E3.0.CO;2-1.

UNESCO: "Indigenous knowledge and implications for the sustainable development agenda. Background paper prepared for the 2016 Global". Education Monitoring Report Education for people and planet: Creating sustainable futures for all. *Global Education Monitoring Report*. 2016, retrieved from: https://unesdoc.unesco.org/ark:/48223/pf0000245623.

Veettil, Bijeesh Kozhikkodan/Kamp, Ulrich: "Global disappearance of Tropical Mountain Glaciers: Observations, causes and challenges". *Geosciences*, 9(5), 2019, p. 196, from DOI 10.3390/geosciences9050196.

Waver, Andrew J./Hillaire-Marcel, Claude: "Global warming and the next Ice Age". *Science*, 304(5669), 2004, pp. 400–2, from DOI 10.1126/science.1096503.

Francisco-Javier Hinojo-Lucena, Inmaculada Aznar-Díaz, María-Pilar Cáceres-Reche, and José-María Romero-Rodríguez

Chapter 16 Mobile learning methodology. Analysis of the use of mobile devices in teaching

Abstract: The application of mobile devices in the classroom is increasingly common in Spanish higher education. The fact of introducing these devices into teaching has acquired its own name: mobile learning methodology. This method consists of using mobile devices (mainly smartphones, tablets) to carry out the teaching-learning process. The purposes of this paper were to analyze the degree of implementation of the mobile learning method in universities in the region of Andalusia (Spain) and determine the reasons why teachers do not apply mobile devices in their teaching. The responses of 316 professors from 10 different universities in Andalusia were analysed. The method used was mixed, a Likert scale and an open questionnaire were used to collect the data. Among the results, the higher percentage of teachers who use mobile devices versus those who do not use them stands out. In addition, the main reasons for not using them were the lack of teacher training on the didactic application of mobile devices and consideration of the possibility that they may distract the student. Finally, the main findings and implications of the study are discussed.

Keywords: Mobile devices, Teacher Education, Higuer Education, Didactic methodology, Program evaluation

Introduction

The use of the mobile device in the university environment is increasingly widespread. In this sense, university professors have begun to include digital elements in their teaching practices (Pozo, López-Belmonte, Moreno-Guerrero, Sola, & Fuentes, 2020). However, there is still a large percentage of teachers who are reluctant to be included. Nevertheless, with the advent of COVID-19, the need to use digital media as the main channel to follow classes normally has become apparent (Huang et al., 2020).

The regular use of the mobile device is part of the mobile learning methodology, which refers to the use of these devices to carry out the teaching and learning process (Díez, Valencia, & Bermúdez, 2017). In addition, mobile learning has begun to be linked to increased student academic performance

(Aznar-Díaz et al., 2018; Ferriols & Oms, 2019; Hu et al., 2019), improving motivation (Nikolopoulou et al., 2020; Valeeva et al., 2019), encouraging cooperative work (Chang, Liu, & Huang, 2017) and developing digital competence (Pozo, López-Belmonte, Rodríguez-García, & López-Núñez, 2020).

This last aspect is essential in today's society, where digital competence is a key factor in developing properly as a citizen in the 21st century (Moreno-Guerrero, Alonso, Ramos, Campos-Soto, & Gómez-García, 2020), since technology is present in most everyday processes. Therefore, the inclusion of the mobile device in the classroom contributes to the personal, professional and social training of students.

In the case of the university qualifications responsible for training future teachers, this takes on a special nuance, where the use of the mobile device acts both as a training element and as a transforming element. In relation to the first case, it contributes to the formation of the students, helping to the understanding and facilitation of the learning contents. While the transforming element refers to the fact that students will apply in the future, in the classrooms where they develop professionally, resources similar to those exposed by the teacher.

In consideration, we have reached a fundamental moment to investigate active methodologies that imply the use of Information and Communication Technologies (ICT), with the purpose of knowing their current state and future projection in a context of uncertainty and change derived from the COVID-19. Furthermore, in recent years there have been several investigations focused on the university environment that have investigated the implementation of mobile learning, highlighting as results: the improvement of relationships between peers and horizontal communication between teacher and student (Soler, Mauri, Lafarga, & Moreno-Guerrero, 2020); development of good teaching practices by almost 40 % of teachers who applied mobile learning in their teaching (Romero-Rodríguez, Aznar-Díaz, Hinojo-Lucena, & Cáceres-Reche, 2020); and in another line those studies that warn of the positive predisposition towards the adaptation of mobile learning (Gómez-Ramírez et al., 2019; Pinto et al., 2020; Seifert et al., 2019; Suárez & Grané, 2019).

On the other hand, some authors have highlighted various reasons given by university professors for not applying ICT in the classroom, including lack of training (Salcines-Talledo et al., 2017; Sánchez-Prieto et al., 2019) and resistance to change (Romero-Rodríguez, Aznar-Díaz, Hinojo-Lucena, & Gómez-García, 2020).

Therefore, this paper proposed as objectives: to analyze the degree of implementation of the mobile learning method in the universities of the region of

Andalusia (Spain); and to determine the reasons why teachers do not apply mobile devices in their teaching. From these objectives were derived as research questions:

- RQ1. What was the percentage of mobile device application in the classroom?
- RQ2. Does the gender or age of teachers influence the application of the mobile device?
- RQ3. What were the main reasons for not applying the mobile device?

Method

A mixed methodology was used, employing both quantitative and qualitative methods (Johnson & Onwuegbuzie, 2004). In this sense, the closed questionnaire was used as a quantitative instrument (Hernández, Fernández, & Baptista, 2016), and the open questionnaire as a qualitative method (Casas, García, & González, 2006; Jansen, 2012). The focus of the study was cross-sectional, as the data were collected at one point in time.

Participants and procedure

The sample was selected on the basis of a convenience sample, focusing on the population of university teachers in the region of Andalusia. An invitation was extended to 1,776 teachers linked to educational areas of knowledge from 10 different public and private universities. Therefore, the significant sample size was established in 316 subjects with a 95 % confidence index and a 5 % of error.

The application of the two instruments was done through Google Forms and was distributed by email along with the data on the purpose of the study and information on the anonymous processing of the data. The informed consent of each of the subjects was a prerequisite for participation in the study.

The sample consisted of 125 men and 191 women aged between 20 and 62 years (M = 43.56; SD = 9.12). Tab. 1 shows the frequency and percentage of the sociodemographic data of the sample.

Instruments

Quantitative data was collected through the Analysis of M-learning practices at the University (APMU) scale (Aznar-Díaz, Romero-Rodríguez, Ramos, & Gómez-García, 2020). This instrument consists of 16 items that evaluate the development of good mobile learning practices through its five dimensions (mobile devices, digital competence, knowledge construction, cooperative work and good use of technology). The response mode is based on a four-level

Tab. 1: Sociodemographic data

	n	%
Gender		
Male	125	39.6
Female	191	60.4
Age		
20–29	25	7.9
30–39	91	28.8
40–49	91	28.8
50–59	82	25.9
60 or more	27	8.5
University		
University of Almeria (UAL)	14	4.4
University of Cadiz (UCA)	32	10.1
University of Cordoba (UCO)	24	7.6
University of Granada (UGR)	105	33.2
University of Huelva (UHU)	23	7.3
University of Jaen (UJAEN)	13	4.1
University of Malaga (UMA)	32	10.1
Pablo de Olavide University (UPO)	6	1.9
University of Seville (US)	63	20
Loyola of Andalusia University (ULOYOLA)	4	1.3

Likert Scale from never to always. Scores on the scale range from 16 to 64 points, where a higher score indicates a higher degree of development of excellent practices with mobile devices. The reliability obtained for this study was good (Cronbach's a =.817). While the open-ended questionnaire picked up only one question which was the open-ended question: "What is the main reason for not applying mobile devices in the classroom?."

Data analysis

The data were analyzed with the SPSS quantitative analysis program, version 24. Mean values and standard deviations on the APMU scale were established for each population subgroup (gender and age). Furthermore, the analysis of the possible existence of significant differences between these two variables was also of interest, applying the T test when there were two groups (gender) and ANOVA with more than two groups (age). Subsequently, a multiple linear

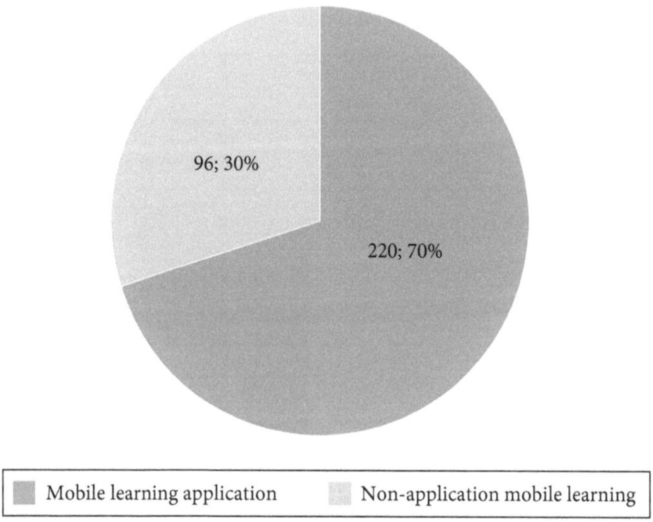

Fig. 1: Percentage of mobile learning application (n = 316)

regression model was carried out to check whether any of these factors directly influenced the development of mobile learning practices.

On the other hand, the qualitative data were categorized and analyzed with the software Atlas.ti. Therefore, a content analysis was used as the main method of analysis of the reasons for not applying mobile learning (Bardín, 1991).

Results

The data indicated a higher percentage of application of mobile learning by university teachers in Andalusia linked to educational knowledge areas, as is the case of Social Sciences where History and Geography are included. Specifically, 70 % used mobile devices to carry out the teaching process in their teaching (Fig. 1).

In relation to the gender variable, the means obtained were similar in both populations. However, women presented a slightly higher average (M = 46.56) than the male population (M = 45.73), although no significant differences were established between the two groups (p =.416) (Tab. 2).

In the age variable something similar happened, the average scores were in similar values. The greatest differences were obtained between the population over 60 years of age (M = 42.94) and the population between 50–59 years of

Tab. 2: Mean and standard deviation of scores and differences based on gender

Gender	n	APMU scale		*t*	*df*	*p*
		M	SD			
Male	94	45.73	7.738	-.815	218	.416
Female	126	46.56	7.124			

Note: *p* calculated through the T test. *df* = degrees of freedom; *n* = 220.

Tab. 3: Mean and standard deviation of scores and differences based on age

Age	n	APMU scale		*F*	*df*	*p*
		M	SD			
20–29	17	45.82	6.858	1.542	4	.191
30–39	60	45.45	6.693			
40–49	65	46.63	7.026			
50–59	61	47.51	7.848			
> 60	17	42.94	9.236			

Note: *p* calculated through ANOVA test; *df* = degrees of freedom; *n* = 220.

Tab. 4: Multiple linear regression analysis

Independent variable	*B*	SE	*T*	*β*	*p*
Gender	.852	1.014	.840	.057	.402
Age	.143	.464	.307	.021	.759

Note: *B* = coefficient; SE = Standard error; *T* = coefficient based on the T of Student; *β* = standardized coefficient; **p*<.05; ***p*<.01; ****p*<.001; *n* = 220.

age (M = 47.51), where the difference was established at 4.57 points. The other scores in the other age ranges were fairly homogeneous between them, resulting in no significant differences (p =.191) (Tab. 3).

In turn, the multiple linear regression model was key to determining whether gender or age influenced the application of the mobile device. The values obtained in the model indicated a good fit, while at the same time it was significant (R^2 =.43; *F*-statistic =*.378; p* =.04). Nevertheless, the data showed that neither gender (p =.402) nor age (p =.759) were significant in the model (Tab. 4).

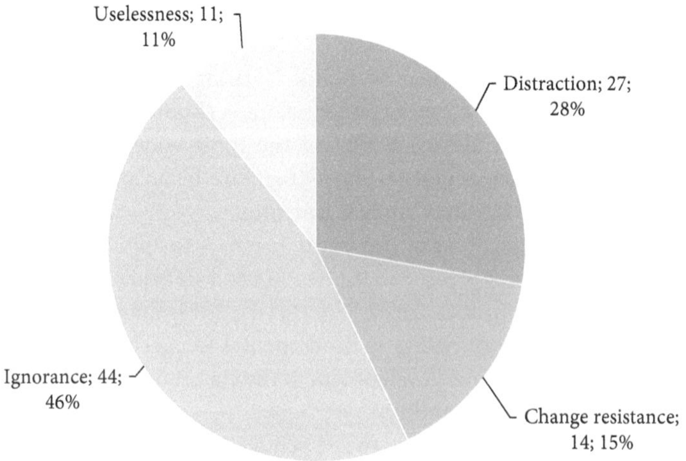

Fig. 2: Reasons for non-application (n = 96)

On the other hand, the main reasons highlighted by teachers for not applying the mobile device were linked to ignorance, distraction, change resistance and uselessness (Fig. 2). Lack of training (ignorance) was the most prominent reason (46 %), followed by the consideration that mobile devices distract students (28 %). Thirdly, there was resistance to change on the part of some teachers who prefer to follow a traditional methodology instead of including ICT (15 %) and finally, the perception of uselessness about the use of mobile devices (11 %).

Discussion

The data showed a high rate of application of mobile learning in the university context, where 70 % of university teachers in the region of Andalusia introduced mobile devices in the classroom. These teachers were linked to different educational areas including the specialties of History and Geography. It is therefore essential to apply ICT resources in areas that are purely theoretical in order to make the learning content enjoyable and to awaken the motivation of the students (Nikolopoulou et al., 2020; Valeeva et al., 2019).

These data highlight a widespread use of ICT (Pozo et al., 2020), the focus of which has been the universities in southern Spain. In this context, there is a collective awareness of the importance of the use of ICT, with digital competence being a key factor in developing adequately in the society of the 21st century (Moreno-Guerrero et al., 2020).

In this sense, the development of mobile learning practices is increasingly common and is growing at an exponential rate (Romero-Rodríguez et al., 2020a), since the mobile device is one of the most widely used devices by the population. Its implementation must therefore be accompanied by good teaching practices that facilitate student learning.

This positive predisposition of university teachers to apply mobile learning has also been previously expressed in studies on analysis of perceptions (Gómez-Ramírez et al., 2019; Pinto et al., 2020; Seifert et al., 2019; Suárez & Grané, 2019). However, the contrast with reality is fundamental to check if the perceptions correspond with the practical development in the classroom, in this case we can say that there is a high applicability.

On the other hand, with respect to the gender variable, the highest average scores were obtained in the female population. Nevertheless, there were no significant differences between women and men. This means that the application of mobile learning does not depend on gender, this fact being confirmed empirically in the multiple linear regression model. Similarly, age did not turn out to be a determining factor either. So neither gender nor age affected the development of mobile learning practices in the university teachers who were studied.

It should be noted that higher average scores were obtained by senior teachers than by novice teachers, which is curious since it is assumed that younger teachers are the ones who usually apply technology in the classroom. In future studies it would be interesting to qualitatively combine this case study.

In relation to the reasons expressed by the teachers who did not apply mobile learning, the lack of training was the main reason derived from the lack of knowledge that the teachers expressed (Salcines-Talledo et al., 2017; Sánchez-Prieto et al., 2019). Thus, active ICT policies are a fundamental commitment of universities so that all their teaching staff are aware of the main educational trends and are competent to carry them out. Another concern was distraction, related to teachers' perception that mobile devices distract students. Therefore, they do not consider it a useful resource for learning, since they perceive more negative consequences than positive ones.

Finally, resistance to change turned out to be another conditioning factor, as previous studies had shown (Romero-Rodríguez et al., 2020b). This resistance on the part of teachers not to change their traditional methodology makes it very difficult to introduce any change in the teaching they give. Therefore, the need for active ICT policies and active methodologies in university training plans is reiterated.

Conclusion

Mobile learning has emerged as an interesting resource for university teaching, which has taken on a leading role during the pandemic experienced by the COVID-19. The introduction of telematic methods to develop training has led teachers to develop or expose their digital competence. Therefore, in the next few years, the development of ICT methodologies will occupy an important place in educational agendas. Therefore, in this work, the objectives were proposed to analyze the degree of implementation of the mobile learning method in the universities of the region of Andalusia and to determine the reasons why teachers did not apply mobile devices. These two objectives have been successfully met, as developed throughout this chapter.

Among the limitations of the study is the cross-sectional nature of the data, which means that these findings cannot be established over time, but are the result of the current moment in which they were collected. Another limitation was the small sample size at certain universities. However, it was decided to keep all cases to ensure the representativeness of all of them.

Much remains to be done on ICT in higher education, but the application data are encouraging, especially in mobile learning. As future lines of research, further research is needed on the percentage of application of methodologies using ICT. At the same time, the concrete practices being implemented will be analysed in order to collect good teaching practices that can serve as a reference for other teachers.

Finally, the COVID-19 pandemic has brought to light certain shortcomings in university education, such as the adaptability of the presential to the virtual. Mobile learning is a method that has come to stay and its implementation will go further in the coming years, being fundamental its inclusion as one more resource within the range of the university teacher.

References

Aznar-Díaz, I./Cáceres-Reche, M.P./Romero-Rodríguez, J.M: "Effect of the mobile learning methodology in university education: Meta-analysis of the research published in WOS and Scopus". *RISTI* 30 2018, pp. 1–16, from DOI 10.17013/risti.30.1–16.

Aznar-Díaz, I./Romero-Rodríguez, J.M./Ramos, M./Gómez-García, G: "Análisis de buenas prácticas docentes de mobile learning en la Universidad: diseño y validación de la escala APMU". *Revista Iberoamericana de Tecnologías del Aprendizaje* 2020, in press.

Bardín, L.: *Análisis de contenido.* Akal: Madrid. 1991.

Casas, J./García, J./González, F: "Guía técnica para la construcción de cuestionarios". *Odiseo. Revista electrónica de Pedagogía* 3(6) 2006, p. 1, retrieved from http://www.odiseo.com.mx/2006/01/casas_garcia_gonzalez-guia.htm.

Chang, W.H./Liu, Y.C./Huang, T.H: "Perceptions of learning effectiveness in Mlearning: scale development and student awareness". *Journal of Computer Assisted Learning* 33 2017, pp. 461–472, from DOI 10.1111/jcal.12192.

Diez, L.F./Valencia, A./Bermudez, J: "Agent-based model for the analysis of technological acceptance of mobile learning". *IEEE Latin America Transactions* 15(6) 2017, pp. 1121–1127, from DOI 10.1109/tla.2017.7932700.

Ferriols, E.Z./Oms, I.B: "Positive effects in university education of using applications based on the mobile-earning methodology". *Revista de estudios empresariales* 2 2019, pp. 92–109, from DOI 10.17561/ree.v2019n2.5.

Gómez-Ramírez, I./Valencia-Arias, A./Duque, L: "Approach to M-learning acceptance among university students: An integrated model of TPB and TAM". *IRRODL*, 20(3) 2019, pp. 141–164, from DOI 10.19173/irrodl.v20i4.4061.

Hernández, R./Fernández, C./Baptista, P.: *Metodología de la investigación (6ª edición)*. McGraw-Hill – Interamericana de México: México. 2016.

Hu, X./Ng, J./Tsang, K.K.Y./Chu, S.K.W: "Integrating mobile learning to learning management system in community college". *Community College Journal of Research and Practice* 2019, from DOI 10.1080/10668926.2019.1640146.

Huang, R./Tlili, A./Chang, T.W./Zhang, X./Nascimbeni, F./Burgos, D: "Disrupted classes, undisrupted learning during COVID-19 outbreak in China: Application of open educational practices and resources". *Smart Learning Environments* 7(1) 2020, p. 19, from DOI 10.1186/s40561-020-00125-8.

Jansen, H: "La lógica de la investigación por encuesta cualitativa y su posición en el campo de los métodos de investigación social". *Paradigmas* 4 2012, pp. 39–72, from DOI 10.2307/j.ctvdf06h7.7.

Johnson, R.B./Onwuegbuzie, A.J.: "Mixed methods research: A research paradigm whose time has come". *Educational Researcher* 33(7) 2004, pp. 14–26, from DOI 10.3102/0013189x033007014.

Moreno-Guerrero, A.J./Alonso, S./Ramos, M./Campos-Soto, M.N./Gómez-García, G.: "Augmented reality as a resource for improving learning in the physical education classroom". *International Journal of Environmental Research and Public Health* 17(10) 2020, p. 3637, from DOI 10.3390/ijerph17103637.

Nikolopoulou, K./Gialamas, V./Lavidas, K.: "Acceptance of mobile phone by university students for their studies: An investigation applying UTAUT2 model". *Education and Information Technologies* 2020, from DOI 10.1007/s10639-020-10157-9.

Pinto, M./Sales, D./Fernández-Pascual, R./Caballero-Mariscal, D.: "Attitudes, perceptions and prospectings on mobile information literacy training: Design and validation of the MOBILE-APP questionnaire". *Journal of Librarianship and Information Science* 52(1) 2020, pp. 208–223.

Pozo, S./López-Belmonte, J./Moreno-Guerrero, A.J./Soler, R.: "Effect of bring-your-own-device program on flipped learning in higher education students". *Sustainability* 12(9) 2020, p. 3729, from DOI 10.3390/su12093729.

Pozo, S./López-Belmonte, J./Rodríguez-García, A.M./López-Núñez, J.A.: "Teachers' digital competence in using and analytically managing information in flipped learning". *Cultura y Educación* 32(2) 2020, pp. 213–241, from DOI 10.1080/11356405.2020.1741876

Romero-Rodríguez, J.M./Aznar-Díaz, I./Hinojo-Lucena, F.J./Cáceres-Reche, M.P.: "Models of good teaching practices for mobile learning in higher education". *Palgrave Communications* 6(80) 2020a, pp. 1–7, from DOI 10.1057/s41599-020-0468-6.

Romero-Rodríguez, J.M./Aznar-Díaz, I./Hinojo-Lucena, F.J./Gómez-García, G. "Mobile learning in higher education: Structural equation model for good teaching practices". *IEEE Access* 8(1) 2020b, pp. 91761–91769, from DOI 10.1109/ACCESS.2020.2994967.

Salcines-Talledo, I./González-Fernández, N./Briones, E.: "Lecturers profiles: Knowledge and professional use of smartphones". *Bordon* 69(2) 2017, pp. 97–114, from DOI 10.13042/Bordon.2017.51445.

Sánchez-Prieto, J.C./Hernández-García, A./García-Peñalvo, F.J./Chaparro-Peláez, J./Olmos-Migueláñez, S.: "Break the walls! Second-order barriers and the acceptance of mLearning by first-year pre-service teachers". *Computers in Human Behavior* 95 2019, pp. 158–167, from DOI 10.1016/j.chb.2019.01.019.

Seifert, T./Hervás-Gómez, C./Toledo-Morales, P.: "Design and validation of the questionnaire on perceptions and attitudes towards learning for mobile devices". *Pixel-Bit* 54 2019, pp. 45–64, from DOI 10.12795/pixelbit.2019.i54.03.

Soler, R./Mauri, M./Lafarga, P./Moreno-Guerrero, A.J.: "How to teach pre-service teachers to make a didactic program? The collaborative learning associated with mobile devices". *International Journal of Environmental Research and Public Health* 12(9) 2020, p. 3755, from DOI 10.3390/su12093755.

Suárez, R./Grané, M: "Mobile creation in communication studies and the challenge of its adoption in higher education". *Catalan Journal of Communication and Cultural Studies* 11(1) 2019, pp. 123–133, from DOI 10.1386/cjcs.11.1.123_1.

Valeeva, N.G./Pavlova, E.B./Zakirova, Y.L.: "M-learning in teaching ESP: Case study of ecology students". *European Journal of Contemporary Education* 8(4) 2019, pp. 920–930, from DOI 10.13187/ejced.2019.4.920.

Janire Castrillo, Iratxe Gillate, Ursula Luna, and Alex Ibañez-Etxeberria

Chapter 17 Developing social and civic competence via apps: The role of historical memory in the initial teacher training[1]

Abstract: The present study is part of a larger research project revolving around the didactic potential of historical memory apps. This paper analyzes and evaluates the implementation of *1936 Project* with 147 students in the initial phase of their primary teacher certification training at the University of the Basque Country UPV/EHU. This initiative, which centers on an approach to historical memory with reference to the Spanish Civil War from a standpoint that recognizes its controversial nature, uses the app *Eibar 1936–1937 Guide*as an educational resource. The main objectives of this study are to evaluate the validity of the didactic proposal for developing the social and civic competence of program participants, increase their interest in learning about history, and enhance their appreciation of innovative methodologies.

Keywords: Civic competence, Apps, Historical memory, Teacher training, Basque Country

Social and civic competence: historical memory as a controversial topic and the use of apps in the classroom

The introduction of a competence approach in the curriculum has brought about a transformation of the teaching and learning process of all school subjects, including the social sciences (Ortuño, Gómez & Ortiz, 2012). The social sciences are directly related to social and civic competence. This relationship means, according to the Recommendation of the European Parliament and Council of Europe of December 18, 2006 *on key competences for lifelong learning*, that persons need to be prepared to participate in civic life through the

1 This research has been funded by the research groups of the University of the Basque Country *GIPyPAC* (IT1193-19) and *Sociedad, Poder y Cultura (XIV-XVIII)* (IT896-16), and by the project *Modelos de aprendizaje en entornos digitales de educación patrimonial* (PID2019-106539RB-I00) of MINECO/FEDER.

knowledge of social and political concepts and structures, as well as through their commitment to active and democratic participation.

Along these lines, and based on the current Spanish Law 8/2013 of December 9th *for the improvement of educational quality*, Merchán and García (2018) contend that social and civic competence "se refieren al uso del conocimiento para la comprensión de fenómenos y problemas sociales, a la interacción con otras personas y, en fin, a la participación en la vida social gracias a la disposición de conocimientos y a ciertas actitudes"[2] (p. 133). Thus, the teaching of social sciences and history needs to contribute to the acquisition of particular kinds of knowledge that are useful in understanding citizenship (i.e., information about institutions or social dynamics), critical social procedures (i.e., confronting a diversity of narratives and fact-checking data), and civic attitudes (e.g., empathy, tolerance) All of these aspects will, in the end, facilitate an understanding of the social context, and enable concrete action within said context (Puig & Morales, 2015).

Our research is being conducted within a regional milieu (i.e., educational administration in the Basque Country) where there is an operational definition of social and civic competence that is very similar to that which was cited above. This definition reads as follows:

> "Conocerse y entenderse a sí mismo, al grupo del que es miembro y al mundo en el que vive, mediante la adquisición, interpretación crítica y utilización de los conocimientos de las ciencias sociales; así como del empleo de metodologías y procedimientos propios de las mismas, para actuar autónomamente desde la responsabilidad como ciudadano en situaciones habituales de la vida; con el fin de colaborar al desarrollo de una sociedad plenamente democrática, solidaria, inclusiva y diversa" (Berritzegune Nagusia, n.d., p. 14)[3].

2 Refers to the use of knowledge for the purpose of understanding social phenomena and problems, for interacting with other persons and, in the final analysis, for participating in social life thanks to the utilization of knowledge and the presence of certain attitudes.

3 Knowing and understanding oneself, the group of which one is a member, and the world in which one lives, by means of the acquisition, critical interpretation, and utilization of knowledge of the social sciences, and by means of the utilization of social scientific methodologies and procedures, all for the purpose of acting autonomously in a civically responsible manner in ordinary life situations. These activities have the purpose of contributing to the development of a fully democratic, inclusive, diverse, and mutually supportive society.

This is why we also are of the opinion that, if schools would hope to contribute to developing this competence, they must teach how to democratically engage with the most pressing social problems – problems that, in the end, are also present in classrooms themselves. In fact, in countries such as the United Kingdom and Ireland, the controversy regarding such issues constitutes part of the core academic curriculum, and associated research has also been conducted (Berg, Graeffe & Holden, 2003). In fact, the Council of Europe has itself promoted the incorporation of controversial issues into school curricula (Council of Europe, 2015). Such inclusion foments a higher degree of emotional investment on the part of students, as well as the acquisition of the skills that enable these selfsame students to act as democratic, responsible, and supportive citizens (López Facal & Santidrián, 2011).

Historical memory is now one among the controversial issues that are subject to social debate. Studies and policies regarding the recovery of discourses of repressed or silenced groups in a number of countries gathered momentum during the 1980s, which saw the fall of non-democratic regimes and the rise of multiculturalism (Kammen, 1995; Klein, 2000). These memories have a high didactic potential for two different reasons. On the one hand, they help engender a more humane historical consciousness based on one's actual personal experience of certain past historical events, and of the suffering that results from violence and repression (Rüsen, 2013; Seixas, 2015). Such a dynamic is possible because what often survives from the aforementioned periods of silencing and repression are individual testimonies. On the other hand, such historical memory involves a highly emotional component, and an engagement with conflicting interpretations and distinct "silences," and this leads to the emergence of a history laden with consciousness and conducive to the development of critical thinking (Martínez & Sánchez, 2018).

In Spain, the expansion of studies of historical memory began in the 1990s. This research revolved around the Spanish Civil War, the subsequent repression under Franco, and the transition to democracy. Historians and social groups have succeeded in recovering a vision of past events that is different from that of the prevailing official history, and have incorporated this vision into collective memory. Currently, in a country in which the highest-ranking officials of a dictatorship were never brought to justice, this vindication of memory is very much a hot topic that is covered extensively in Spain's mass media. This recovery of memory has in turn led to different viewpoints regarding the specific steps that have been taken to engender the recovery. Ever since approval of Organic Law 52/2007 of December 26th, *which recognizes and expands rights and establishes remedies for those who have suffered persecution or violence during the civil war and dictatorship,* historical memory has also led to tensions because of the policies that have been enacted toward the end of providing reparation to victims, creating spaces of memory, removing existing memorials, etc.

In their respective analyses of the educational treatment of this controversial topic within Spain, both Valls (2007) and Fuertes (2018) contend that the didactic proposals for working on the Spanish Civil War in classrooms have important shortcomings resulting from the Civil War not being approached in a truly democratic way. These researchers point out that this might be the result of the difficulties inherent in the ongoing historiographical debates regarding the interpretation of particular events. Furthermore, recent research has drawn attention to the widespread ignorance among Spanish students of the most rudimentary facts regarding the Civil War (Arias et al., 2019), and has stressed the need to delve more deeply into this topic of recent Spanish history.

Historical memory has hardly been dealt with at all within the context of primary education (Domínguez & Santiago, 2014). Research conducted within the realm of secondary education by Delgado-Algarra (2014) has concluded that, in general, prevailing educational practices neither encourages democratic action on the part of students nor places a great deal of importance on citizenship education (when this latter is understood as a means of social transformation). However, on occasion, there are individual teachers who do have a personal commitment to such issues, and this is something that is conducive to democratic debate in the classroom, with students being called on to defend their positions, and with testimonies being utilized to work on the topic of historical memory (Delgado-Algarra & Estepa, 2016).

Along with this new competence-based approach that needs to be incorporated into the teaching of history, and the new topics proposed for inclusion in classroom curricula, these past 20 years have also witnessed the emergence of digital technology and its educational potential, and of mobile learning (Brazuelo & Gallego; Hylén, 2012; Kukulska-Hulme, 2007; Sharples, Taylor & Vavoula, 2006). Apps have been a resource that has generated the most interest as regards their use in education, and research has been conducted on their application and evaluation. This has led to the proposal of tools to be used for an educational analysis of apps (Kortabitarte et al., 2017), which have been applied in research within specific sectors (Gillate et al., 2017; Ibáñez-Etxeberria & Kortabitarte, 2016; Luna, Ibáñez-Etxeberria & Rivero, 2019; Luna, Rivero & Vicent, 2019). In addition – and this has been something sorely lacking – the use of apps in real-life educational contexts has been evaluated (Ibáñez-Etxeberria et al., 2020a). Studies have also been conducted that indicate the different educational contexts within which apps are used in Spain (Luna, Ibáñez-Etxeberria

& Kortabitarte, 2019) and comparing their use with that of other nations (Luna et al., 2020).

One of the benefits of using apps is that they allow easy and efficient access to the oral testimonies of those who have lived through particular events. This bringing of app users into direct contact with aspects of history that might have previously been perceived as abstract and distant has an emotional impact on individual users. In addition, these testimonies come to constitute primary sources and, as such, are valid for recreating the historical method. This is something that is essentially important for learning about history in a meaningful way (Santacana, Pibernat & Hernàndez, 1998). Another of the benefits of using apps is that they increase student motivation (Faria et al., 2019; Kortabitarte et al., 2017; Kortabitarte et al., 2018) by offering content adapted to the interest of each individual user, as well as a particular learning context that also is conducive to the acquisition of digital competence (Ibañez-Etxeberria et al., 2019).

Notwithstanding the powerful emergence of apps in the classroom as support tools, they have not yet been employed to maximum effect within a formal educational context. This is due to a number of different factors, such as insufficient education of teachers and a lack of proper resources in the schools themselves. For this reason, it is critically important that knowledge related to mobile learning be integrated into the initial training of teachers (Ally, Grimus & Ebner, 2014; Baran, 2014). This becomes especially important when we take into account the fact that, as Kortabitarte and her colleagues point out (Kortabitarte et al., 2018), apps have certain deficiencies when used as educational tools, thus making teacher mediation indispensable for making them productive classroom resources.

We entirely agree with Gómez, Ortuño and Miralles (2018) that a change in teaching methodologies must play a critical role in reforming the more traditional model of teaching history. In this regard,, the teaching of procedures characteristic of the social sciences and of strategies with respect to controversial situations will – along with the use of digital technologies – help generate an educational ecosystem where such pathways of change can be effectively introduced. In accordance with these premises, a program-analysis research study has been conducted on the educational potential of apps that specifically focuses on the subject of historical memory (Gillate et al., 2020; Vicent et al., 2020). In this new research phase, we are attempting to integrate apps utilized within real educational settings with students. For this purpose, we have chosen the app *Eibar 1936–1937 Guide*, which focuses on the Spanish Civil War within the Basque Country. This research has in turn been integrated into *1936 Project* a didactic framework that allows us to evaluate the educational results

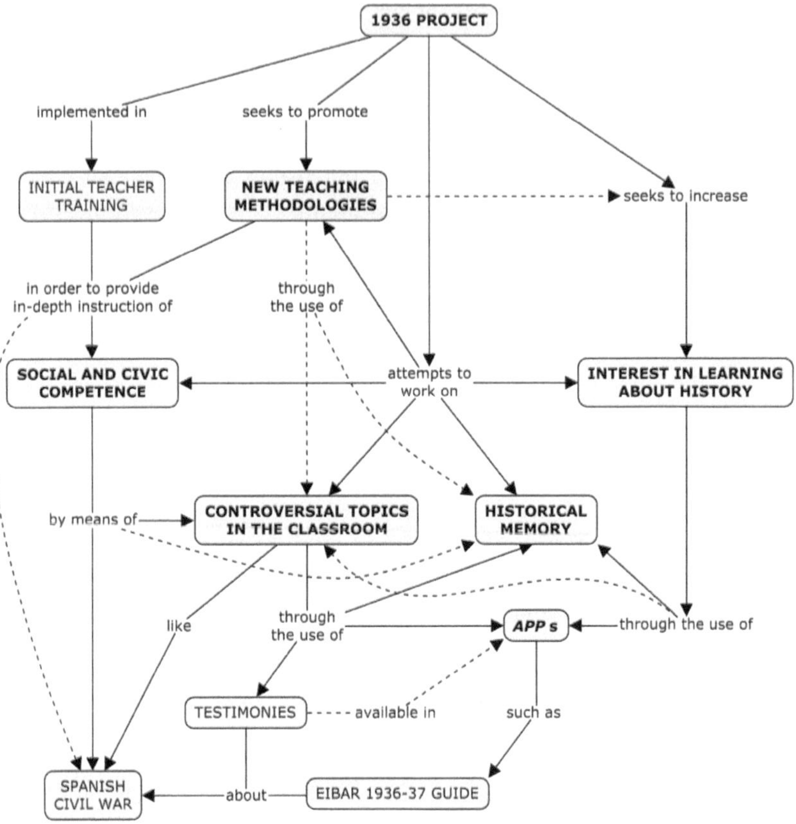

Fig. 1: Conceptual map of the theoretical framework upon which 1936 Project is based. Source: Authors

of the proposal from different points of view, and that we will describe in the next section.

1936 project and the *Eibar 1936–1937 guide* app

1936 Project seeks to train future teachers in the integration of controversial topics in social science classes, as well as in the utilization of mobile devices and apps for teaching and learning processes. Specifically, the Spanish Civil War and the recovery of historical memory were both addressed through the use of the app *Eibar 1936–1937 Guide* as the main educational resource.

Tab. 1: Structure of sessions and tasks for 1936 Project

Session	Activities
1	-Presentation of project. -Theory: Apps in education. -Download of app.
2	-Theory: The Spanish Civil War and historical memory.
3	-Questionnaire on basic concepts of the Spanish Civil War.
4	-Viewing of the documentary "Gernikaren Egiak" (The truth about Gernika). -Analysis of testimonies on the app.
5–6	-Suggestions regarding a memory space capable of being recovered in terms of historical memory.
7–8	-Theory: How to deal with controversial subjects in the classroom -Written reflection.

Source: Authors

Throughout the course of the project, instruction was first provided on theoretical content regarding the possibilities of utilizing apps in primary education. This initial information was presented from a procedural point of view. In addition, work was also done, from a perspective of social and civic competence, on the Spanish Civil War, historical memory, and on approaching controversial topics in the classroom. A variety of materials and educational formats were combined in the project, such as structured teacher presentations, the viewing of videos, and use of the app *Eibar 1936–1939 Guide* as a source for students to look for information in the texts and testimonies of that resource.

Eibar 1936–1937 Guide is an app launched in 2016 by the local government of Eibar, a municipal district in the Basque province of Guipuzcoa, located in northern Spain. The app offers three differently themed guided tours on the Spanish Civil War as experienced in Eibar (Vicent et al., 2020). The first of these, *A tour of Eibar*, passes through a number of points of interest in the municipality, and allows users to view scenes where battles were fought, thus facilitating comprehension of key aspects of the war. The second tour is of a museum in Eibar dedicated to preserving the memory of the local impact of the Spanish Civil War. The third tour, titled *Eibar 1931–1945*, traverses the historic center of the town of Eibar in order to inform users of the most important historical events associated with that period, from the proclamation of the Second Republic in 1931 until the Spanish Civil War, and the subsequent reconstruction of the municipality following the war's conclusion.

Even though the app has no explicit teaching objectives (Vicent et al., 2020), it does have an implicit didactic and interpretative component. This component is conveyed by presenting information in an interesting way through the prism of real existing circumstances and local conditions during the era portrayed. Among the app's strengths are its presentation of a wealth of information on the war and its course within the Basque Country and Eibar. This presentation is multilayered, and can be easily navigated by the user. The app includes historical photographs, videos of survivor testimonies, and utilizes GPS technology. It also allows users to select points of interest in order to create their own personalized tours.

The objective of the present study is to implement *1936 Project* with the *Eibar 1936–1937 Guide* app within real-life educational settings, and to evaluate the impact of the app's utilization on students' acquisition of social and civic competencies and their level of interest in history, and on changes in the methodology employed by instructors. Specifically, our threefold purpose of this study is as follows:

1. To analyze if there has been a statistically significant increase in average development of social and civic competence following students' participation in *1936 Project.*
2. To determine if there has been a statistically significant increase in students' interest in learning about history following their involvement in *1936 Project.*
3. To known the assessment made by students about new teaching methodologies before and after the implementation of the *1936 Project.*

Method

Design

The design of this study is quasi-experimental, with data collected prior to and following the intervention of interest, but without a control group.

Participants

A total of 147 students (68.7 % women) participated in this study. These students were in their third year of study for the Primary Education Degree at the University of the Basque Country UPV/EHU. Participating students were enrolled in *1936 Project* as part of their class "Didactics in Social Sciences II" in April-May 2020.

Tab. 2: Variables and dimensions

Variable	Dimension
1. Social and civic competence	Knowledge of the past: Spanish Civil War. Analysis of conflict situation: historical memory.
2. Interest	Interest in studying history.
3. Teaching methodologies	Traditional methodologies. New methodologies.

Source: Authors

Measurement tool

A self-completion questionnaire consisting of 23 items was created. Possible responses were structured according to the Likert scale (ranging from 1=very little to 5=extremely well). The objective of this questionnaire was to evaluation participants' perceptions regarding the development of *social and civic competence, interest in studying history,* and *the use of different teaching methodologies* (Tab. 2).

The variable *social and civic competency* was conceptualized in accordance with the document *Guidelines for drafting the report on the acquisition of Basic Competences in Secondary Education.* These guidelines were prepared by the Basque government's Department of Education (Berritzegune Nagusia, n.d.). Although numerous components are associated with this competence, the following two indicators were chosen to measure development of the competence in this particular study:

1. Knowledge of the characteristics of the society in which one lives, and especially its recent past, formation, organization, and functioning.
2. Analysis of a conflict situation in the setting where one lives, and an investigation of said conflict's background, as well as its social, cultural, and historical causes.

These two components were operationalized through the use of ten items whose suitability was evaluated by means of both exploratory factor analysis and analysis of internal consistency. Both the KMO test (.80) as well as Bartlett's test of sphericity ($\chi^2(45)=684.97$; $p<.001$) indicated suitability of the data for factor analysis. The two-factor solution that was found explained 60.44 % of variability. The factor "knowledge of the Spanish Civil War," which was associated with the first variable identified above, comprised six items whose levels of saturation exceeded 0.50, and its reliability was very high (α=.88). The factor of historical memory, which was associated with the second variable, comprised four items with high levels of saturation and high levels of internal consistency (α=.82). The reliability of the variable as a whole was .86.

The second variable is *interest in studying history.* Increase in said variable was evaluated by means of the topic that was studied (i.e., the Spanish Civil War and historical memory) and the use of new technologies such as apps. The measurement utilized was modeled after a questionnaire devised by Rey, Hidalgo and Espinosa (1989). The data indicated suitability for factor analysis (KMO=.72; $\chi 2(10)$=377.18; $p<.001$), identified an underlying structure of five of the items that explains 63.47 % of variability, and that also has a high level of reliability (α=.86).

The third variable measures students' evaluations of the utilization of *teaching methodologies* (KMO=.78; $\chi 2(28)$=356.27; $p<.001$) and is distributed along two different dimensions: *traditional methodologies*, comprising two items (α=.59) and *new methodologies,* comprising six items (α=.81). All together, the eight items explain 59.18 % of total variance, and have a high reliability (α=.80).

Procedure

After participants signed the informed consent form, the questionnaire was presented under routine teaching conditions on two occasions: prior to the intervention and after eight sessions (12 hours) of specific education.

Data analysis

It was initially established that there were no atypical values or patterns of non-random missing data. Afterward, descriptive statistics were calculated for each of the global variables, as well as for each of the individual items (i.e., distribution, central tendency, dispersion, and form). Compliance with the assumption of normality was established. Two different kinds of tests were utilized to study differences in means before and after the activity: Student's t-test for related samples in the case of variables, and non-parametric sign tests in the case of the items. The results are presented along with calculations of effect size. The program IBM SPSS Statistics 26 was utilized to calculate the results.

Results

Social and civic competence

Analysis of differences between means revealed a statistically significant increase in the measurement of the variable *social and civic competence* following intervention, both globally and within component sub-dimensions (Tab. 3), and with very high associated effect sizes.

For the purposes of illustrating in greater detail some of the indicators of acquisition of the key components of social and civic competence, we will now present the data obtained from the most significant items on the questionnaire, we will now present the data obtained from the most significant items on the questionnaire, regarding the two dimensions that form the mentioned variable.

On the first dimension, *knowledge of the past: the Spanish Civil War,* the first item refers to students' self-reported knowledge regarding this topic. The results (Zs(147)=9,27; p<.001; d=2.379) indicate statistically significant differences of high magnitude between pre- and post-intervention scores, with the latter being higher (Tab. 4).

Item 5 asks students about their knowledge of the consequences of the Spanish Civil War. The results (Zs(147)=6,22; p<.001; d=.781) indicate a statistically significant increase as well, with a high effect size (Tab. 5).

Within the second dimension, *analysis of a conflict situation: historical memory,* the results obtained are similar to those for item 8 (Zs(145)=9.68; p<.001; d=2.68) (regarding current reasons for talking about historical memory) and item 10 (Zs(147)=5.303; p<.001; d=.972) (regarding the importance accorded the recovery of the memory of the losing side in the Spanish Civil War). In both cases, there were statistically significant increases – with

Tab. 3: Differences in means prior to and following intervention

Variable	Time of Measurement	n	M	SD	t	p	d
Social and civic competence.	PRE	142	3.18	0.62	18.02	<.001	1.55
	POST	142	4.06	0.46			
Knowledge of the past: the Spanish Civil	PRE	145	2.81	0.76	16.05	<.001	1.36
War.	POST	145	3.71	0.60			
Analysis of conflict	PRE	145	3.72	0.71	13.32	<.001	1.12
situation: historical memory.	POST	145	4.58	0.48			

Source: Authors

Tab. 4: Responses to item "I know the history of the Civil War"

Response	PRE		POST	
	n	%	n	%
Very little	12	8.2	2	1.4
A little	38	25.9	2	1.4
Enough	77	52.4	66	44.9
A lot	18	12.2	74	50.3
Extremely well	2	1.4	3	2

Source: Authors.

Tab. 5: Responses to the item "I know what the consequences of the Civil War were"

Response	PRE		POST	
	n	%	n	%
Very little	8	5.4	0	0
A little	13	8.8	5	3.4
Enough	53	36.1	32	21.8
A lot	65	44.2	87	59.2
Extremely well	8	5.4	23	15.6

Source: Authors

Tab. 6: Responses to the item "I know the reasons why people talk about historical memory nowadays"

Response	PRE		POST	
	n	%	n	%
Very little	4	2,7	0	0
A little	25	17.1	0	0
Enough	51	34.9	7	4.8
A lot	53	36.3	59	40.4
Extremely well	13	8.9	80	54.8

Source: Authors

Tab. 7: Responses to the items "I think that it is important to recover the memory of the losing side in the Civil War"

Response	PRE		POST	
	n	%	n	%
Very little	0	0	0	0
A little	0	0	0	0
Enough	29	19.7	8	5.4
A lot	58	39.5	39	26.5
Extremely well	60	40.8	100	68

Source: Authors

associated high effect sizes – in the scores obtained following the intervention (Tabs. 6 and 7).

Interest in studying history

Similarly, for the purpose of determining if there has been any increase in *interest in studying history,* a statistical analysis of differences between means prior to and following the group intervention was conducted. The results reported in Tab. 8 indicate that there is a statistically significant positive difference of moderate magnitude in interest reported post-intervention.

More specifically, the results indicated statistically significant differences of low magnitude in changes related to the responses to item 15 (Zs(144)=2.195; p<.001; d=.373) which concerns interest in the history of the Spanish Civil War (Tab. 9), and a moderate effect size in item (Zs(146)=4.19; p<.001; d=.69) which concerns the study of historical memory (Tab. 10).

Tab. 8: Differences in means before and after activity

Variable	Time of Measurement	n	M	SD	t	p	d
Interest in studying history	PRE	142	3.87	0.72	5.621	<.001	0.48
	POST	142	4.15	0.68			

Source: Authors' own

Tab. 9: Responses to item "I am interested in the history of the Civil War"

Response	PRE		POST	
	n	%	n	%
Very little	2	1.4	0	0
A little	9	6.2	6	4.1
Enough	28	19.3	30	20.5
A lot	65	44.8	58	39.7
Extremely well	41	28.3	52	35.6

Source: Authors

Tab. 10: Responses to the item "I am interested in studying historical memory"

Response	PRE		POST	
	n	%	n	%
Very little	1	.7	0	0
A little	7	4.8	2	1.4
Enough	31	21.2	15	10.2
A lot	71	48.6	78	53.1
Extremely well	36	24.7	52	35.4

Source: Authors

Tab. 11: Differences between means before and after activity

Variable	Time of Measurement	n	M	SD	t	p	d
Traditional methodologies	PRE	144	3.66	0.67	4.38	<.001	0.37
	POST		3.88	0.61			
Interactive methodologies	PRE	143	3.86	0.53	4.50	<.001	0.40
	POST		4.06	0.52			

Source: Authors

Teaching methodologies

The variable *teaching methodologies* comprises two distinct dimensions: *traditional methodologies* and *new methodologies*. The results obtained indicate statistically significant increases of a moderate magnitude in both of these dimensions in post-intervention scores (Tab. 11).

As regards the dimension *new teaching methodologies*, the specific changes in item 12 (Zs(147)=2.151; p=.001; d=.36) as regards studying history using apps can be seen in Tab. 12. Changes with respect to item 17, the value of testimonies in learning history (Zs(147)=3.674; p<.001;d=.635) are reported in Tab. 13.

Discussion and conclusions

Controversial topics constitute a part of classroom curricula in a number of different countries (Berg et al., 2003). European governmental bodies have recommended that such topics be addressed in schools (Council of Europe,

Tab. 12: Responses to item "Studying history by using apps helps me learn"

Response	PRE		POST	
	xn	%	n	%
Very little	3	2	4	2.7
A little	16	10.9	9	6.1
Enough	56	38.1	47	32
A lot	59	40.1	68	46.3
Extremely well	13	8.8	19	12.9

Source: Authors

Tab. 13: Responses to the item "Learning about history through testimonies leads to greater understanding"

Response	PRE		POST	
	n	%	n	%
Very little	0	0	2	1.4
A little	4	2.7	1	.7
Enough	16	10.9	9	6.1
A lot	58	39.5	42	28.6
Extremely well	69	46.9	93	63.3

Source: Authors

2015) because it is considered an excellent means of developing social and civic competence. The results of the present study constitute further evidence of the positive results obtained by implementation of *1936 Project* with respect to the attainment of *social and civic competence,* with statistically significant differences and high effect sizes in both of the dimensions analyzed here: *knowledge of the past: the Spanish Civil War* and *analysis of conflict situations: historical memory.*

In fact, the results here are consistent with those of previous studies reporting an increase in participants' development of *social and civic competence* (Berritzegune Nagusia, n.d.). On the one hand, participants indicate greater knowledge of the characteristics of the societies in which they live, and of its past, with respect to the Spanish Civil War and its consequences. What this means is that, during the course of their didactic experience, participants have acquired knowledge that is crucial for not only understanding social phenomena and problems, but for understanding and exercising citizenship (Puig & Morales, 2015; Merchán & García, 2018). On the other hand, participants remedied their deficient knowledge of the Civil War (Arias et al., 2019; Fuertes, 2018; Valls, 2007) and proved themselves more capable of analyzing the continuing repercussions of this traumatic event in contemporary society. In this way, there is an increase in both the level of understanding of the reasons for historical memory currently proffered, as well as of the importance accorded to the recovery of the historical memory of those who lost the war. Otherwise stated, there appears to be a greater understanding of the phenomenon, as well as an increase in awareness regarding it.

These results, which indicate that historical memory has a great deal of potential in working with *social and civic competence*, lead us to consider the need to reverse the current trend in Spanish schools, and to insist that controversial topics be addressed in classrooms (Delgado-Algarra, 2014; Delgado-Algarra & Estepa, 2016). This reversal of course would accord historical memory the prominence that it deserves because of its educational potential, given that it is a field that has scarcely been utilized, especially in primary education (Arias et al., 2019; Domínguez & Santiago, 2014). In addition, this new method reaffirms the need to reorient educational proposals within the social sciences and history from conventional approaches based on the transmission of content to alternative approaches that are aimed instead at addressing dilemmas of the current social reality, and that are congenial to the learning mechanisms utilized by students.

Another of the indications of the validity of *1936 Project*, and one related to the second objective of this study, is the increased interest of participants in the study of history – and more specifically, in the history of the Spanish Civil War and its memory. Albeit of modest dimensions, this result does lend support to the contentions of López Facal and Santidrián (2011) and of the Council of Europe (2015), to the effect that addressing controversial subjects in the classroom helps to engender a higher degree of emotional investment in the learning process on the part of students.

The present study has focused on historical memory, a hot topic in Spain that receives a great deal of attention in the nation's mass media, and that serves as a catalyst for social debate. For these reasons, it is a subject that is directly related to the real world. It is this relationship that creates the possibility that learning that occurs with reference to this subject could become especially meaningful. All the more so when the teaching resources used focus on working within a local and immediate context – in our particular study, the Basque Country and Eibar – thanks to the app *Eibar 1936–1937 Guide*. We understand that this topic, given that it is so closely tied to the daily life of citizens, contributes to understanding the world and to engendering citizens who are tolerant, engaged, and critical. This is because, as Martínez and Sánchez (2018) contend, such attitudes involve an analysis of differing viewpoints.

This increased interest may in turn have something to do with the teaching methodology utilized and, most especially, with the integration of the app *Eibar 1936–1937 Guide* as the primary didactic resource. Such a conclusion would be in line with the findings of Kortabitarte et al. (2017, 2018) and Faria et al. (2019), who have reported that apps contributed to increased motivation in the classroom. In this regard, it is noteworthy that future teachers who are pursuing university studies came to view apps more positively after witnessing them in use in the classroom. This is shown by the higher post-intervention scores of the item regarding the studying of history using apps as a means to facilitate learning. This more positive evaluation may also be related to concomitant improvements in the digital competence of participants (Ibañez-Etxeberria et al., 2019).

Another of the positive aspects of *1936 Project* from the standpoint of teaching methodology has been the integration of the testimonies that are provided in the app. In these testimonies, various survivors of the Spanish Civil War in Eibar provide first-person accounts. These accounts include a strong emotional element that add value to the material being studied, which may in turn lead to a higher degree of historical understanding (Rüsen, 2013; Seixas, 2015). We thus see an increase in scores on the item regarding whether or not testimonies engender a greater understanding of history. This is one of the most noteworthy contributions of *Eibar 1936–1937 Guide*, because it makes it possible to work with primary sources within classrooms themselves, thus allowing a re-creation of the historical method (Santacana, Pibernat & Hernàndez, 1998), something that would otherwise be difficult to accomplish.

It should nevertheless be remembered that apps are generally not designed to be used in formal education (Gillate et al., 2017) and that teacher intervention is necessary in order to adapt them so that they can be utilized as suitable educational resources (Kortabitarte et al., 2018). The specific training that this requires highlights the need for orientation in primary teacher training programs regarding the use of apps (Ally et al., 2014; Baran, 2014). In this way, *1936 Project* has served as food for thought for future teachers with regard to the evaluation of the strengths and weaknesses of these tools, and also with respect to the need to introduce both general and specific pedagogical adaptations in these apps in order to make them optimally suitable to the educational level of the students that will be used with.

We therefore conclude that the work on historical memory and the use of the app *Eibar 1936–1937 Guide* proposed in *1936 Project* has been useful in allowing participants in said program to develop abilities associated with social and civic competence; to increase their interest in studying history; and to appreciate the value of the new teaching methodologies. In addition, from the standpoint of the education of future teachers, the initiative has been useful in two different ways. On the one hand, it has led to the acquisition of knowledge of how to work on developing social and civic competence by dealing with controversial topics. On the other hand, it has served as a means of utilizing new teaching approaches and new educational resources, including those based on the use of mobile devices.

References

Ally, Mohamed/Grimus, Margarete/Ebner, Martin: "Preparing teachers for a mobile world, to improve access to education". *Prospects*, 44, 2014, pp. 43–59, retrieved 4.6.2020, from DOI 10.1007/s11125-014-9293-2.

Arias, Laura/Egea, Alejandro/Sánchez, Raquel/Domínguez, Jesús/García, Francisco Javier/Miralles, Pedro: "Forgotten history or history not taught? The students of Spanish High School and their lack of knowledge about the Civil War". *Revista Complutense de Educacion*, 30(2), 2019, pp. 461–478, retrieved 6.6.2020, from DOI 10.5209/RCED.57625.

Baran, Evrim: "A review of research on mobile learning in teacher education". *Educational Technology & Society*, 17(4), 2014, pp. 17–32.

Berg, Wolfgang/Graeffe, Leena/Holden, Cathie: *Teaching Controversial Issues: A European Perspective.* London Metropolitan University: London 2003.

Berritzegune Nagusia-Gobierno Vasco (n.d.): *Orientaciones para elaborar el informe sobre el grado de adquisición de las Competencias Básicas Educación Secundaria.* Retrieved 03.01.2020, from https://www.berrigasteiz.com/site_argitalpenak/docs/300_competencias/3002017002c_konpetentziak_eskuratze_maila_bh_c.pdf.

Brazuelo, Francisco/Gallego, Domingo. J.: *Mobile learning: Los dispositivos móviles como recurso educativo.* Eduforma: Sevilla 2011.

Council of Europe: *Living with Controversy Teaching. Controversial Issues through Education for Democratic Citizenship and Human Rights (EDC/HRE)*, 2015. Retrieved 25.5.2020, from https://rm.coe.int/16806948b6.

Delgado Algarra, Emilio: *Educación para la ciudadanía en la enseñanza de las Ciencias Sociales y vinculación con las dimensiones de la memoria. Estudio de caso.* (Universidad de Huelva) (doctoral thesis) 2014.

Delgado-Algarra, Emilio/Estepa Giménez, Jesús: "Ciudadanía y memoria histórica en la enseñanza de la historia: análisis de la metodología didáctica en un estudio de caso en ESO". *Revista de Investigación Educativa*, 34(2), 2016, pp. 521–534, retrieved 9.2.2020, from DOI 10.6018/rie.34.2.224891.

Domínguez Almansa, Andrés/Santiago, Borja: "La respuesta está en el viento: reflexión para una nueva docencia de la guerra civil". In Lourenzo Fernández/Aurora Artiaga (eds.): *Otras miradas sobre golpe, guerra y dictadura. Historia para un pasado.* La Catarata: Madrid 2014, pp. 300–322.

Faria, Cláudia/Guilherme, Elsa/Pintassilgo, Joaquim/Mogarro, Maria João/Pinho, Ana Sofia/Baptista, Mónica/Chagas, Isabel/Galvão, Cecília: "The Portuguese Maritime Voyages of Discovery: The exploration of the history of a city with an *App* as an educational resource". *Digital Education Review*, 36, 2019, pp. 85–99, retrieved 11.5.2020, from http://greav.ub.edu/der/.

Fuertes, Carlos: "La dictadura franquista en los manuales escolares recientes: una revisión crítica". *Revista Historia Autónoma*, 12, 2018, pp. 279–297.

Gillate, Iratxe/Luna, Ursula/Castrillo, Janire/Ibáñez-Etxeberria, Alex: "Historical memory in heritage education *Apps*. A resource for building social and civic competence". In Emilio Delgado Algarra/José María Cuenca (eds.): *Handbook of Research on Citizenship and Heritage Education.* IGI-Global: Pennsylvania 2020, pp. 285–310.

Gillate, Iratxe/Vicent, Naiara/Gómez-Redondo, Carmen/Marín- Cepeda, Sofía: "Características y dimensión educativa en *apps* de educación patrimonial. Análisis a partir del método OEPE". *Revista de Estudios*

Pedagógicos, 43(4), 2017, pp. 115–136, retrieved 26.5.2019, from DOI 10.4067/S0718-07052017000400006.

Gómez, Cosme J./Ortuño, Jorge/Miralles, Pedro: *Enseñar ciencias sociales con métodos activos de aprendizaje. Reflexiones y propuestas a través de la indagación*. Octaedro: Barcelona 2018.

Hylén, Jan: *Turning on Mobile Learning in Europe*. UNESCO: Paris 2012.

Ibáñez-Etxeberria, Alex/Gómez-Carrasco, Cosme/Fontal, Olaia/García-Ceballos, Silvia: "Virtual environments and augmented reality applied to heritage education. An evaluative study". *Applied Sciences*, 10(7), 2020a, pp. 23–52.

Ibáñez-Etxeberria, Alex/Kortabitarte, Aroia: "*Apps*, Redes Sociales y dispositivos móviles en educación patrimonial", 2016, retrieved 8.8.2019 from http://www.culturaydeporte.gob.es/planesnacionales/dam/jcr:4cc8d5a2-3084-409e-a134-b46b62f7126e/appsy-socialmedia-en-educacionpatrimonial.pdf.

Ibáñez-Etxeberria, Alex/Kortabitarte, Aroia/Castro, Pablo de/Gillate, Iratxe: "Competencia digital mediante *apps* de temática patrimonial en el marco DigComp". *Revista Electrónica Interuniversitaria de Formación del Profesorado*, 22(1), 2020b, pp. 13–27, retrieved 26.3.2020, from DOI 10.6018/reifop.22.1.356231.

Kammen, Michael: "Frames of remmbrance: The dynamics of collective memory". *History and Theory*, 34(3), 1995, pp. 245–261, retrieved 27.7.2020, from www.jstor.org/stable/2505624.

Klein, Kerwin L.: "On the emergence of memory in historical discourse". *Representations*, 69(69), 2000, pp. 127–150.

Kortabitarte, Aroia/Gillate, Iratxe/Luna, Ursula/Ibáñez-Etxeberria, Alex: "Las aplicaciones móviles como recursos de apoyo en el aula de Ciencias Sociales: estudio exploratorio con el *app* Architecture gothique/romane en Educación Secundaria". *ENSAYOS, Revista de la Facultad de Educación de Albacete*, 33(1), 2018, pp. 65–79, retrieved 28.11.2019 from DOI 10.18239/ensayos.v33i1.1743.

Kortabitarte, Aroia/Ibañez-Etxeberria, Alex/Luna, Ursula/Vicent, Naiara/Gillate, Iratxe/Molero, Begoña/Kintana, Jurgi: "Dimensiones para la evaluación de aprendizajes en APPs sobre patrimonio". *PULSO. Revista de Educación*, 40, 2017, pp. 17–33.

Kukulska-Hulme, Agnes: "Mobile usability in educational contexts: What have we learnt?" *The International Review of Research in Open and Distributed Learning*, 8(2), 2007, retrieved 4.1.2019 from DOI 10.19173/irrodl.v8i2.356.

López Facal, Ramón/Santidrián, Victor M.: "Los «conflictos sociales candentes» en el aula". *Íber Didáctica de las Ciencias Sociales, Geografía e Historia*, 69, 2011, pp. 8–20.

Luna, Ursula/Canto Herrera, Pedro J./Zapata González, Alfredo/Ibáñez-Etxeberria, Alex: "*Apps*, educación y patrimonio en México. Análisis de situación y estudio comparativo con el caso español". *Aula Abierta*, 49(1), 2020, pp. 9–16.

Luna, Ursula/Ibañez-Etxeberria, Alex/Kortabitarte, Aroia: "Herramientas digitales innovadoras en espacios educativos. El uso de las *apps* para la enseñanza-aprendizaje del patrimonio en España". In Juan Calvillo (coord.): *Escenarios, heterogeneidad y nuevas perspectivas para la comprensión en la educación*. Universidad Estatal de Oriente: Puebla, México 2019, pp. 90–106.

Luna, Ursula/Ibañez-Etxeberria, Alex/Rivero, Pilar: "El patrimonio aumentado. 8 *apps* de Realidad Aumentada para la enseñanza-aprendizaje del patrimonio". *Revista Interuniversitaria de Formación del Profesorado*, 94(33.1), 2019, pp. 43–62.

Luna, Ursula/Rivero, Pilar/Vicent, Naiara: "Augmented reality in heritage *Apps*: Current trends in Europe". *Applied Sciences*, 9(2756), 2019, pp. 1–15, retrieved 24.1.2020, from DOI 10.3390/*app*9132756.

Martínez, Rosendo/Sánchez, María: "La enseñanza de la historia reciente en la adquisición de competencias para una ciudadanía democrática". In Pedro Miralles/Cosme J. Gómez (coord.): *La educación histórica ante el reto de las competencias*. Octaedro: Barcelona 2018, pp. 139–148.

Merchán, Francisco J./García, Francisco F.: "La enseñanza de la historia y la adquisición de competencias sociales y cívicas". In Pedro Miralles/Cosme J. Gómez (coord.): *La educación histórica ante el reto de las competencias*. Octaedro: Barcelona 2018, pp. 127–137.

Organic Law 52/2007 of December 26th, *which recognizes and expands rights and establishes remedies for those who have suffered persecution or violence during the civil war and dictatorship*/Ley Orgánica 52/2007, de 26 de diciembre (BOE nº310 del 27 de diciembre), por la que se reconocen y amplían derechos y se establecen medidas en favor de quienes padecieron persecución o violencia durante la guerra civil y la dictadura.

Organic Law 8/2013 of December 9th *for the improvement of educational quality*/ Ley Orgánica 8/2013, de 9 de diciembre (BOE nº295 del 10 de diciembre), para la mejora de la calidad educativa (LOMCE).

Ortuño, Jorge/Gómez, Cosme J./Ortiz, Eva: "La evaluación de la competencia educativa social y ciudadana desde la didáctica de las ciencias sociales. Un estado de la cuestión". *Didáctica de las ciencias experimentales y sociales*, 26, 2012, pp. 53–72, retrieved 20.3.2020, from DOI: 10.7203/DCES.26.1931.

Puig, María/Morales, J. Antonio: "La formación de ciudadanos: conceptualización y desarrollo de la competencia social y cívica". *Educación XXI*, 18(1), 2015, pp. 259–282, retrieved 20.3.2020, from DOI: 10.5944/educxx1.18.1.12332.

Recommendation of the European Parliament and Council of Europe of December 18, 2006 *on key competences for lifelong learning*/Recomendación del Parlamento Europeo y del Consejo, de 18 de diciembre de 2006 (*Diario Oficial de la Unión Europea*, 30 de diciembre de 2006), sobre las competencias clave para el aprendizaje permanente (2006/962/CE). Retrieved 25.5.2020, from https://eur-lex.europa.eu/LexUriServ/LexUriServ.do?uri=OJ:L:2006:394:0010:0018:ES:PDF.

Rey, José María/Hidalgo, Eugenio/Espinosa, Carlos: La motivación en la escuela. Cuestionarios para su análisis. Málaga: Ágora 1989.

Rüsen, Jörn: "Forming historical consciousness towards a humanistic history didactics". *Antíteses*, 5(10), 2013, pp. 519–536.

Seixas, Peter: "A history/memory matrix for history education". *Public History Weekly*, 4(6), 2015, retrieved 2.5.2019, from DOI 10.1515/phw-2016-5370.

Sharples, Mike/Taylor, Josie/Vavoula, Giasemi: "A theory of learning for the mobile age". In Richard Andrews/Caroline Haythornthwaite (eds.): *The Sage Handbook of Elearning Research*. Sage: London 2006, pp. 221–247.

Valls, Rafael: "La guerra civil española y la dictadura franquista: las dificultades del tratamiento escolar de un tema potencialmente conflictivo". Enseñanza de las ciencias sociales, 6, 2007, pp. 61–73.

Vicent, Naiara/Gillate, Iratxe/Campos-López, Teresa/Castrillo, Janire: "Potencial educativo de *apps* de Memoria Histórica. Estudio de casos". *Aula Abierta*, 49(1), 2020, pp. 35–43.

Attachment: questionnaire used within the research

1. I know the history of the Civil War
2. I know the history of the Civil War in my town
3. I know the causes of the Civil War
4. I have information about the development of the Civil War in my town
5. I know the consecuences of the Civil War
6. I know which were the consecuences of the Civil War in my town
7. I know the meaning of "Historical Memory"
8. I know which are the reasons to speak about "Historical Memory" nowadays

9. Learning on "Historical Memory" helps me to understand the current society
10. It is important to recover the memory of those who lost the Civil War
11. I am interested in history
12. Working on history through apps helps me to learn
13. I make an effort when doing exercises and history works
14. Working on history through textbooks helps me to learn
15. I am interested in the Civil War
16. Using multiple resources helps me learning history (books, technology, videos...)
17. History is better understood through testimonies
18. I find it interesting to study the development of the Civil War in my town
19. I am keen to work on history through videos
20. Searching information with technological devices helps me to learn history
21. Working on history with technological devices helps me to learn (ex. power point)
22. I am interested in historical memory
23. I like learning my town's history

José Monteagudo-Fernández, Juan Ramón Moreno-Vera, Raimundo A. Rodríguez-Pérez, and Ramón Cózar-Gutiérrez

Chapter 18 Motivation strategies for gamified flipped classrooms in Social Sciences education[1]

Abstract: This study examines different motivation strategies for gamified processes in the classroom and shows the perception that students and trainee teachers have regarding to gamification and flipped-classroom as motivating resources. We analyse existing approaches in gamification and flipped-classroom studies for Social Sciences education, focusing our research on the different strategies (games, videogames, quizzes, investigation, etc.) that improve students' motivation. The literature shows a very high valuation of motivation in gamified and flipped-classroom implemented processes. This fact can be attributed to an increase of student participation in the construction of their own conceptual knowledge, using ITC resources and videos, the implementation of a reward evaluation system and the use of cooperative work strategies during the lessons.

Keywords : Education, Social studies, Methodology, Gamification, Flipped-classroom

Introduction

Current teaching cannot ignore the relevance of ITC. But it is torn between mere entertainment and the need for students to learn, acquire knowledge, skills and values. Technology advances very fast, much more than teaching in any discipline, its regulatory laws, teacher training or teaching materials. Therefore, it can be a dangerous aid if not applied carefully. The playful part can end up engulfing academic rigor (Jiménez-Alcázar, 2018). So teachers require skills and continuous updating. All technology per se is not good, if it is not adapted to the educational context in which it must be applied. Even if applied well, it can lead to negative or disappointing results. If the student learns the same or less than with traditional methodologies, educational innovation derived from

---

1 This study is the result of the research Projects PGC2018-094491-B-C33 and PID2019-106539RB-100, funded by the Spanish Ministry of Science, Innovation and Universities, co-funded by the ERDF and the project entitled "Teaching competences and active learning methods. An evaluative research with the teaching staff in social sciences" (20638/JLI/18), funded by the Seneca Foundation.

technology loses meaning. Not by using a modern device or app will there be an educational improvement.

In the field of social science teaching, the reference disciplines (history, geography, archeology, art history) have long considered technology as a mere complementary element. That it served to illustrate, adorn, but it was still a more modern support than paper or simple oral discourse. However, in recent years the progress made has been considerable. And they have tried to respond to the usual shortcomings of this basic subject: to train critical and responsible citizens (Cabero, Torres & Hermosilla, 2019). For this reason, it is necessary that new technologies help to become aware of the past and present problems of the world in which they live: environment, economy, society, politics, heritage, etc. To know them, they must be worked in depth, not only in the classroom, but also outside of it. This is not only achieved with flipped class or games, but must be combined with debates, simulations, case studies, projects, challenges or school outings.

Therefore, new technologies should be the support to put together an educational planning that promotes the selection of content. Not all programming can be studied in detail, it is better to do it with some relevant topics. The student's boredom towards theoretical contents and away from their daily life is a challenge. But it is not solved only with videogames or tablets, it also requires familiarizing the student with the initiation processes in social research. Documenting about a topic, collecting primary and secondary sources, developing hypotheses, contrasting them and discussing them to obtain conclusions. This is how the utility and dynamism of the social sciences, forgers of identity and active citizenship, will be seen. Undoubtedly, new technologies make teaching more dynamic, connecting with students who are digital natives and making it easier for them to seek and expand information on various topics (Bowser, Hansen, He, Boston, Reid, Gunnar & Preece, 2013).

An approach to ITC, games and active methodologies teaching and learning Social Sciences

The appearance and development of numerous digital technologies and new methodological concepts in recent years at the service of the teacher in the classroom, has generated a good number of change processes in educational approaches, notably expanding the pedagogical ecosystem (Adell & Castañeda, 2013). ICTs offer many advantages to promote student learning, and the analysis of how to use them in an appropriate, effective and meaningful way is quite beneficial. For this reason, there are multiple published studies that address the possibilities of ICTs from simple approaches, focused only on the mere introduction of technological tools in the educational context, to more complete ones, in which a more correct integration of technology is achieved, being it focused on facilitating the acquisition of competences and the fulfilment of learning objectives, under the pedagogical and curricular domain of technological tools and the disciplines in which it is used.

In this sense, one of the proposals for the effective integration of technology at the service of teaching disciplinary content is the TPACK model (Technological Pedagogical Content Knowledge) popularized by Mishra and Koehler (2006), in which the technological dimension is added to the Pedagogical Content Knowledge model, previously formulated by Shulman (1986). The TPACK model suggests the need to generate interrelationships and interactions between the disciplinary, pedagogical and technological areas, in order to produce a satisfactory result in student learning and in the collaboration between teachers and students. Only when the three dimensions converge, the objective of the model is achieved: the technological pedagogical content knowledge. Despite the fact that empirical research based on the TPACK model in Social Sciences has been approached with less intensity than in other disciplines (Willemark, 2017), some educational interventions have been carried out, especially in initial teacher training (Cózar, Zagalaz & Saez, 2015; Gómez-Trigueros, 2016; Colomer, Sáiz & Bel, 2018; Gómez-Trigueros & Moreno-Vera, 2018), which reflect the potential of the model to promote effective and motivating teaching.

The generalization of technology in the teaching of Social Sciences has led to access to an overabundance of information, devices, learning scenarios and resources that differ from traditional ones (textbooks). The Web, with its progressive updates, has generalized access to knowledge and the appearance of a wide range of content "creators" who share resources, proposals and guidelines, fostering the construction of active, meaningful and student-centered learning models. Rivero and Mur (2015) have established three groups of Web 2.0 resources that stand out for their didactic utility in the Social Sciences classroom: cooperative/collaborative writing applications, such as wikis and blogs; material repositories that can be built by students (YouTube, Instagram, Picassa, Google Drive, etc.); and spaces for communication between users, such as virtual worlds or social networks. Miralles, Gómez and Monteagudo (2019) point out as main advantages of the use of ICTs for the teaching of Geography, History and other Social Sciences, its possibilities to promote self-regulated and meaningful learning, collaboration, both between students and between teachers and students, and the fact that education can be more individualized.

Among the proposals for ICT resources, the use of videogames is awakening an extraordinary interest in the literature. Games, which have traditionally been used as a tool for socialization and cooperative learning among children, have begun to be incorporated into educational environments as pedagogical resources, since it is believed that they provide the required social context to promote the acquisition of learning, motivation, engagement, retention and critical thinking, among others (Talan, Doğan & Batdı, 2020).

The Game-Based Learning methodology has a proven track record and has been used in several subjects and educational levels to improve the learning experience, maintaining the balance between content, playfulness and application in the real world. With the advent of videogames and their popularity among children and adolescents due to the enormous amount of leisure time they dedicate to these activities, Digital Game-Based Learning (Prensky, 2001) has focused its attention on the introduction of these media in the classrooms to unify serious learning and interactive entertainment. Van Eck (2006) establishes three aspects on the use of this approach in the classroom: (a) the use of commercial titles consisting of engaging and motivating games that can be used for educational purposes; (b) the so-called Serious Games, those specially developed to educate, train and inform; and (c) games built by the students themselves, which develop problem solving, programming and game design skills. The digital learning games can be designed to promote learning or the development of cognitive skills through precise curricular plans that do not lose sight of the pedagogical approach, or through simulations that allow students to practice their skills in a virtual environment (Erhel & Jamet, 2013).

The initial resistance of Social Sciences teachers to introducing videogames in the classroom has vanished in the last decade thanks to the publication of multiple experiences and investigations in which the didactic potential of this resource has been evidenced. Works such as those of Cuenca and Martín (2010) or Cózar and Sáez (2016) emphasize its benefits to address knowledge of a historical-social nature and work on cultural and social references in a contextualized, attractive and dynamic way, through highly motivating activities for the students.

In recent years, the strong irruption of mobile devices in both homes and schools, together with technological advances in the development of applications, has allowed the generalization of 3D simulations, virtual worlds and games in which digital scenarios are reconstructed, which facilitate the adaptation of these resources by teachers for teaching and learning of specific curricular contents, at all educational levels. In this line, the possibilities of Augmented Reality (AR) and Virtual Reality (VR) stand out. In 2017, the Observatory of Educational Innovation of Tecnológico de Monterrey (2017) dedicated one of its *EduTrends* reports to these emerging technologies, highlighting among their benefits: the reinforcement of attention, due to its sensory richness and immersive nature; the improvement of short- and long-term memory; an efficient learning pace; the cognitive development derived from the practical experimentation of the theory; the increase in motivation; and the possibility of individualizing the learning experience and adapting it to different cognitive styles and ways of learning. In addition, it favors the introduction of emerging pedagogical trends in the classroom, such as: "learning by doing"; gamification; learning based on experimentation; mobile, hybrid or blended learning; and expanded education.

Leaving aside the research carried out from the field of archaeology and virtual museography – a consolidated line and with a great trajectory –, there are still not many publications in the teaching of Social Sciences on the use of these resources in the classroom, although some of them have already advanced their positive effects at different educational levels. Villena, Cózar, González-Calero and López (2019) analyse the benefits of VR in the teaching of History on students in the fourth year of Primary Education (9–10 years old) compared to a traditional methodology based on a coursebook. The results show statistically significant differences in favor of those students who used VR, both in academic performance and in motivation. Regarding AR, the works of Cózar, De Moya, Hernández and Hernández (2015), López and Miralles (2018), and Sáez, Cózar, González-Calero and Gómez (2020), highlight its advantages in the development of the creativity, collaborative activities, innovation, active approaches, interest and motivation of students.

Nowadays, both technologies are increasingly accessible, which will allow the enrichment of the means and resources available to teachers in the short term.

Advantages and disadvantages of learning in a flipped-classroom context

The integration of ICTs in the educational field is not always easy, especially since we almost always focus on learning to use technology and not on how we can integrate it properly in the classroom for a concrete content and using an appropriate methodology.

In this sense, flipped-classroom is a new methodological option that allows the effective incorporation of ICT in the classroom and it achieve significant teaching renewal. The flipped-classroom model, devised by Bergman and Sams (2012), can be defined as a teaching-learning method in which the theoretical lesson is given outside the classroom and homework is done in classroom. This means a change in the traditional dynamics of classrooms thanks to the use of new technologies. Teachers provide students with the theoretical material in an audio-visual format, mainly explanatory videos that have great didactic potential due to the motivation that audio-visual material has in students (Casanova & Serrano, 2017), with the option of incorporating complementary materials. Students must consult it outside the classroom where, when, how and with whoever they want. This means that students must organize themselves to view the material and take responsibility for their learning process. Once in the classroom, the teaching task is to review or clarify doubts about the material and organize students in cooperative groups to investigate or deepen the previously seen materials through simulations, debates, short researches, problem solving, projects, etc. The interaction generated allows the teacher to obtain real-time information on the progress and deficiencies of the students, being able to influence, correct or clarify the most dubious or problematic questions, thus implementing an authentic continuous and formative assessment.

There are several works that refer to the advantages and disadvantages of using the flipped-classroom in education. Tourón and Santiago (2015) talked about the innovation and potential improvement of educational quality that this model implies, contributing, as main benefits among others, the following:

- Allows teachers to dedicate more time to pay attention to individual differences.
- It is an opportunity for teachers to share information and knowledge with each other, with students, with families and with the community.
- Provides students with the possibility of re-accessing, as many times as necessary, the best content generated or provided by their teachers.
- Create a collaborative learning environment in the classroom.
- It involves families from the beginning of the learning process.

In the same work, Tourón and Santiago cited several case studies that reflected that classes, from different educational levels (from primary school to university), experienced gains in performance and satisfaction by teachers and students with the use of flipped-classroom. Regarding teachers, it was shown that teachers who adapt the flipped-classroom do not usually want to return to more traditional models, while their students improve their academic results and their satisfaction with the teaching received and their learning.

Alarcón (2016), following the ideas presented by Luque (2015), collected in her work the benefits and difficulties of the method for both teachers and students:

- Advantages for students:

- They have more control over their own learning.
- Short videos allow them to follow their learning pace.
- Focus learning and collaboration.
- It allows them to participate actively and builds confidence in their abilities.
- Free access to internet content that makes it easy for them to catch up at any time.
- Receive help from parents for access.
- Flipped-classroom provides more free time and avoids recovery work.

- Disadvantages for students:

- Not taking responsibility for their learning.
- The digital divide increases if you do not have access to the internet or it is limited.
- Flipped-classroom does not prepare them to pass standardized tests.
- Many hours in front of the computer could cause dispersion.
- Outdated technology in educational centres.
- Workload if many videos and questionnaires are sent.

- Advantages for teachers:

- They dedicate more time to pay attention to diversity.
- To clarify student questions immediately.
- They identify errors and can correct it individually.
- They share information and knowledge between teachers, students and families.
- They speak the same language as the students.
- They have greater flexibility to take advantage of time.

- Disadvantages for teachers:

- Little motivation and involvement to attract students.
- Take responsibility for creating the material.
- Not giving up control, not trusting the student.
- Do not make students protagonist in their learning.
- To believe that they are the sole holder of the information.
- Time and technical capacity.

The method seems to have more advantages than disadvantages, but the biggest problem is that it can marginalize students with limited resources (Alarcón, 2016). Something that arises clearly during this COVID-19 isolation crisis as demonstrate research from Van Lancker and Parolin (2020) or Beaunoyer, Dupére and Guitton (2020).

Another problem found in the use of the inverted classroom is to use it as an exclusive method, since it does not solve anything by itself, as Vélez and González (2017) argued, being necessary for each teacher to combine different methods until finding the balance.

Santiago and Bergmann (2018) reported other demonstrable advantages of the method:

- Shorten and streamline homework.
- Improves discipline problems in the classroom by promoting a more active and participatory environment.
- Helps diversity outreach of student learning styles.

In the research they carried out, reported in the same publication, Santiago and Bergmann (2018) showed evidence of the improvements promoted by the implementation of flipped-classroom according to the opinion of the students. In this sense, the method is very well received by students because it allows them to learn and work more and better, both with their schoolmates and teacher. They also argued that it allows them to work at their own pace and gives them more freedom in accessing the materials that better suit their way of learning.

This way, a more personalized teaching is obtained, the attention, interest and motivation of the educational agents are improved and a new impetus is given to the formative and continuous facets of the assessment.

In spite of these works, it remains a strategy with relatively little empirical research in academia (O'Flaherty & Phillips, 2015), above all in Social Sciences, a field still very linked to traditional teaching & learning methodologies. But, from researches in others educative fields, we can know that, like other didactic actions focused on active student learning, flipped-classroom improves traditional formulas (Steen-Utheimm & Foldnes, 2018) increases student participation (Gilboy, Heinerichs & Pazzaglia, 2015) and, at the same time, allows for research and critical activity. Studies show that a direct consequence of this methodology is the increase of motivation and a greater procedural learning (Abeysekera & Dawson, 2015; Burke & Fedorek, 2017). On the other hand, there are studies (Yilmaz, 2017) detecting that, although flipped-classroom increases motivation, there are also decreases as students become used to this way of working, so it is important to program activities correctly to permit class work encompasses different methodological strategies.

It is possible that, today, following the Gartner Curve described by Adell and Castañeda (2012) about the expectations of emerging pedagogies, we are in phase number 4, where more evidence emerges of how technology can benefit education and we begin to better understand how it works.

Placing this phase in the case of flipped-classroom, Akçayir and Akçayir's research (2018) reviews the main advantages and disadvantages of flipped-classroom through 71 research articles in the Social Sciences Citation Index of the Web of Knowledge, and concludes that this methodological strategy improves student learning, although its weak points are found in the daily classroom activities and also in the confusion it generates in students not knowing directly the learning objectives of the subject to be studied (Shibukawa & Taguchi, 2019). For Sergis, Sampson and Pelliccione (2018) this improvement in learning is due to the theory of self-determination that increases student motivation, especially those students with low grades who prefer a new methodological model, such as flipped-classroom, although as Cai et al. (2019) stress, the success of this strategy is always linked to good planning on the part of teachers in terms of content, ease of use and access to materials.

After following different investigations, Feliu and Cózar (2018) collected the benefits of the implementation of flipped-classroom in Social Sciences subject, which were increased motivation, improved learning results and, in general, a good assessment by the students as a way of approaching certain contents, those of the social sciences, that students find rote and boring (Monteagudo, Gómez & Miralles, 2017).

Other interesting researches about flipped-classroom in Social Sciences are those of Gómez-Carrasco, Monteagudo-Fernández, Moreno-Vera and Sainz-Gómez (2019, 2020), where they combined the flipped-classroom with gamification. They showed a very positive impact on motivation, the learning achieved and the strategies applied in the program. The learning results were satisfactory, although lower than the students perceived. The limitations of these researches tell us that learning about theoretical conceptualization of capacity proposals was too superficial, so we must teach, at the same time, to design activities, technical skills and a greater theoretical framework.

Does really gamification motivate our students? Different strategies to develop good gamified lessons

In 2015, a paper from Dicheva, Dichev, Agre and Angelova (2015) mapped the publications about gamification in the field of Education. Their conclusions were clear: this strategy is highly developed in topics as marketing, business, corporation management, human resources or wellness initiatives but there were very few investigations about empirical results for gamified contexts in Education.

After five years of research, there still are some questions to answer in the concern of gamification: is it a good strategy to develop long-term lessons and conceptual knowledge? Does gamification allow to work inside other strategies as Project-based Education? Teachers gamify through ITC or it is not necessary? How does it work better: individually or in teams? Is gamification a good tool to evaluate content instead of traditional exams?

Scientific conclusions in these topics are still very preliminary. For sure, gamification is gaining presence in classrooms in the last 5 years, as way to engage student's attention and to make more fluent and dynamic lessons, but empirical results evaluating the good or bad implementation of gamification in the schools need to be took.

From 2012 onwards, one of the subjects that centered the research in educational gamification is motivation. Investigators, teachers and projects asked themselves if gamification helped to increase student's motivation during the lessons (Mekler, Brühlmann, Opwis & Tuch, 2013). Hursen and Bas (2013) obtained very good results about motivation from the students and, concluded as well, that even parents had positive opinions on the usage of gamification in the classroom.

Regarding to psychological perspectives, Sailer, Hense, Mandl and Klevers (2013) advanced that using game elements in a no-game context can foster motivation in students. A system by points, badges, leaderboards can help to rank the success and motivate pupils to improve their own skills. Progress bar, performance graphs, quests or avatars are able to motivate students in a competition among each other, as indicated by gamification taxonomy by Robinson and Bellotti (2013).

The psychological advantages of using gamification in a school context are resume in six perspectives (McClelland, 2009):

(1) Traits: different individual characters in students can evoke different motivations. For example: good students find in gamification a new context to compete and rank themselves, but even low-profile students can use gamification to succeed in a new learning context not based on discipline but in playing games.
(2) Behaviourist learning: students learning in a gamified classroom can be motivated by the immediate feedback. Motivation comes from the reward system implemented in the strategy. Positive or negative reinforcement increase motivation.
(3) Cognitive: motivation depends on the specific goals proposed by the teacher and the outcome expectancies of the students. Gamification will

improve the motivation in the classroom if the strategy provides a clear and achievable target to reach.

(4) Self-determination: the way that gamification could help to motivation, in this approach, is related to the student self needs of competence, autonomy and social relatedness (Ryan & Deci, 2000a). Intrinsic motivation is defined, in this perspective, as the desire to execute an action for an own sake. The feeling of competence could motivate a student using gamification.

(5) Interest: collective or individual preferences of contents constitute an important motivational variable in gamification. This motivation is based on affective and cognitive interest. If gamification can link the strategy and an interest content, the students will be motivated to learn it in a different way. Regarding to this perspective, centring motivation on student interests and environment is especially recommended to be used while implementing a significant education, through nearby and daily tools in students' life (TV programs, internet, sports, videogames, films, etc.)

(6) Emotion: in this case, the link between gamification and positive feelings (as fun, pleasure, happiness, etc.) increases motivation among the students. So, allowing a dynamic strategy as gamification in our classroom could improve emotions as freedom and decrease the sensation of fear or envy. This is why motivation is especially important when used as evaluation instrument.

Once identified the psychological perspectives able to allow motivation while using gamification in a classroom, the discussion moves to establish the better approach to design our gamification activities: do I need to reward the efforts in the game to maintain motivation or is it better to use a game that suppose a challenge for the students learning without using prizes?

Most of the empirical investigations made in the last years about gamification reach the same conclusion: teaching in a gamified context increases student's motivation (Bowser, Hansen, He, Boston, Reid, Gunnar & Preece, 2013; Mekler, Brühlmann, Opwis & Tuch, 2013; Richter, Raban & Rafaeli, 2015; Gómez, Monteagudo, Moreno-Vera & Sáinz, 2019). But according to the different psychological perspectives of motivation, which kind of motivation is increasing in that lessons?

Research from Gómez, Monteagudo, Moreno-Vera and Sáinz (2019; 2020) enlight, partially, the question: in their study, four groups of higher education used a methodology based on flipped-classroom and gamification duringa complete semester. Students were asked about their learning perceptions at the end of the semester and, in all the groups students answer that they felt more motivated using this strategy than in "traditional lessons" (they meant by "traditional" master classes). The key question arose when investigators compared the evaluation results between the gamified topic and traditional topics, the conclusion was clear: similar qualifications in both methodologies. So, where was the problem? The students preferred gamification model and felt more motivated but, at the same time, learning results have a little improvement with an active strategy.

Attending to this "contradiction," gamification became a subject of controversy for some researchers (Wu, 2012): obviously, the goal of a game is to entertain, but we have to redirect our goals when applying games to education, in this case, the goal of a gamified activity is to learn contents, not just to entertain.

Some investigators as Hanus and Fox (2015) or Banfield and Wilkerson (2014) studied the effect of intrinsic motivation and how this kind of motivation is, at the same time, related to efficacy in knowledge results. As Ryan and Deci (2000b) indicated, the intrinsically motivated students are more involved during the lessons and activities as their goal is to get the learning objectives. So the target for gamification it is not just to be used as a game in classroom to entertain (extrinsic motivation), if not, to be used as a strategy to reach learning contents (intrinsic motivation).

Conclusions

New technologies are not a target themselves, but a support element. That helps to pass several phases. ICT do not arise for educational purposes, but they are a first-rate didactic resource to speed up the teaching process and give examples. When they are given an educational bias, they promote the acquisition of knowledge. And finally, they would lead to promote empowerment and participation (Cambil & Romero, 2016). Therefore ICT, in this complex process, can favor the acquisition of a collective critical consciousness. Among so much information available, the student must know how to manage it, assume it and compare it to obtain their own conclusions.

This shifts the traditional role of teachers towards students. Teachers remain as mediators of learning, facilitators of resources and ideas, but they do not present complete or closed results. The danger that this implies is that the student seeks only fun, something little related to the learning that we want to achieve. So much autonomy, so much work outside the classroom, at home and with computers or smartphones, must be continuously supervised (Alarcón, 2016; Santiago & Bergmann, 2018). And, above all, it must be combined with face-to-face activities, on-site knowledge of nearby environments, discussions and corrections.

New technologies, applied to the educational field, are necessary and very relevant. But by themselves they are nothing more than a medium, a format, also subject to continuous change (Casanova & Serrano, 2017). Every computer program, web or videogame will be out of date in a short time. The updating of teachers and students must be almost permanent, adapting these technologies to the programming and needs of the school and its context (age, social and cultural level of the families). Flipped-classroom and videogames have taken an increasingly prominent position in social science teaching and it is a phenomenon that has come to stay. But they cannot completely eliminate face-to-face or more traditional methods, which help to establish skills such as the reading habit or knowing how to work in a team. Nor should values such as tolerance, respect for others, ethics or care for heritage and the environment be forgotten. These are the great challenges of teaching social studies, regardless of the technology or the means available.

References

Abeysekera, Lakmal/Dawson, Phillip: "Motivation and cognitive load in the flipped classroom: Definition, rationale and a call for research". *Higher Education Research & Development* 34, 2015, pp. 1–14, retrieved 10.9.2020, from https://doi.org/10.1080/07294360.2014.934336.

Adell, Jordi/Castañeda, Linda: "Tecnologías emergentes, ¿pedagogías emergentes?" In: Hernández, José/Pennesi, Massimo/Sobrino, Diego/ Vázquez, Azucena (coords.): *Tendencias emergentes en educación con TIC*. Barcelona: Asociación Espiral, Educación y Tecnología. 2012, pp. 13–32.

Adell, Jordi/Castañeda, Linda: "El ecosistema pedagógico de los PLEs" In Castañeda, Linda/Adell, Jordi (Eds.): *Entornos personales de aprendizaje: claves para el ecosistema educativo en red*. Marfil: Alcoy. 2013, pp. 29–52.

Akçayir, Gökçe/Akçayir, Murat: "The flipped classroom: A review of its advantages and challenges". *Computers & Education* 126, 2018, pp. 334–345, retrieved 2.6.2020, from https://doi.org/10.1016/j.compedu.2018.07.021.

Alarcón-Martínez, Virginia: "Metodología de Flipped Classroom en humanidades y ciencias sociales". In: Bernard-Monferrer, Estela/Mut-Camacho, Magdalena (Eds.): *Aula Virtual: Contenidos y Elementos*. McGraw-Hill: Madrid. 2016, pp. 1–18.

Banfield, James/Wilkerson, Brad: "Increasing student intrinsic motivation and self-efficacy through gamification pedagogy". *Contemporary Issues in*

Education Research 7(4), 2014, pp. 291–298, retrieved 2.6.2020, from https://doi.org/10.19030/cier.v7i4.8843.

Beaunoyer, Elisabeth/Dupére, Sophie/Guitton, Mathew J.: "COVID-19 and digital inequalities: Reciprocal impacts and mitigation strategies". *Computers in Human Behaviour* 111, 2020, retrieved 12.6.2020, from https://doi.org/10.1016/j.chb.2020.106424.

Bergmann, Jonathan/Sams, Aaron: *Flip Your Classroom: Reach Every Student in Every Class Every Day.* International Society for Technology in Education: Eugene, OR. 2012.

Bowser, Anne/Hansen, Dereck/He, Yurong/Boston, Carol/Reid, Matthew/Gunnell, Logan/Preece, Jennifer: "Using gamification to inspire new citizen science volunteers". In: *Gamification'13 Proceedings of the First International Conference on Gameful Design, Research and Applications.* University of Ontario, Institute of Technology: Ontario. 2013, pp. 18–25, retrieved 2.9.2020, from https://doi.org/10.1145/2583008.2583011.

Burke, Alison/Fedorek, Brian: "Does flipping promote engagement? A comparison of a traditional, online and flipped class". *Active Learning in Higher Education* 18, 2017, pp. 11–24, retrieved 15.5.2020, from https://doi.org/10.1177/1469787417693487.

Cabero, Julio/Torres, Luisa/Hermosilla, José Manuel: "Las TIC y la creación de una ciudadanía crítica e-digital". *Education in the knowledge society (EKS)* 20, 2019, retrieved 2.6.2020, from http://dx.doi.org/10.14201/eks2019_20_a22.

Cai, Jin/Yang, Harrison H./Gong, Di/MacLeod, Jason/Zhu, Sha: "Understanding the continued use of flipped classroom instruction: A personal beliefs model in Chinese higher education". *Journal of Computing in Higher Education* 31, 2019, pp. 137–155, retrieved 21.6.2020, from https://doi.org/10.1007/s12528-018-9196-y.

Cambil, María Encarnación/Romero, Guadalupe: "Las TIC, las TAC y Redes 3.0 para la enseñanza de las Ciencias Sociales". In: Liceras, Ángel/Romero, Guadalupe (coords.): *Didáctica de las ciencias sociales. Fundamentos, contextos y propuestas*. Pirámide: Madrid. 2016, pp. 271–295.

Casanova, Óscar/Serano, Rosa: "Flipped-classroom en ciencias sociales". *Íber: Didáctica de las Ciencias Sociales, Geografía e Historia* 88, 2017, pp. 47–52.

Colomer Rubio, Juan Carlos/Sáiz Serrano, Jorge/Bel Martínez, Juan Carlos: "Competencia digital en futuros docentes de Ciencias Sociales en Educación Primaria: análisis desde el modelo TPACK". *Educatio Siglo XXI* 36(1), 2018, pp. 107–128. https://doi.org/10.6018/j/324191.

Cózar Gutiérrez, Ramón/De Moya, María V./Hernández, José A./Hernández, Juan R.: "Tecnologías emergentes para la enseñanza de las Ciencias Sociales. Una experiencia con el uso de Realidad Aumentada en la formación inicial de maestros". *Digital Education Review* 27, 2005, pp. 138–153.

Cózar Gutiérrez, Ramón/Sáez López, José Manuel: "Game-based learning and gamification in initial teacher training in the social sciences: An experiment with MinecraftEdu". *International Journal of Educational Technology in Higher Education* 13(1), 2016, retrieved 2.6.2020, from https://doi.org/10.1186/s41239-016-0003-4.

Cózar Gutiérrez, Ramón/Zagalaz, Juan/Sáez López, José Manuel: "Creando contenidos curriculares digitales de Ciencias Sociales para Educación Primaria. Una experiencia TPACK para futuros docentes". *Educatio Siglo XXI* 33(3), 2015, pp. 147–168.

Cuenca, José María/Martín, Myriam J.: "Virtual games in social science education". *Computers & Education* 55(3), 2010, pp. 1336–1345, retrieved 2.6.2020, from https://doi.org/10.1016/j.compedu.2010.05.028.

Dicheva, Darina/Dichev, Christo/Agre, Gennady/Angelova, Galia: "Gamification in education: A systemic mapping study". *Educational Technology & Society* 18(3), 2015, pp. 75–88.

Erhel, Séverine/Jamet, Eric: "Digital game-based learning: Impact of instructions and feedback on motivation and learning effectiveness". *Computers & Education* 67, 2013, pp. 156–167, retrieved 23.7.2020, from https://doi.org/10.1016/j.compedu.2013.02.019.

Feliu, Maria/Cózar, Ramón: "Nuevos recursos TIC para la enseñanza de la historia en educación obligatoria". In: Miralles, Pedro/Gómez, Cosme J. (coords.): *La educación histórica ante el reto de las competencias. Métodos, recursos y enfoques de enseñanza*. Octaedro: Barcelona. 2018, pp. 39–50.

Gilboy, Mary Beth/Heinerichs, Scott/Pazzaglia, Gina: "Enhacing student engagement using the flipped classroom". *Journal of Nutrition Education and Behavior* 47, 2015, pp. 109–114, retrieved 2.6.2020, from https://doi.org/10.1016/j.jneb.2014.08.008.

Gómez-Carrasco, Cosme J./Monteagudo-Fernández, José/Moreno-Vera, Juan R./Sáinz-Gómez, Marta: "Effects of a gamification and flipped-classroom program for teachers in training on motivation and learning perception". *Education Sciences* 9(4), 2019, 299, retrieved 2.6.2020, from https://doi.org/10.3390/educsci9040299.

Gómez-Carrasco, Cosme J./Monteagudo-Fernández, José/Moreno-Vera, Juan R./Sáinz-Gómez, Marta: "Evaluation of a gamification and flipped-classroom program used in teacher training: Perception of learning and outcome". *PLoS ONE* 15(7), 2020, retrieved 2.9.2020, from https://doi.org/10.1371/journal.pone.0236083.

Gómez-Trigueros, Isabel María: "La inclusión de las tecnologías en la formación inicial del profesorado: una intervención de aula a través del modelo TPACK". *Tendencias Pedagógicas* 28, 2016, pp. 133–152, retrieved 2.6.2020, from https://doi.org/10.15366/tp2016.28.010.

Gómez-Trigueros, Isabel María/Moreno-Vera, Juan Ramón: "Nuevas didácticas geográficas: el modelo TPACK, los MOOCs y Google Earth en el aula". *EDMETIC* 7(2), 2018, pp. 146–165, retrieved 2.6.2020, from https://doi.org/10.21071/edmetic.v7i2.9547.

Hanus, Michael D./Fox, Jesse (2015): "Corrigedum to assessing the effects of gamification in the class-room: A longitudinal study on intrinsic motivation, social comparison, satisfaction, effort and academic performance". *Computers & Education* 80, 2015, pp. 152–161, retrieved 2.6.2020, from https://doi.org/10.1016/j.compedu.2018.09.019.

Hursen, Cigdem/Bas, Cizem: "Use of gamification applications in Science Education". *International Journal of Emerging Technologies in Learning* 14(1), 2019, pp. 4–23.

Jiménez-Alcázar, Juan Francisco: "La historia no fue así: reflexiones sobre el fenómeno de la historia contrafáctica en los videojuegos históricos". *Clío. History and History Teaching* 44, 2018.

López-García, Alejandro/Miralles Martínez, Pedro: "La realidad aumentada en la formación del profesorado. Una experiencia en las prácticas del Máster de Profesorado de Enseñanza Secundaria". *Campus Virtuales* 7(2), 2018, pp. 39–46.

Luque Ana: *Flipped Classroom en el Colegio Singuerlin. Experiencia del modelo en 4.º de primaria: Valoraciones.* Universitat Internacional de Catalunya: Barcelona. 2015.

McClelland, David C.: *Human Motivation.* Cambridge University Press: Cambridge. 2009.

Mekler, Elisa, D./Brühlmann, Florian/Opwis, Klaus/Tuch, Alexandre N: "Do points, levels and leaderboards harm instrinc motivation?: An empirical analysis of common gamification elements". In: *Gamification'13 Proceedings of the First International Conference on Gameful Design, Research and Applications.* University of Ontario, Institute of Technology: Ontario. 2013, pp. 66–73, retrieved 2.6.2020, from https://doi.org/10.1145/2583008.2583017.

Miralles Martínez, Pedro/Gómez Carrasco, Cosme J./Monteagudo Fernández, José: "Percepciones sobre el uso de recursos TIC y «Mass-Media» para la enseñanza de la Historia. Un estudio comparativo en futuros docentes de España-Inglaterra". *Educación XX1* 22(2), 2019, pp. 187–211.

Mishra, Punya/Koehler, Matthew: "Technological pedagogical content knowledge: A framework for teacher knowledge". *Teachers College Record* 108(6), 2006, pp. 1017–1054.

Monteagudo Fernández, José/Gómez Carrasco, Cosme J./Miralles Martínez, Pedro: "Evaluación del diseño e implementación de la metodología flipped-classroom en la formación del profesorado de ciencias sociales". *RED. Revista de Educación a Distancia* 55, 2017, retrieved 2.6.2020, from http://dx.doi.org/10.6018/red/55/7.

O'Flaherty, Jacqueline/Phillips, Craig: "The use of flipped classrooms in higher education: A ccoping review". *The Internet and Higher Education* 25, 2015, pp. 85–95, retrieved 22.6.2020, from https://doi.org/10.1016/j.iheduc.2015.02.002.

Prensky, Marc: *Digital Game-Based Learning*. McGraw-Hill: New York. 2001.

Richter, Ganit/Raban, Daphne/Rafaeli, Sheizaf: "Studying gamification: the effect of rewards and incentives on motivation". In: Reiners, Torsten/Wood, Lincoln C. (eds.): *Gamification in Education and Business*. Springer International Publishing: New York. 2015, pp. 21–46.

Rivero, Pilar/Mur, Lorenzo: "Aprender ciencias sociales en la web 2.0". *Íber: Didáctica de las Ciencias Sociales, Geografía e Historia* 80, 2015, pp. 30–37.

Robinson, David/Bellotti, Victoria: "Preliminary taxonomy of gamification elements for varying anticipated commitment". In: *Proceedings of the CHI 2013*. CHI: Paris. 2013.

Ryan, Richard M./Deci, Edward L.: "Intrinsic and extrinsic motivations: Classic definitions and new directions". *Contemporary Educational Psychology* 25(1), 2000a, pp. 54–67.

Ryan, Richard M./Deci, Edward L.: "Self-determination theory and the facilitation of intrinsic motivation, social development and well-being". *American Psychologist* 55(1), 2000b, pp. 68–78.

Sáez-López, José M./Cózar-Gutiérrez, Ramón/González-Calero, José A./Gómez Carrasco, Cosme J.: "Augmented reality in higher education: An evaluation program in initial teacher training". *Education Sciences* 10(2), 26, 2020, retrieved 16.7.2020, from https://doi.org/10.3390/educsci10020026.

Sailer, Michael/Hense, Jan/Mandl, Heinz/Klevers, Markus: "Pyschological perspectives on motivation through Gamification". *Interaction Design and Architecture(s) Journal* 19, 2013, pp. 28–37.

Santiago, Raúl/Bergmann, Jonathan: *Aprender al revés. Flipped learning 3.0 y metodologías activas en el aula*. Paidós Educación: Barcelona. 2018.

Sergis, Stylianos/Sampson, Demetrios G./Pelliccione, Lina: "Investigating the impact of flipped-classroom on students' learning experiences: A self-determination theory approach". *Computers in Human Behaviour* 78, 2018, pp. 368–378, retrieved 2.6.2020, from https://doi.org/10.1016/j.chb.2017.08.011.

Shibukawa, Sachika/Taguchi, Mana: "Exploring the difficulty on students' preparation and the effective instruction in the flipped classroom. A case study in a physiology class". *Journal of Computing in Higher Education* 31, 2019, pp. 311–339, retrieved 2.6.2020, from https://doi.org/10.1007/s12528-019-09220-3.

Shulman, Lee S.: "Those who understand: Knowledge growth in teaching". *Educational Researcher* 15(2), 1986, pp. 4–14.

Steen-Utheimm, Anna Therese/Foldnes, Njal: "A qualitative investigation of student engagement in a flipped classroom". *Teaching in Higher Education* 23, 2018, pp. 307–324, retrieved 2.6.2020, from https://doi.org/10.1080/13562517.2017.1379481.

Talan, Tarik/Doğan, Yunus/Batdı, Veli: "Efficiency of digital and non-digital educational games: A comparative meta-analysis and a meta-thematic analysis". *Journal of Research on Technology in Education* 52, 2020, retrieved 2.6.2020, from https://doi.org/10.1080/15391523.2020.1743798.

Tecnológico de Monterrey: *EduTrends: Realidad aumentada y virtual.* Observatorio de Innovación Educativa Tecnológico de Monterrey: Monterrey. 2017.

Tourón, Javier/Santiago, Raúl: "El modelo *Flipped Learning* y el desarrollo del talento en la escuela". *Revista de Educación* 368, 2015, pp. 197–208, retrieved 2.6.2020, from https://doi.org/10.4438/1988-592X-RE-2015-368-288.

Van Eck, Rick: "Digital game-based learning: It's not just the digital natives who are restless". *EDUCAUSE Review* 41(2), 2006, pp. 16–30.

Van Lancker, Wim/Parolin, Zachary: "COVID-19, school closures, and child poverty: A social crisis in the making". *The Lancet* 5(5), 243–244, retrieved 2.6.2020, fromhttps://doi.org/10.1016/S2468-2667(20)30084-0.

Vélez Alonso, Francisco J./González Martínez, Carlos: "Flipped classroom en las aulas de ciencias sociales". In: Martínez-Medina, Ramón/García-Morís Roberto/García Ruiz, Carmen R. (coords.): *Investigación en didáctica de las ciencias sociales. Retos preguntas y líneas de investigación*. Universidad de Córdoba-AUPDCS: Córdoba. 2017, pp. 810–816.

Villena, Rafael/Cózar-Gutiérrez, Ramón/González-Calero, José Antonio/López, Isabel: "Strolling through a city of the Roman Empire: An analysis of the potential of virtual reality to teach history in Primary Education". *Interactive Learning Environments*, 2019, retrieved 2.6.2020, from https://doi.org/10.1080/10494820.2019.1674886.

Willermark, Sara: "Technological pedagogical and content knowledge: A review of empirical studies published from 2011 to 2016". *Journal of Educational Computing Research* 56(3), 2017, pp. 315–343.

Wu, Michael: *The Gamification Backlash + Two Long Term Business Strategies.* 2012, retrieved May 25, 2020, from http://lithosphere.lithium.com/t5/science-of-social-blog/The-Gamification-Backlash-Two-Long-Term-Business-Strategies/ba-p/30891.

Yilmaz, Ramazan: "Exploring the role of e-learning readiness on student satisfaction and motivation in flipped classroom". *Computers in Human Behaviour* 70, 2017, pp. 251–260, retrieved 2.6.2020, from https://doi.org/10.1016/j.chb.2016.12.085.

www.ingramcontent.com/pod-product-compliance
Ingram Content Group UK Ltd.
Pitfield, Milton Keynes, MK11 3LW, UK
UKHW041824100926
8556IPUK00002B/17

9 783631 818978